Advanced Composite Materials for Structure Strengthening and Resilience Improving

Advanced Composite Materials for Structure Strengthening and Resilience Improving

Guest Editors

Xinghuai Huang
Yeshou Xu

MDPI

Basel • Beijing • Wuhan • Barcelona • Belgrade • Novi Sad • Cluj • Manchester

Guest Editors
Xinghuai Huang
School of Civil Engineering
Southeast University
Nanjing
China

Yeshou Xu
School of Civil Engineering
Southeast University
Nanjing
China

Editorial Office
MDPI AG
Grosspeteranlage 5
4052 Basel, Switzerland

This is a reprint of the Special Issue, published open access by the journal *Buildings* (ISSN 2075-5309), freely accessible at: https://www.mdpi.com/journal/buildings/special_issues/structure_strengthening.

For citation purposes, cite each article independently as indicated on the article page online and as indicated below:

Lastname, A.A.; Lastname, B.B. Article Title. *Journal Name* **Year**, *Volume Number*, Page Range.

ISBN 978-3-7258-6926-8 (Hbk)
ISBN 978-3-7258-6927-5 (PDF)
https://doi.org/10.3390/books978-3-7258-6927-5

Contents

About the Editors

Xinghuai Huang

Xinghuai Huang is affiliated with the School of Civil Engineering at Southeast University in Nanjing, China. His research interests center on the mechanics of composite materials, structural strengthening and retrofitting, and the experimental and numerical simulation of reinforced concrete structures. He is actively involved in national-level research projects investigating the durability and long-term performance of FRP-strengthened concrete elements under complex environmental conditions. Dr. Huang has published numerous articles in peer-reviewed journals and contributes to the development of technical standards for the application of advanced composites in civil engineering.

Yeshou Xu

Yeshou Xu is a researcher at the School of Civil Engineering, Southeast University, Nanjing, China. His scholarly work focuses on the design of resilient structures, smart composite materials, and innovative manufacturing technologies for construction applications. He participates in key projects aimed at integrating sensor technology and composite materials for structural health monitoring and adaptive strengthening systems. Dr. Xu has achieved recognition for his contributions to advancing the understanding of smart material behavior in civil infrastructure and for his efforts in translating laboratory research into practical engineering solutions.

Preface

This Reprint focuses on the application of advanced composite materials for the strengthening and resilience improvement of engineering structures. Its subject encompasses the mechanics, design, manufacturing, and experimental or numerical analysis of composite materials and structural elements. The scope extends to smart composites and their role in creating more durable and adaptable infrastructure. The primary aim is to consolidate and highlight recent scientific and technological advancements in this critical field, providing a comprehensive resource for researchers and practitioners.

The motivation for this work stems from the growing need to enhance the longevity, safety, and performance of existing and new structures in the face of aging infrastructure, increased environmental demands, and the pursuit of sustainable development. By compiling cutting-edge research, this Reprint seeks to address challenges related to structural maintenance, interconnectivity, resilience, and energy efficiency. It is addressed to academics, engineers, graduate students, and industry professionals involved in civil engineering, materials science, and structural design who are dedicated to pioneering solutions for the built environment of the future.

Xinghuai Huang and Yeshou Xu
Guest Editors

buildings

MDPI

Editorial

Recent Developments of Advanced Composite Materials for Structural Strength and Resilience Improvements

Xinghuai Huang and Yeshou Xu *

China-Pakistan Belt and Road Joint Laboratory on Smart Disaster Prevention of Major Infrastructures, Southeast University, Nanjing 211189, China; huangxh@seu.edu.cn
* Correspondence: xuyeshou@163.com

With the increasing demand for building structures with diversified functions driven by social development, the necessity to cope with more extreme and complex service environments (involving earthquakes, marine corrosion, high temperatures, and droughts) is imposing higher strength and toughness requirements on building structures. Advanced composite materials are characterized by a high strength stiffness, and special functions and represent a key breakthrough for enhancing the performance and resilience of structural systems. With the development of modern automated manufacturing processes, further progress has been made in the mechanics, technology, and analysis of composite materials and structural elements. Composite materials cover a wide array of materials, ranging from the more commonly used forms (such as glass fibers) to more advanced materials (such as carbon fibers and magnetorheological elastomers). New ideas and methods for the application and integration of composite materials have made significant contributions to enhancing the safety, reliability, and durability of structures under multiple scenarios. For instance, magnetorheological fluids, which are used to make magnetorheological dampers, possess the characteristics of adjustability, a fast response speed, small volume, and a light weight. Xu [1] used ultrasonic and mechanical stirring–grafting techniques to coat multiwalled carbon nanotubes (MWNTs) onto carbonyl iron (CI), proposed the single- and double-chain micromechanical model, and compared it with shear yield stress tests to verify the effectiveness and accuracy of this model. In addition, a new piece of equipment that utilizes the synergistic effects of vibration isolation and damping devices, magnetorheological dampers, and viscoelastic dampers was proposed [2–6], adopting a multi-state control concept to regulate the input current of magnetorheological dampers. By comparing the time-domain and frequency-domain responses of damped and undamped structures, it was concluded that these devices can significantly reduce the dynamic response of the platform. However, research on composite materials still faces challenges. On one hand, the preparation techniques and formulation optimization of composite materials have not yet been fully developed. Whether they concern the fiber orientation in fiber-reinforced composites, the deposition sequence in laminated composite materials, or the nanoparticle reinforcement in matrix composite materials, even minor parameter deviations may lead to significant fluctuations in material properties. On the other hand, the integration of composite materials with traditional structures requires practical verification to promote the large-scale application of advanced composite materials in various complex scenarios. These factors have motivated research in the field of advanced composite materials for the purposes of strengthening structures and improving resilience, focusing on areas such as the mechanics of composite materials used to strengthen structures; design methods using composite materials to improve strength and resilience; technologies for the manufacturing

https://doi.org/10.3390/buildings15224097

of composite materials; experimental and numerical studies on composite materials and devices; experimental and numerical studies on the strengthening of structures using composite materials; and research on and applications of smart composites.

This Special Issue, "Advanced Composite Materials for Structural Strength and Resilience Improvements", provides an opportunity to share experiences and knowledge regarding the preparation of advanced composite materials and their applications in complex scenarios. It has collected eleven original research papers and one review paper. These papers cover research on the preparation and mechanical properties of composite materials, design methods using composite materials to enhance structural toughness in complex scenarios, and experimental and numerical studies on the application of composite materials to improve structural performance.

Among the studies on the preparation and mechanical properties of composites published in this Special Issue, Li [7] studied the influences of strain amplitude, number of cycles, and heat treatment on the fatigue performance of shape memory alloy (SMA) bars. The experimental and simulation results showed the energy dissipation capacity and small residual strain of the SMA bars under low-cyclic-fatigue loading, implying that SMA bars can be used as self-centering energy-dissipating elements in seismic design.

Xu [8] developed a higher-order fractional derivative model modified with the internal variable theory and temperature–frequency equivalent principle (ITHF) to describe the dynamic behaviors of viscoelastic dampers with varying frequencies and displacement amplitudes. Based on a test on a plate-shear-type viscoelastic damper at room temperature under sinusoidal displacement excitations, the higher-order fractional derivative model and the temperature–frequency equivalent principle were employed to characterize the influence of frequency and temperature, and the internal variable theory considering the evolution of the internal/microscale structure is introduced to capture the displacement affection.

Among the studies on design methods using composite materials to enhance structural toughness in complex scenarios published here, Luo [9] used silicomanganese slag powder to replace grounded blast furnace slag to generate alkali-activated silicomanganese slag concrete (AASSC) with excellent performance. The problem of ferromanganese-silicon slag being difficult to handle and prone to causing environmental pollution can be solved by making use of it in AASSC. The experimental results indicate that the AASSC with a 10% silicomanganese slag substitution and 2% steel fibers retains 80% of its compressive strength, making it suitable for practical needs. The AASSC with a 60% silicomanganese slag substitution also can applied to construction to achieve a massive substitution of GBFS.

Silwal [10] studied the application of unidirectional carbon FRP (CFRP) sheets to steel I-beams. Finite element analysis (FEA) was utilized to investigate the effect of various corrosion levels and the number of circular web openings on the yield and ultimate load capacities of the beams. The simulation results showed that thinning the bottom flange reduced both the yield and ultimate load capacities, with a nearly perfect linear reduction in ultimate load for each 2.5% reduction in thickness. Three configurations that involved applying CFRP sheets to both flanges and the web effectively restored the strength of SBWOs when an adequate number of CFRP layers were used.

Li [11] designed the layout of glass–polypropylene (PP) laminate layers to enhance their stiffness and used them as load-bearing members in civil engineering. A bend model of CSIP thin-wall box-beams under uniform loading and the deflection curve equation under shearing deformation were built. The FE result validated the theoretical method as being applicable for the design of thin-wall box-beams made from composite materials.

Rageh [12] proposed the use of two bonding materials, epoxy (EP) and geopolymer (GPP), with different ratios of short glass fibers (SGFs) to strengthen the bond between glass fiber-reinforced polymer (GFRP) and RC concrete surfaces. Failure load experiments were conducted on seven RC beams. These beams had the same cross-section and length and were bonded with adhesive materials (EP and GPP) mixed with SGF at different proportions. Experiments and simulations demonstrated that the addition of bonding materials plays a significant role in enhancing the failure load-bearing capacity of beams.

Malakopoulos and Salifoglou [13] investigated the performance of mortars composed of Portland limestone cement, calcium carbonate, butyl stearate, and oleic acid, aiming to reduce the carbon footprint and enhance the durability of concrete structures. Measurements of electrical resistivity, chloride migration, porosity, water permeability, drying shrinkage, thermal expansion, and compressive strength were conducted to assess durability. The results indicate that incorporating calcium carbonate, butyl stearate, and oleic acid into mortar mixtures provides enhanced durability.

Among the experimental and numerical studies on the application of composite materials for the purposes of improving structural performance published in this Special Issue, Zhang [14] developed a guided-rail tension device (GR) and a novel guided-rail isolation rubber bearing (GR&RB). The GR is designed to enhance the tensile strength of rubber bearings, while the GR&RB is designed to mitigate the risk of overturning due to tension in rubber isolation bearings. The experimental results suggest that the inclusion of GRs has a limited effect on the horizontal mechanical attributes of rubber isolation bearings but does enhance their tensile strength. The numerical analysis suggests that the utilization of GR&RB can mitigate the tensile stress levels of rubber isolators.

Wu [15] proposed a new type of steel shear-connection horizontal joint to ensure reasonable force transmission and structural integrity. The influence of this new type of joint on mechanical behavior was investigated through simulations, and a bilinear constitutive model based on the M-θ relationship of the new steel shear-connection joint was further advanced and deduced. The results show that the proposed new joint can provide stable shear capacity and superior energy dissipation capacity.

Schmidt [16] researched the response of a novel response-controlled prestressing anchor system used for CFRP NSMR strengthening. An initial assessment of an in situ precast concrete bridge was performed to provide a basis for the most desirable anchor system placement. Five laboratory tests were performed on a novel ductile response-controlled anchor system used for prestressed CFRP NSMR strengthening. These results were used as a basis for further implementation in a pilot project where an in situ cast concrete bridge was strengthened with the developed system.

Li [17] proposed a new nonlinear dynamic model of cylindrical reticulated shells with initial damage to investigate the effect of initial damage accurately. The nonlinear vibration differential equations with damage were built and the nonlinear natural vibration frequency with initial damage was derivatized. After that, a bifurcation problem with initial damage was studied by using the Flouquet Index, and the dynamic stability state at the equilibrium point was analyzed in depth. The theoretical results indicate that as the damage accumulates to 0.618, the nonlinear natural vibration frequency drops to zero, leading to the failure of local stability and even the loss of overall stability in the cylindrical reticulated shell. This study provided a theoretical foundation for the future investigation of overall stability with initial damage.

Huang [18] conducted a review on advanced composite materials. They provided a comprehensive overview of the types and characteristics of advanced composite materials, explored new methods and technologies, and discussed the challenges and opportunities for future work.

Although this Special Issue only includes 12 papers, new advanced composite materials applied in structural strengthening and resilience improvement have been presented as much as possible. Given that this is still a challenging topic, new thoughts, perspectives, ideas, and methods, as well as new advances in materials and devices, are still sought to promote research on advanced composite materials that can strengthen structures and improve resilience.

We are striving to enhance the influence of the journal *Buildings* and its leading position in the field of advanced composite materials used for structural strengthening and resilience improvement. We are grateful for the contributions of the authors who were vital to making this Special Issue successful.

Author Contributions: All authors listed have made substantial, direct, and intellectual contributions to the work and approved it for publication. All authors have read and agreed to the published version of the manuscript.

Data Availability Statement: The datasets generated and/or analyzed during the current study are available from the corresponding author upon reasonable request.

Acknowledgments: We would like to acknowledge all the authors, reviewers, editors, and publishers who have supported this Special Issue.

Conflicts of Interest: The authors declare no conflicts of interest.

References

1. Guo, Y.Q.; Sun, C.L.; Xu, Z.D.; Jing, X. Preparation and tests of MR fluids with CI particles coated with MWNTs. *Front. Mater.* **2018**, *5*, 50. [CrossRef]
2. Xu, Z.D.; Xu, F.H.; Chen, X. Intelligent vibration isolation and mitigation of a platform by using MR and VE devices. *J. Aerosp. Eng. ASCE* **2016**, *29*, 04016010. [CrossRef]
3. Xu, Z.D.; Xu, F.H.; Chen, X. Vibration suppression on a platform by using vibration isolation and mitigation devices. *Nonlinear Dyn.* **2016**, *83*, 1341–1353. [CrossRef]
4. Xu, Z.D.; Dong, Y.R.; Chen, S.; Guo, Y.Q.; Li, Q.Q.; Xu, Y.S. Development of hybrid test system for three-dimensional viscoelastic damping frame structures based on Matlab-OpenSees combined programming. *Soil Dyn. Earthq. Eng.* **2021**, *144*, 106681. [CrossRef]
5. Xu, Z.D.; Shen, Y.P.; Guo, Y.Q. Semi-active control of structures incorporated with magnetorheological dampers using neural networks. *Smart Mater. Struct.* **2003**, *12*, 80–87. [CrossRef]
6. Xu, Z.D.; Jia, D.H.; Zhang, X.C. Performance tests and mathematical model considering magnetic saturation for magnetorheological damper. *J. Intell. Mater. Syst. Struct.* **2012**, *23*, 1331–1349. [CrossRef]
7. Li, L.; Zhao, X.; Cheng, J. Research on fatigue performance of shape-memory alloy bars under low cyclic loading. *Buildings* **2023**, *13*, 1553. [CrossRef]
8. Xu, Y.; He, Q.; Guo, Y.Q.; Huang, X.H.; Dong, Y.R.; Hu, Z.W.; Kim, J. Experimental and theoretical investigation of viscoelastic damper by applying fractional derivative method and internal variable theory. *Buildings* **2023**, *13*, 239. [CrossRef]
9. Luo, B.; Wang, D.; Mohamed, E. Study on mechanical properties and durability of alkali-activated silicomanganese slag concrete (AASSC). *Buildings* **2022**, *12*, 1621. [CrossRef]
10. Silwal, P.K.; Parvin, A.; Alhusban, M. Numerical investigation on strengthening of steel beams for corrosion damage or web openings using carbon fiber reinforced polymer sheets. *Buildings* **2024**, *14*, 1069. [CrossRef]
11. Li, L.; Uddin, N.; Zhao, X.; Tian, L. Mechanical property research for CSIP thin-wall box-beams. *Buildings* **2023**, *13*, 1822. [CrossRef]
12. Rageh, B.O.; El-Mandouh, M.A.; Elmasry, A.H.; Attia, M.M. Flexural behavior of RC beams strengthened with GFRP laminate and retrofitting with novelty of adhesive material. *Buildings* **2022**, *12*, 1444. [CrossRef]
13. Malakopoulos, A.; Salifoglou, A. Assessment of durability indicators for service life prediction of portland limestone Cementitious systems produced with permeability-reducing admixtures. *Buildings* **2022**, *12*, 1712. [CrossRef]
14. Zhang, L.; Lan, X.; Wu, K.; Yu, W. Tests and seismic response analysis of guided-rail-type anti-tensile rubber bearing. *Buildings* **2024**, *14*, 992. [CrossRef]
15. Wu, X.; Wang, Y.; Ji, S.; Liu, M.; Wang, D. Numerical study on mechanical behaviors of new type of steel shear-connection horizontal joint in prefabricated shear wall structure. *Buildings* **2023**, *13*, 3000. [CrossRef]
16. Schmidt, J.W.; Sørensen, J.D.; Christensen, C.O. In situ concrete bridge strengthening using ductile activated NSMR CFRP system. *Buildings* **2022**, *12*, 2244. [CrossRef]

17. Li, L.; Li, G.; Uddin, N.; Tian, L.; Zhu, Z.; Bai, C.; Shen, C. Nonlinear dynamic stability of cylindrical reticulated shells with initial damage. *Buildings* **2023**, *13*, 2852. [CrossRef]
18. Huang, X.; Su, S.; Xu, Z.; Miao, Q.; Li, W.; Wang, L. Advanced composite materials for structure strengthening and resilience improvement. *Buildings* **2023**, *13*, 2406. [CrossRef]

buildings

MDPI

Article

Numerical Investigation on Strengthening of Steel Beams for Corrosion Damage or Web Openings Using Carbon Fiber Reinforced Polymer Sheets

Prabin Kumar Silwal [1], Azadeh Parvin [1,*] and Mohannad Alhusban [2]

[1] Department of Civil and Environmental Engineering, The University of Toledo, Toledo, OH 43606, USA; prabin.silwal@rockets.utoledo.edu
[2] Crawford, Murphy & Tilly, Inc., St. Louis, MO 63102, USA; malhusban@cmtengr.com
* Correspondence: azadeh.parvin@utoledo.edu

Abstract: Fiber-reinforced polymers (FRPs) have been widely used to strengthen steel structures, which could suffer from corrosion or the introduction of web openings, for utilities such as ductwork, plumbing, electrical conduits, and HVAC systems. The present numerical study involves the application of unidirectional carbon FRP (CFRP) sheets to steel I-beams, damaged due to corrosion or web openings, to regain their lost load-carrying capacity. Finite element analysis (FEA) was utilized to develop and validate three beam models against existing experimentally tested specimens. Subsequently, a parametric study was conducted investigating the effect of various corrosion levels and the number of circular web openings on the yield and ultimate load capacities of the beams. The optimum number of CFRP layers needed to strengthen corroded beams was determined and six CFRP strengthening scenarios were adopted to determine the best configurations to retrofit steel beams with openings (SBWOs). The results revealed that corrosion, introduced by thinning the bottom flange, reduced both yield and ultimate load capacities, with a nearly perfect linear reduction in ultimate load for each 2.5% thickness loss. The optimum number of CFRP layers depended on the level of corrosion damage. Furthermore, while maintaining a constant total opening area, beams with a greater number of smaller circular web openings demonstrated higher yield and ultimate load capacities than those with fewer larger openings. Out of the six adopted CFRP strengthening scenarios, three configurations that involved applying CFRP sheets to both flanges and the web effectively restored the strength of SBWOs, when adequate CFRP layers were used.

Keywords: steel beam; strengthening; corrosion; fiber-reinforced polymer; CFRP sheets; web openings; numerical analysis; finite element analysis; ANSYS-MAPDL

1. Introduction

The strengthening and repairing of existing structures are indispensable so that they can function safely, while maintaining serviceability throughout their service life [1]. In steel structures, design errors, environmental conditions, loss of material due to corrosion, sudden unanticipated loadings such as earthquake, etc., warrant strengthening and repair. The conventional strengthening and repair of steel structures involves the addition of steel plates or stiffeners to existing structures. However, it requires heavy tools and equipment to carry steel plates and the welding of steel plates would, ultimately, increase the vulnerability to corrosion and fatigue damages, while also increasing the self-weight of the structure [1,2]. As an alternative to conventional techniques, the use of fiber-reinforced polymers (FRPs) is growing in popularity in the field of structural engineering, along with other areas of civil and mechanical engineering. Being light weight, having a high strength-to-weight ratio, corrosion resistance, durability, and high stiffness are among the desirable properties of FRPs [3,4] that make them an excellent choice for structural strengthening applications.

Many experiments have been conducted to assess the performance of FRP-strengthened steel structures, whose findings point towards their improved performance. It is also observed that the performance of FRP-strengthened structures depends on different parameters such as fiber type, thickness, mechanical properties, fiber orientation, number of layers, etc. [5]. Among the various kinds of available FRP, based on the fiber type, carbon FRP (CFRP) is most widely used because of its higher strength and stiffness. The use of CFRP plates or laminates has been widely studied for the flexural strengthening of steel beams [6,7]. However, the application of CFRP sheets using the wet lay-up technique has not been as popular as laminate plates [8].

Corrosion and delamination in steel structures often lead to the thinning of the flanges in beams and are, therefore, two of the major reasons for the need to strengthen and replace steel structures. Corrosion can gradually weaken steel structures over time, affecting its load-bearing capacity and overall stability. Recent experiments have focused on the durability of FRPs and the FRP–steel bond under different environmental conditions, as well as the effects of corrosion on its durability. An experimental study of CFRP–steel double-strap joints studied under an accelerated corrosion environment showed chances of deterioration in CFRP, resulting in its rupture [9]. Additionally, a study on the influence of corrosion on the bond behavior of CFRP–steel joints showed that some degree of surface irregularity could maximize the bond behavior [10]; however, alternate cycles of dry–wet conditions could cause significant negative impacts on the bond behavior [11].

Additionally, web openings are typically introduced in steel beams to serve a variety of functional and operational demands, such as passage for services (ductwork, plumbing, electrical conduits, etc.), HVAC systems, and other architectural considerations, to achieve specific aesthetic goals. Previous attempts have been made to investigate different types and sizes of web openings, different spacing-to-diameter ratios, the optimal location of openings, different opening ratios and web post widths, and their effects on the overall load-carrying capacity and failure modes of the steel beams [12–16]. The overall load-carrying capacity and stiffness were found to decrease with the increase in the area of each web opening (or increase in the total web opening area), based on those studies. An increase in von Mises stress and shear stress was also observed with an increase in opening area [17]. In this study, the effect of a varying number of web openings of different sizes, with the total web opening area (sum of area of all openings) remaining constant, was investigated. Moreover, the studies that have been made towards the strengthening of steel beams with web openings are limited in number and the strengthening typically involves conventional techniques and materials. Compared to ordinary (homogeneous) beams with web openings, an increase in strength was observed in hybrid beams (made up of beams with web and flange plates of different strengths) [18]. Additionally, a study on strengthening techniques involving the use of steel stiffeners around web openings [19] and the application of steel flange cover plates at the flange bottom [20] demonstrated their effectiveness. The use of CFRP plates to strengthen simply supported steel I-beams with rectangular web openings was also studied experimentally and improvements in load-carrying capacity and stiffness were observed [21]. However, rectangular openings in steel I-beams may introduce higher stress concentrations around their perimeter, particularly near the corners, compared to circular ones, which could potentially limit the FRP's efficacy. In addition, given the complex behavior of steel I-beams with openings, which introduces challenges for reinforcement, the use of FRP sheets emerges as a more feasible option compared to FRP plates. Therefore, the use of FRP sheets in steel I-beams with circular instead of rectangular web openings and the optimization of strengthening layouts need to be investigated.

Based on the existing literature, studies on the efficacy of CFRP sheets on strengthening steel I-beams, damaged due to corrosion and cut-out web openings, are limited. Therefore, the present FEA study investigates the effect of corrosion on the flexural strength of as-built steel beams and the optimum number of CFRP layers required in restoring their strength. This study further examines the strengthening of steel beams with web openings introduced at the flexure-dominated region of the beam, near the mid-span, using CFRP sheets. Specifically, six different strengthening layouts, each considering various numbers of CFRP layers, were adopted to identify the optimal configurations.

2. Description of the Experiment Used for the Validation of FE Models

This section describes the experimental study by Bastani et al. [8] used in the validation of the developed FE models. Three FE models have been validated with their respective experimental tests. The experiments were performed on simply supported steel I-beams under four-point bending loading. The load was applied using a simply supported spreader beam with a span of 500 mm and transferred to the specimen using two steel plates [8]. The effective and total span lengths of the specimens were 1500 mm and 2000 mm, respectively. The first developed FE model was validated against the tested undamaged un-strengthened control steel beam (CV-00), as shown in Figure 1. The second FE beam model was developed based on specimen CD-20, in which the bottom flange was thinned out to simulate corrosion effects. The maximum thinning was 20% of the thickness of the bottom flange. The specimen was thinned out in a circular profile, such that the total length of damage was 100 mm, as shown in Figure 2. Lastly, the third tested beam taken for validation was RB-20-2L (Figure 3), representing the rehabilitated version of the CD-20 beam by using two layers of CFRP sheets at the bottom face of the bottom flange. Unidirectional CFRP sheets were oriented along the longitudinal axis of the beam. The CFRP sheets were soaked in epoxy resin and then attached to the bottom flange using a wet lay-up method. CFRP cross-wraps were used to prevent debonding between steel and longitudinal CFRP sheets. The material properties of steel and CFRP sheets used in the development of FE models are presented in Table 1.

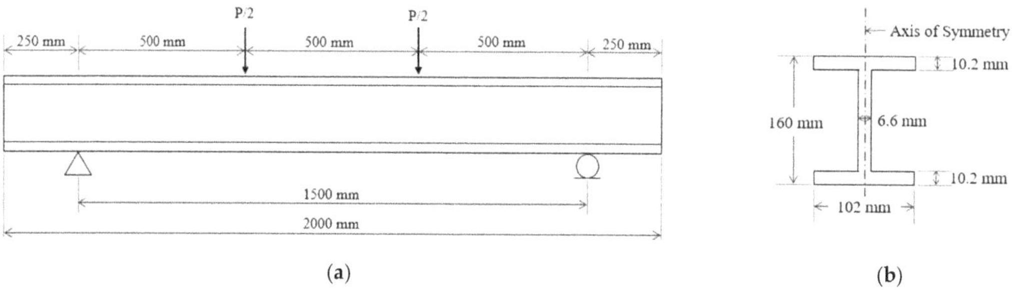

Figure 1. Beam CV-00 details and dimensions: (**a**) elevation; (**b**) cross section. (Data from Bastani et al., 2019 [8]).

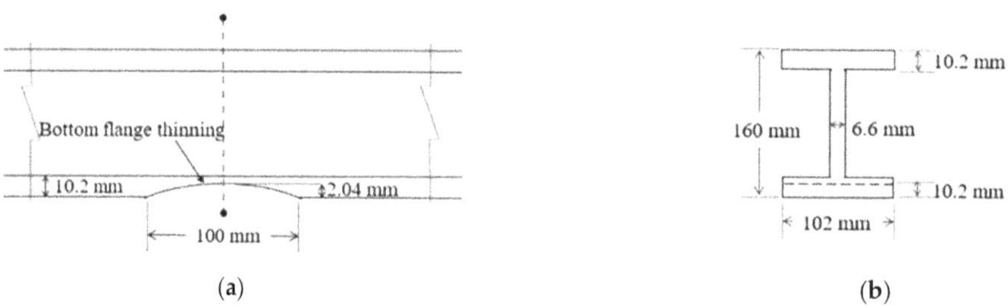

Figure 2. Beam CD-20 with 20% thinning: (**a**) elevation; (**b**) cross section. (Data from Bastani et al., 2019 [8]).

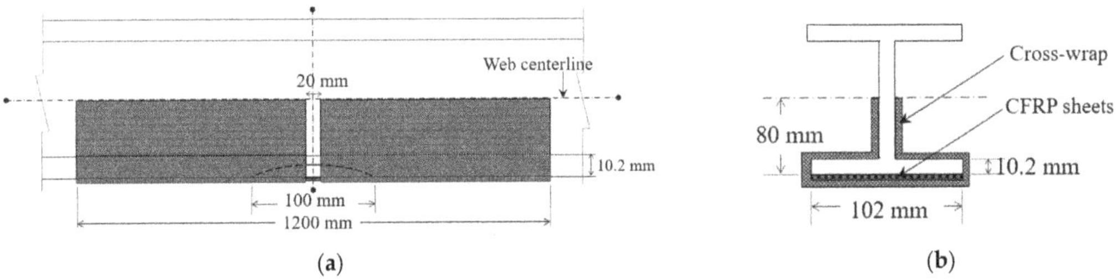

Figure 3. CFRP-strengthened beam (RB-20-2L): (**a**) elevation; (**b**) cross section. (Data from Bastani et al., 2019 [8]).

Table 1. Material properties of steel and CFRP sheets (Data from Bastani et al., 2019 [8]).

Material	Mechanical Properties	Symbol	Unit	Value
Steel	Yield stress	fy	MPa	379
	Ultimate stress	fu	MPa	484
	Yield strain	εy	%	0.2
	Ultimate strain	εu	%	35
	Elastic Modulus	E	GPa	205
	Poisson's ratio	μ	-	0.3
CFRP	Sheet thickness	-	mm	0.26
	Tensile strength in longitudinal direction (0°)	-	MPa	1780
	Tensile modulus in longitudinal direction (0°)	-	GPa	152
	Tensile elongation in longitudinal direction (90°)	-	%	1.75
	Tensile strength in transverse direction (90°)	-	MPa	60
	Tensile modulus in transverse direction (90°)	-	GPa	9
	Tensile elongation in transverse direction (90°)	-	%	0.5
	In-plane shear strength	-	MPa	100
	In-plane shear modulus	-	GPa	12

3. Finite Element Modeling Procedure

The numerical models were developed using a nonlinear FEA software program, ANSYS 2020 R2, which was used to model steel beams. FE models were developed in accordance with the geometry, material properties, loading, and boundary conditions of the experimental setup [8]. The steel beams, loading and supporting plates, and CFRP sheets were modeled using element type 'SOLID185', which is an eight-node 3D element having plasticity, hyperelasticity, stress stiffening, creep, large deflection, and large strain capabilities [22]. Both the elastic and plastic properties were assigned for steel.

To define linear elasticity, the modulus of elastic moduli and Poisson's ratio were assigned. Strain-hardening in steel after yielding was considered using ANSYS's multilinear-isotropic material model in which the yield and ultimate stress and strain values were assigned (Table 1). The resulting stress–strain curve for the steel exhibits a bilinear behavior. For the steel plates, only linear elastic behavior was assumed, with the modulus of elasticity 'E' and Poisson's ratio of the steel being assigned. On the other hand, CFRP sheets were modeled as a linear–elastic material up to its rupture using ANSYS's linear–orthotropic material model. The material properties of CFRP sheets are provided in Table 1. To define the linear–orthotropic material model in ANSYS, only the elastic moduli in three (x, y, and z) directions and Poisson's ratios and shear moduli in all three (xy, yz, and xz) planes were defined. The elastic modulus in the three directions (E_X, E_Y, and E_Z) were assigned as 152, 9, and 9 GPa, respectively. Poisson's ratios (μ_{XY}, μ_{YZ}, and μ_{XZ}) were all input as 0.3 and the shear moduli in all three planes (G_{XY}, G_{YZ}, and G_{XZ}) was assigned as 12 GPa. The tensile and shear strength of CFRP and the tensile elongations were used to identify when the failure is reached in the models.

CFRP layers were modeled as distinct solid layers and a perfect bond at the interlaminar interface between the layers was ensured by merging the coincident nodes. In addition, since debonding did not occur in test specimens, steel and CFRP sheets were connected in FE models, assuming a perfect bond by merging only the coincident nodes between the two surfaces, as in the experiment. This simplified method yields similar outcomes to the bonded surface-to-surface contact method that assumes no relative motion or slippage between the two materials at their interface. Similarly, the bond behavior between the steel beam and loading and supporting plates was assumed to be perfect.

The use of finer mesh improves the convergence of FE results and yields more accurate predictions [8]. Based on a mesh sensitivity analysis, the largest element size selected was 5 mm. When compared with the experimental results, the yield, ultimate loads, and failure mode of the FE models agreed well with experimental results. Additionally, the CFRP sheet used in this test was very thin (0.26 mm thickness). Therefore, the selection of 5 mm avoids an element aspect ratio exceeding ANSYS's warning limit of 20, to eliminate the distortion of irregular or thin elements.

Due to symmetry (refer to Figure 1b), only the symmetrical half of the beam was modeled. The developed FE model of the control beam (CV-00) is shown in Figure 4a. The experiment was carried out using simply supported boundary conditions. Therefore, restraints were assigned at the support locations in the model. To model a hinged support, restraints were assigned for all three directions, X, Y, and Z, and to model roller support, restraints were assigned only in the Y and Z directions for the line of nodes located at the location of supports. Likewise, the nodes along the plane of symmetry were constrained using roller supports in the perpendicular direction (Z), as shown in Figure 4b.

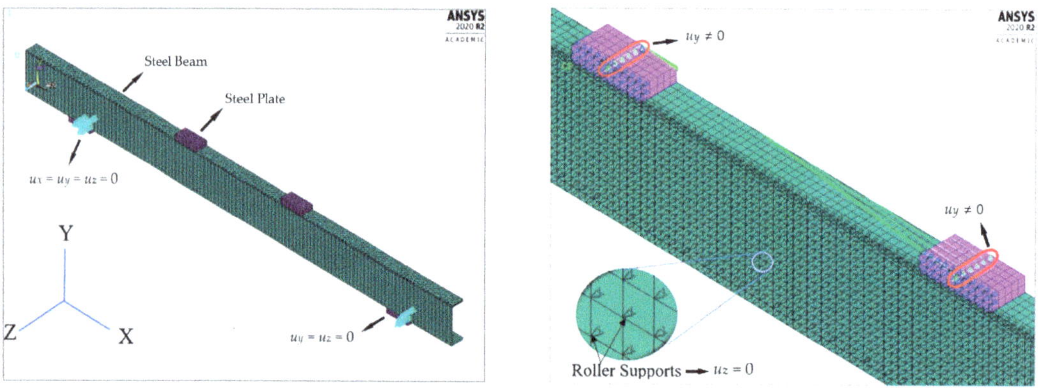

(a) (b)

Figure 4. (**a**) FE model of control beam (CV-00); (**b**) roller supports perpendicular to axis of symmetry.

In the experimental study [8], the loading of the beams was carried out monotonically, using the displacement control method. The loads were applied to the steel plates at the top flange of the beams to avoid any stress concentration and to ensure a smooth load transfer between the nodes of the plates and the beam. To achieve convergence and to avoid premature failure of the beams, a displacement-controlled analysis was implemented by gradually applying a vertical displacement on the top loading plates. To ensure the same displacement would be applied at the corresponding loading locations, the nodes at the location of application of point loads were tied together in the Y direction using the "coupling" feature of the ANSYS preprocessor, so that all the coupled nodes would displace vertically by the same amount.

4. Validation of Developed FE Beam Models

The developed finite element models were validated against their experimental counterparts. The models were validated both at the yield and failure stages. Figures 5–7 show load versus mid-span displacement curves of the validated beam models. The FEA and experimental results are compared to each other and are in good agreement, with less than a 4% difference, as shown in Table 2.

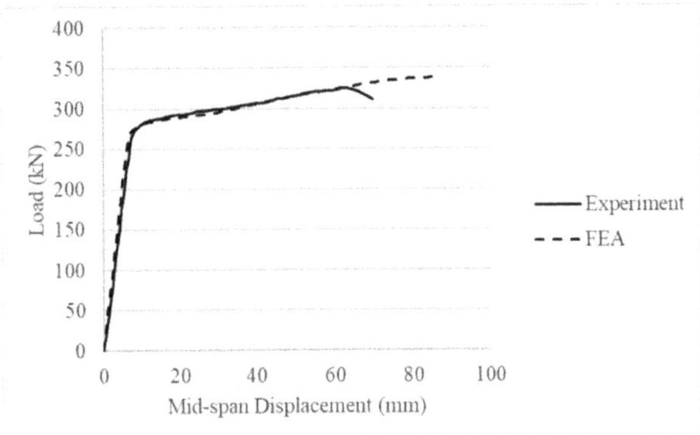

Figure 5. Load versus mid-span displacement curve for CV-00.

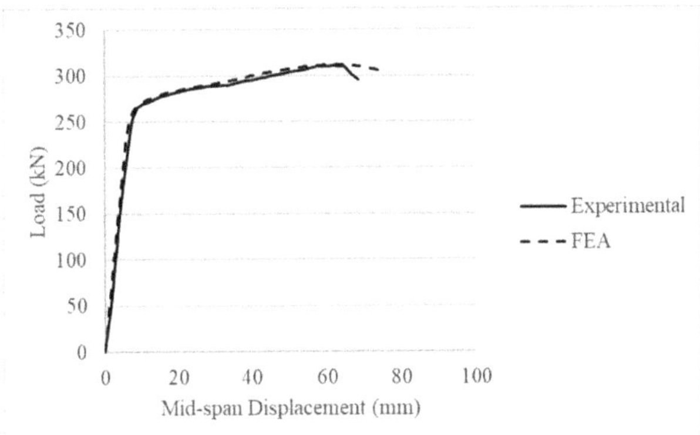

Figure 6. Load versus mid-span displacement curve for CD-20.

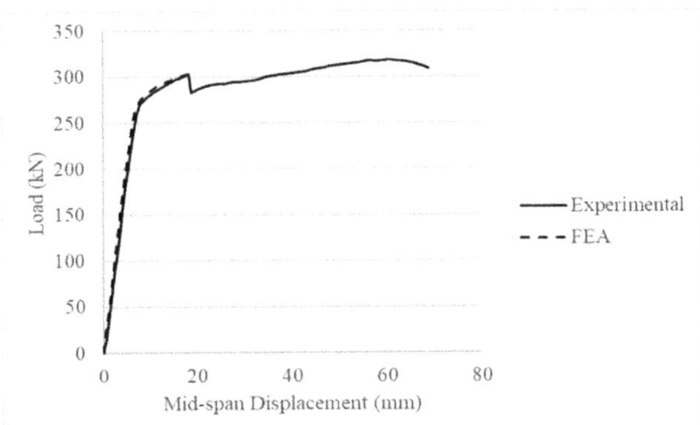

Figure 7. Load versus mid-span displacement curve for RB-20-2L.

Table 2. Comparison of experimental and FEA results.

	Yield Load (kN)			Ultimate Load (kN)			CFRP Rupture Load (kN)		
Beam ID	Experiment	FEA	Difference (%)	Experiment	FEA	Difference (%)	Experiment	FEA	Difference (%)
CV-00	273	272	0.4	325	338	4	-	-	-
CD-20	255	256	0.4	312	312	0	-	-	-
RB-20-2L	270	271	0.4	-	-	-	303	304	0.3

As indicated in the previous section, CV-00 is the control (undamaged) beam and '00' represents no (0%) damage; CD-20 is the control damaged beam and '20' represents 20% damage (maximum 20% thickness loss of the bottom flange); RB-20-2L represents the rehabilitated beam, which had a maximum 20% thickness loss with 2 layers of CFRP sheets [8].

FE Beam Models Failure Behavior

The yield load was obtained in the experiment [8] when the beam specimen reached the yield strain value of 0.2%, which was recorded using the strain gauge mounted at the mid-span of bottom flange. In the developed FE models, the beam was considered to have yielded after the strain value of 0.002 was reached at the flange gauge location in the experiment. The corresponding load at the same time step was the yield load of the FE beam models. The ultimate load in the experimental study was taken to be the maximum load that the beam could carry before the load started to drop as the applied displacement increased. In the cases of the developed FE models of the control undamaged beam, CV-00, and the control damaged beam, CD-20, the load versus mid-span deflection data points were extracted from ANSYS and plotted using a spreadsheet. The load was observed to reach a certain maximum value (the ultimate load) and was then gradually decreased, as the mid-span displacement increased. Likewise, the observed load versus mid-span deflection of the CFRP-strengthened beam, RB-20-2L, in the experiment could be divided into five segments [8]. Among those segments, the sudden vertical drop is of interest and is caused by the rupture of CFRP sheets. This sudden vertical drop is due to the brittle nature of CFRP sheets, whose rupture is abrupt and without warning [8]. After the rupture of the CFRP sheets, the load versus deflection curve follows a similar trend to that of the un-strengthened beam [8]. In the developed FE model of the strengthened beam, RB-20-2L, the CFRP rupture point was regarded as the failure point and only the behavior up to the CFRP rupture point was validated, as shown in Figure 7. Modeling the behavior of FRP-strengthened steel beams after a load drop in ANSYS is highly challenging due to the abrupt change in load distribution and the potential for localized damage or failure. The CFRP rupture point was identified in the FEA when the maximum stress in the CFRP sheet reached its tensile strength of 1780 MPa.

5. Parametric Study Results

After the validation of FE models, various scenarios such as corrosion damage and web opening cutout were considered, to study the behavior of steel beams and the efficacy of CFRP sheets in restoring the load capacities of the altered beams. To strengthen the beams subjected to corrosion damage, CFRP sheets were applied to their bottom flanges, whereas six CFRP strengthening configurations were studied in detail to determine the efficient strengthening configuration to strengthen steel beams with web openings (SBWOs). The subsequent section presents the parametric study carried out on the validated FE model of the control beam, 'CV00', which is the undamaged steel beam. Other beams, 'CD-20' and 'RB-20-2L', were validated to enhance the accuracy of developed FE models.

5.1. Effect of Corrosion and CFRP Strengthening

The effect of corrosion in steel beams was incorporated in the FE model by reducing the thickness of the bottom flange. The study was conducted in five distinct levels of corrosion damage, i.e., 2.5%, 5%, 10%, 15%, and 20% thinning of the bottom flange, corresponding to 0.26 mm, 0.52 mm, 1.04 mm, 1.56 mm, and 2.08 mm thinning, respectively. The thinning was introduced to 80% of the beam effective span (refer to Figure 8) to emulate non-localized corrosion cases and maximize the effects due to thinning.

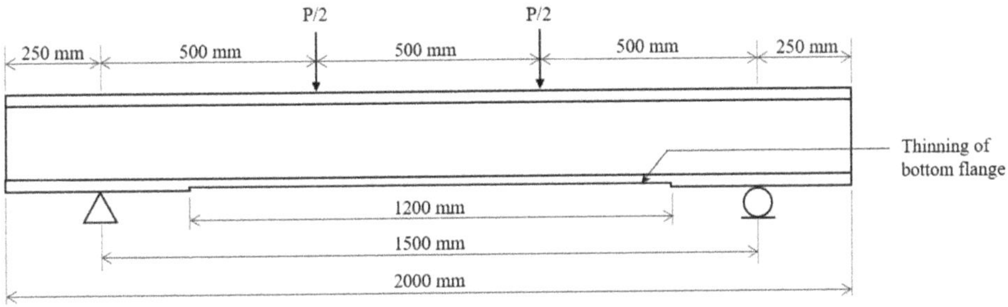

Figure 8. Beam subjected to thinning at the bottom flange.

The FE beam models are named to indicate the damage level and number of layers of CFRP sheets used for rehabilitation. The model ID is divided into two parts, using a dashed line if the model is a strengthened beam. The first and second parts specify the type of damage and number of layers of CFRP sheets, respectively. In the first part, the letter 'D' indicates that the beam is damaged. The number following the letter 'D' indicates the level of corrosion damage. In the second part, the first letter 'R' indicates that the beam has been rehabilitated and the remaining letters 'nL' refers to the number of layers of CFRP sheets used for strengthening. The beams without rehabilitation do not have the second part in their IDs.

The effect of thickness loss of the bottom flange on the yield and ultimate load of the beam is tabulated in Table 3. The load versus mid-span displacement diagram is shown in Figure 9. As the level of bottom flange damage (thickness loss) increased, the reduction in the yield and the ultimate load also increased. Both the reduction in yield and ultimate loads followed a linear trend, as shown in Figure 10.

Table 3. Effect of bottom flange thinning.

Beam ID	Yield Load (kN)	Reduction in Yield Load Capacity (%)	Ultimate Load (kN)	Reduction in Ultimate Load Capacity (%)
D2.5	272	0	333	1.5
D5	268	1.5	328	3.0
D10	261	4.0	317	6.2
D15	256	5.9	307	9.2
D20	249	8.5	296	12.4

Note: reductions in the load capacities are obtained by comparing the undamaged beam (CV-00) to the damaged ones.

To strengthen the damaged beams, successive layers of CFRP sheets were attached to the bottom face of the bottom flange, covering the damaged (thinned out) area, as shown in Figure 11. The unidirectional CFRP sheets were oriented along the longitudinal axis of the beam.

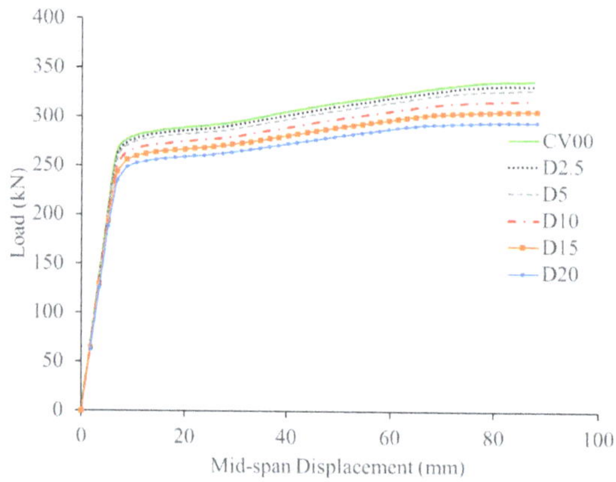

Figure 9. Load versus mid-span displacement behavior of corroded steel beams.

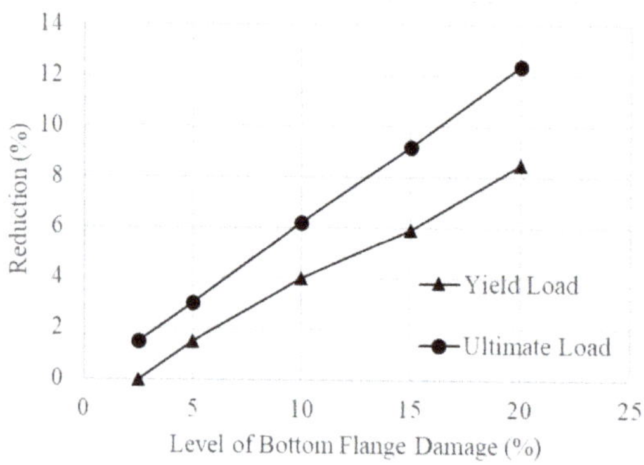

Figure 10. Reduction in load capacities of corroded steel beams.

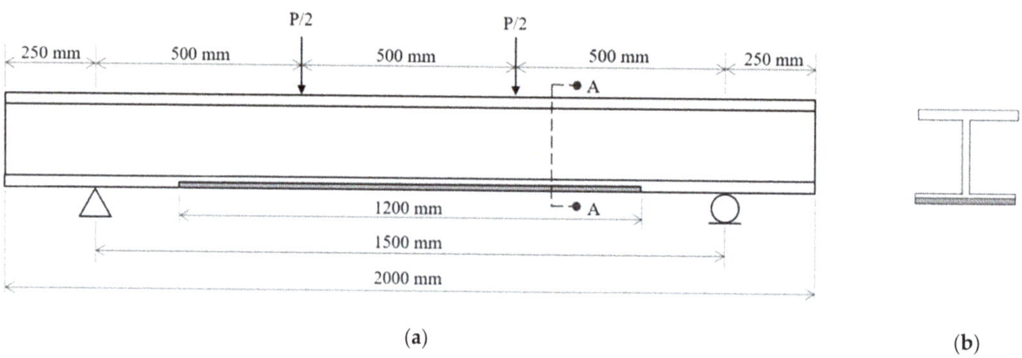

Figure 11. CFRP strengthening in beams subjected to thinning of bottom flange: (**a**) elevation; (**b**) section A-A.

Seventeen FE models were developed for different levels of corrosion damage. The initial number of layers of CFRP sheets for each level of corrosion damage was based on the number of layers required by the previous level of corrosion damage and the layers were added successively until both the target yield and ultimate capacities were reached. The observed yield and the failure loads for each FE model are shown in Table 4. The failure load was reached when the CFRP sheets began to rupture. Debonding was prevented in the experiment using cross-wraps [8]. Therefore, the debonding of the CFRP sheets was not allowed in the FE models.

Table 4. Yield and failure loads after CFRP strengthening in FE beam models subjected to thinning of the bottom flange.

	Yield Load (kN)		Failure Load (kN)	
Beam ID	**Target Value**	**FEA Results**	**Target Value**	**FEA Results**
D2.5-R1L D2.5-R2L D2.5-R3L	273	275 279 282	325	302 319 336
D5-R1L D5-R2L D5-R3L	273	272 276 279	325	299 316 331
D10-R3L D10-R4L	273	273 276	325	325 342
D15-R3L D15-R4L D15-R5L D15-R6L	273	266 270 273 276	325	317 333 350 365
D20-R4L D20-R5L D20-R6L D20-R7L D20-R8L	273	251 267 270 273 276	325	325 342 357 373 387

The study focused on the optimum number of layers of CFRP sheets needed to restore the yield and the ultimate load capacities of the damaged beam to the original undamaged state (to that of CV-00). By doing so, an improvement in the serviceability requirements in terms of reduction in the mid-span deflection due to the application of CFRP was also observed.

For the beam with damage level of 2.5% (D2.5 beam), one layer of CFRP strengthening would be sufficient to restore the yield load capacity; whereas, three layers were required to restore the ultimate load capacity to that of the undamaged control beam 'CV00' (Table 4). Similarly, for the 'D5' beam, two layers of CFRP sheets were sufficient to restore the yield load capacity; however, three layers were needed to restore the ultimate load capacity. For the 'D10' beam, three layers of CFRP sheets were just sufficient to reach the yield and ultimate load capacities and four layers were needed to exceed those capacities. For the 'D15' beam, four layers were sufficient to restore the ultimate load capacity, whereas six layers were needed to exceed the target yield load capacity. For the 'D20' beam, five layers were needed to exceed the target ultimate load. However, eight layers were needed to exceed the target yield load.

It is also observed that the failure load of the 'D2.5-R1L', 'D2.5-R2L', 'D5-R1L', and 'D5-R2L' strengthened beam models (Table 4), which correspond to CFRP rupture, is less than the ultimate load of the damaged beams 'D2.5' and 'D5' (Table 3). It may give the impression that the strengthening worked to the opposite effect. However, the load versus mid-span deflection curves of these beams shows a considerable improvement in the stiffness of the inelastic regime and the reduction in the mid-span deflection, due to strengthening. Consequently, the lower number of CFRP sheet layers in these beams was subjected to higher levels of stresses, due to which they ruptured at lower load levels.

Figure 12 shows the load versus mid-span deflection curves for all FE beam models. It is observed that a higher number of CFRP layers resulted in a reduced mid-span deflection and improved stiffness and the adequate number of layers of CFRP sheets that depends upon the level of damage, as presented in Table 4, could fully restore the yield and ultimate load capacities of corroded beams to the level of the undamaged beam.

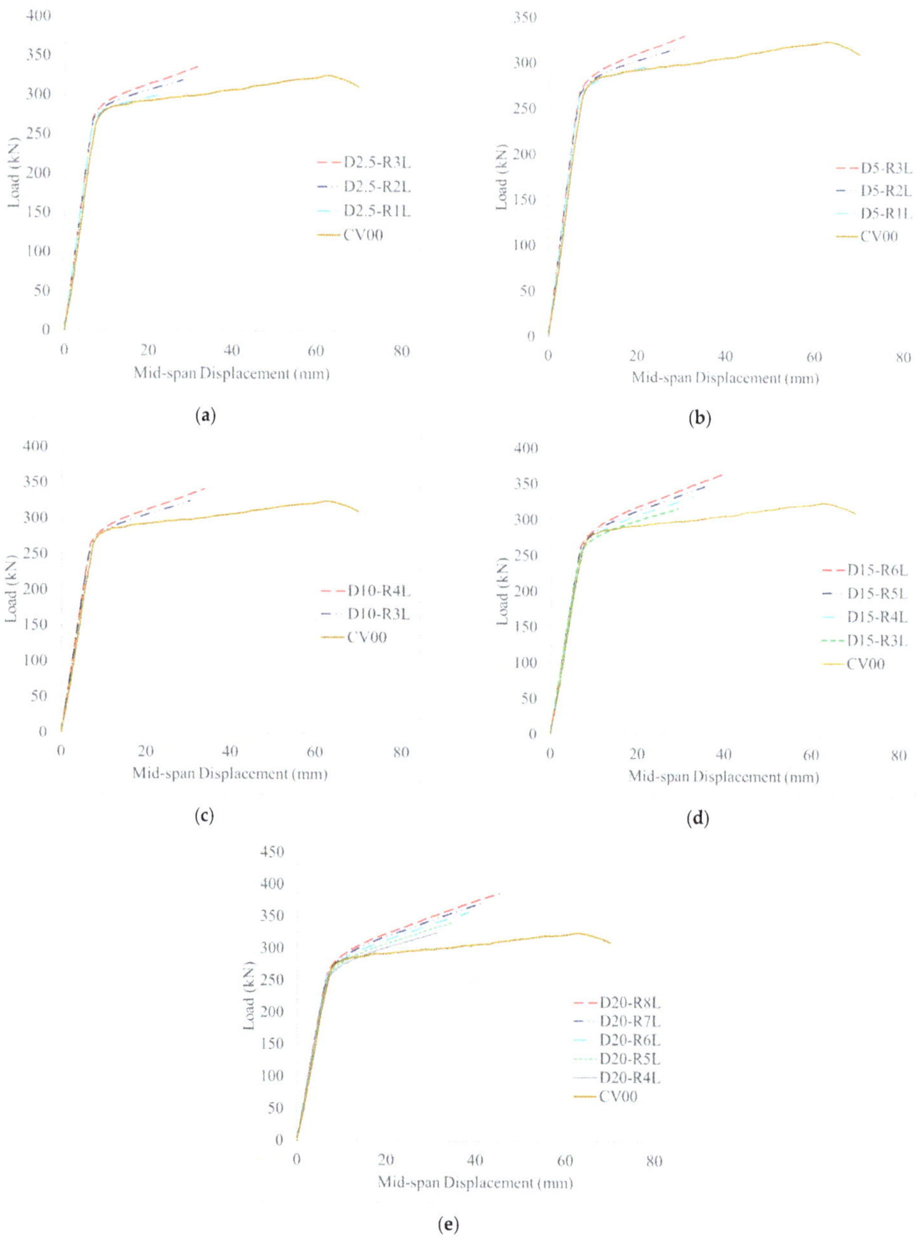

Figure 12. Load versus mid-span displacement behavior of undamaged (CV-00) and strengthened beams with: (**a**) 2.5% damage; (**b**) 5% damage; (**c**) 10% damage; (**d**) 15% damage; and (**e**) 20% damage.

5.2. Effect of Web Openings

Past studies [12,13,16] have shown that the performance of steel beams with circular web openings is superior to that of other opening shapes. For the same opening area, steel beams with circular web openings had a higher ultimate load capacity and lower stress concentrations around their edges. Therefore, circular openings were considered in the present study. Since the bending stresses are at a maximum at the extremities and the shear stress is the highest at the neutral axis of the cross-section, the flanges and the web should be able to resist high bending stresses and high shear, respectively Therefore, it is prudent to provide openings in the web (where shear stress is at a maximum) in the regions of low shear forces. One study [14] concluded that the openings are to be provided in the bending predominant region instead of in the shear predominant region, and the optimal location of web openings is the middle two-thirds of the beam's span, which is a bending predominant region.

Under the four-point bending load, the middle-third span of the beam is subjected to pure bending and there is no shear force. Therefore, openings were introduced at this location of the beam, as shown in Figure 13. Design guidelines across the globe have provisions and limits on the dimension and location of openings, spacing between openings, etc. Therefore, an offset from point load 'C' and clear spacing between openings 'S', in this study, were chosen based on the SCI P355 [23], which suggests generalized design methods for discrete or closely spaced web openings.

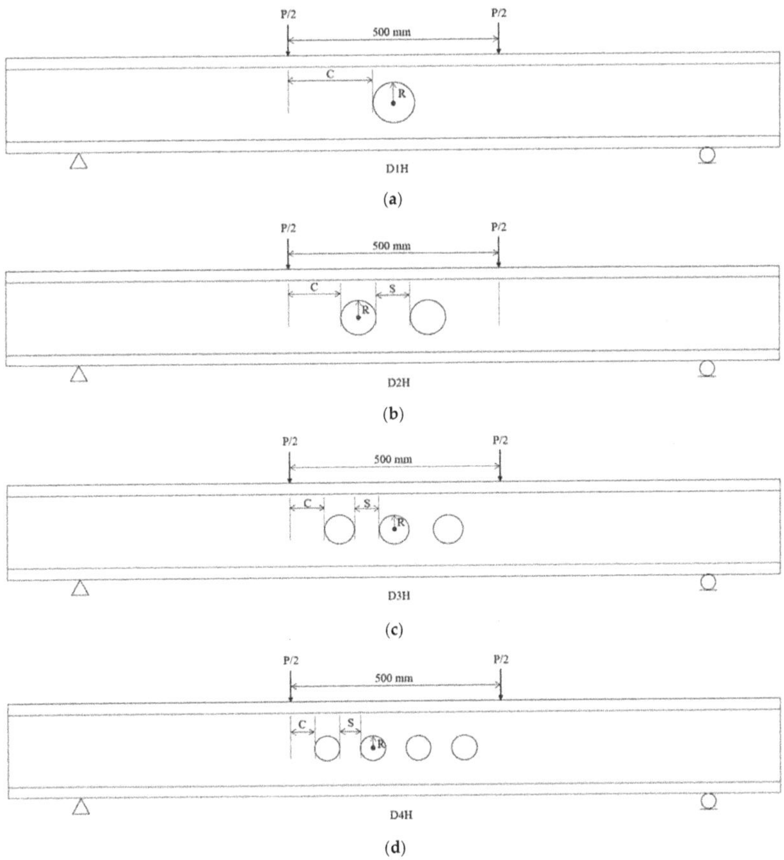

Figure 13. Steel beams with circular web openings (SBWOs) with: (**a**) one opening; (**b**) two openings; (**c**) three openings; and (**d**) four openings.

In total, four FE beam models with different sets of web openings were studied, with the number of openings ranging from one to four, as shown in Figure 13. The total area of openings in all the beams was kept nearly constant. The details and dimensions of the beam models are described in Table 5.

Table 5. Details and dimensions of web openings shown in Figure 13.

Beam ID	'C' (mm)	'S' (mm)	'R' (mm)	'N'	'Ae' (mm^2)	'A' (mm^2)	Area Ratio	Opening Ratio
D1H	193.5	-	56.5	1	10029	10029	1.000	0.71
D2H	80	180	40	2	5027	10053	1.002	0.50
D3H	80	71	33	3	3421	10264	1.023	0.41
D4H	80.5	37	28.5	4	2552	10207	1.018	0.36

Note: 'C' = offset from point load; 'S' = clear spacing between openings; 'R' = radius of each opening; 'N' = number of openings; 'Ae' = area of each opening; 'A' = total area of openings; 'Area Ratio' = Total area of openings of each model (A) divided by total opening's area of D1H; and 'Opening Ratio' = depth (diameter) of opening divided by depth of beam.

The nomenclature of the beams is such that the first character 'D' indicates the damage induced by the web opening and the remaining characters, 'nH', indicate 'n' number of circular web openings or holes in the beam. The load versus mid-span deflection curves for the FE models are shown in Figure 14. The maximum load attained is the ultimate failure load, after which, the load started to drop upon further increase in applied displacement.

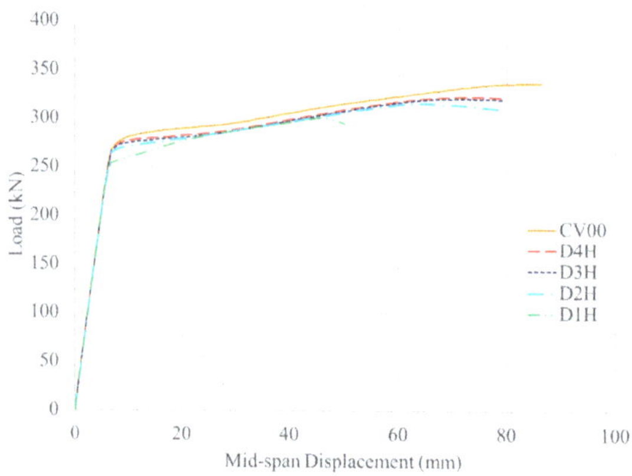

Figure 14. Load versus mid-span displacement behavior of SBWOs.

To obtain the reduction in the load capacities, as tabulated in Table 6, the beam models with openings (D1H, D2H, D3H, and D4H) were compared to that of the undamaged beam (CV-00).

Table 6. Effects of introduction of web openings.

Beam ID	Reduction in Yield Load Capacity (%)	Reduction in Ultimate Load Capacity (%)
D1H	6.6	10.9
D2H	1.1	6.5
D3H	0	5
D4H	0	4.4

For SBWOs with an opening ratio <0.5 (D3H and D4H), the maximum reduction in ultimate load capacity was 5% and there was no effect on the yield load capacity. However, for model D1H with an opening ratio >0.5, the reduction in the ultimate and yield load capacities was the highest. In general, it was observed that beams with a higher number of circular web openings and smaller sizes had higher load capacities, as compared to those with a smaller number of bigger openings, if the total area of the openings was kept constant.

Based on previous studies [15,21], commonly observed global failure modes in SBWOs are flexural, Vierendeel mechanism, and web-post buckling. The flexural failure mode, also known as the flexural mechanism, is predominant in the region with high bending moments, as it could lead to yielding of T-regions below and above the web openings and the consequent formation of plastic hinges. The Vierendeel mechanism is a failure due to the formation of plastic hinges at the corners of the T-regions above and below the openings, causing them to deform. It is observed in the region with high shear forces. Web-post buckling generally occurs when the openings are closely spaced without sufficient clear spacing between the adjacent openings. Since the openings were introduced in the middle-third of the beam span, where it was subjected to high bending moments and low shear, the flexural failure mode was anticipated in all four beam models, D1H, D2H, D3H, and D4H. The stress diagrams of the beam models at the ultimate load are shown in Figure 15. The stress contour shows that both the top and bottom Ts at the opening locations had yielded as the average stress in those T-regions were higher than the yield stress. Therefore, it was confirmed that the governing mode of failure was the flexural failure mode.

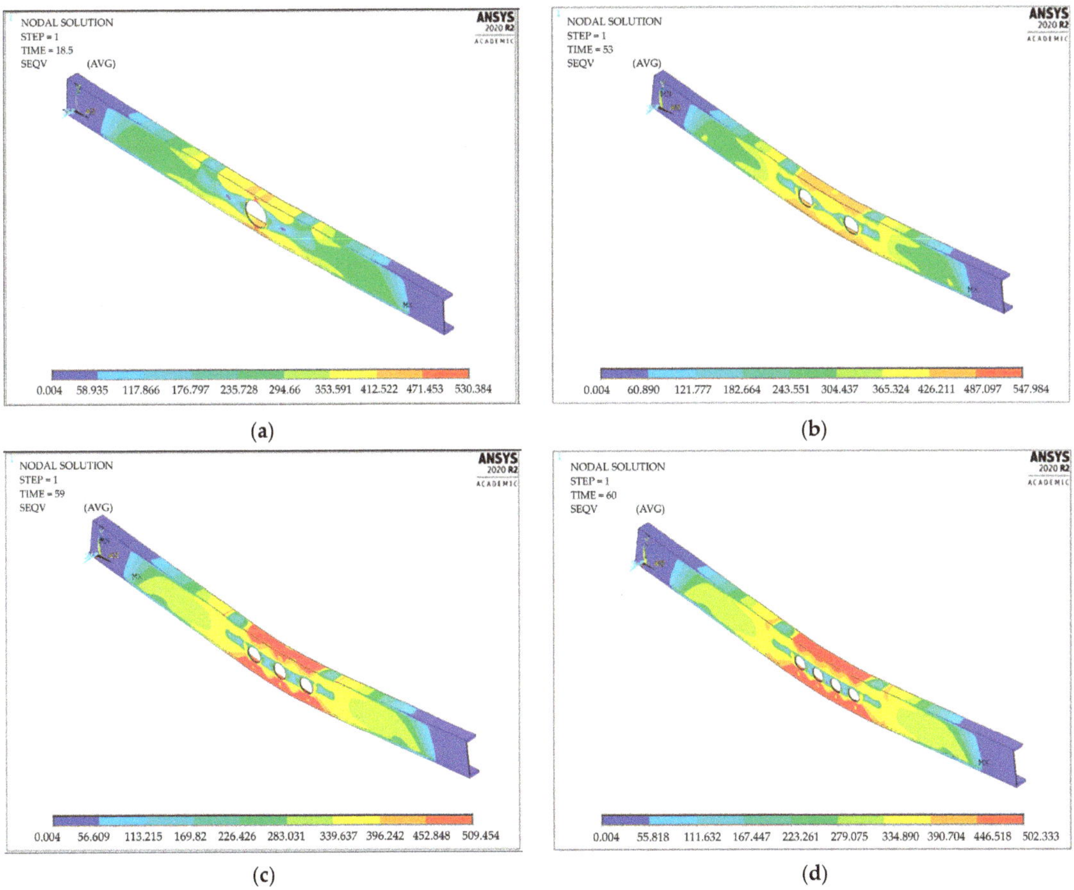

Figure 15. von Mises stress distribution at ultimate stages for SBWOs: (**a**) D1H; (**b**) D2H; (**c**) D3H; and (**d**) D4H.

The maximum ultimate load reduction of 10.9% (refer to Table 6) occurred in model D1H, which had an opening ratio of 0.71 (refer to Table 5). The area of the web opening was equal to 4.8% of the total area of the web, along the span of 1500 mm.

Most available design guidelines, including SCI P355 [23], allow for a maximum opening ratio of 0.8. Therefore, another model—D1H*, with maximum permissible opening ratio of 0.8—was developed. Figure 16 shows the dimensions and location of the opening. The load versus mid-span deflection curve for D1H* is shown in Figure 17, which is compared against the steel beam without an opening (CV00).

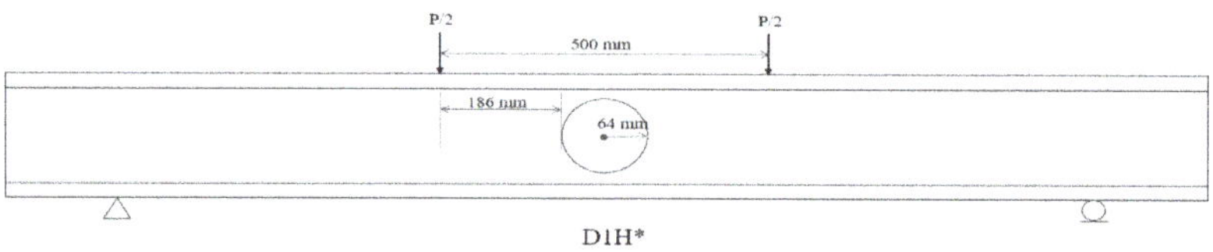

Figure 16. Steel beam with maximum opening ratio.

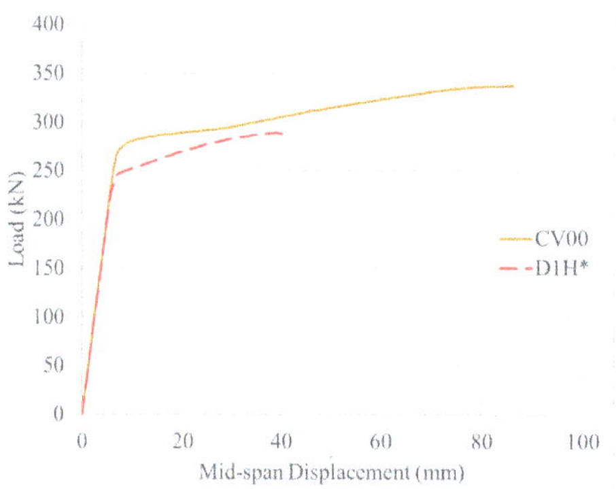

Figure 17. Load versus mid-span deflection curve of beam model D1H*.

Compared to the control undamaged beam (CV00), reductions of 9.6% and 14.5% in the yield and ultimate load capacities were observed for D1H*. The stress distribution diagram of D1H* at the ultimate stage is shown in Figure 18. The stress on the top and bottom Ts exceeded yield stress and were close to ultimate strength of steel. Therefore, the governing failure mode for D1H* was the flexural mechanism.

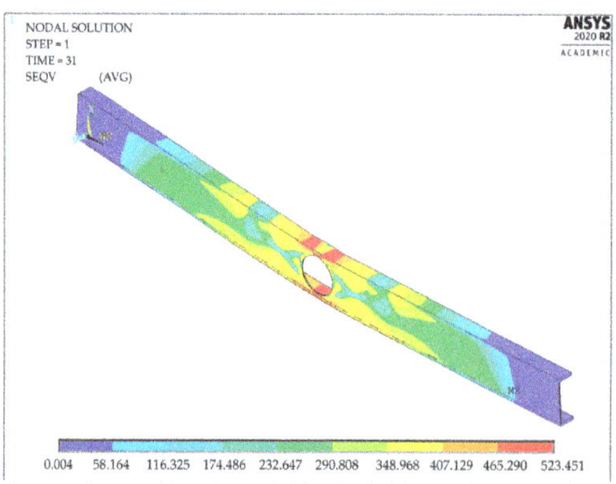

Figure 18. von Mises stress distribution of D1H* at ultimate load stage.

5.3. CFRP Strengthening of Steel Beams with Openings

The use of FRP is generally more efficient in strengthening weak beams compared to strong beams. Therefore, the beam model D1H*, with the highest loss of load capacity due to opening, was selected for CFRP strengthening. Six different CFRP strengthening configurations (schemes) were examined. The mode of application of CFRP sheets for each scheme is shown in Figure 19, where the shaded blocks represent CFRP sheets. The first strengthening layout was selected based on the failure of the D1H* beam, which failed in the flexural mechanism, with high stresses observed at the bottom and top flanges. Consequently, each strengthening arrangement was selected based on the predicted failure or limitation of the previous configuration. The aim of adopting various strengthening arrangements is to determine the most efficient strengthening configuration to strengthen steel beams with web openings. The specifics of each strengthening scenario are discussed in detail in the subsequent paragraphs. It is noted that all CFRP sheets were oriented along the longitudinal axis of the beam, including the ones applied to the beam's web.

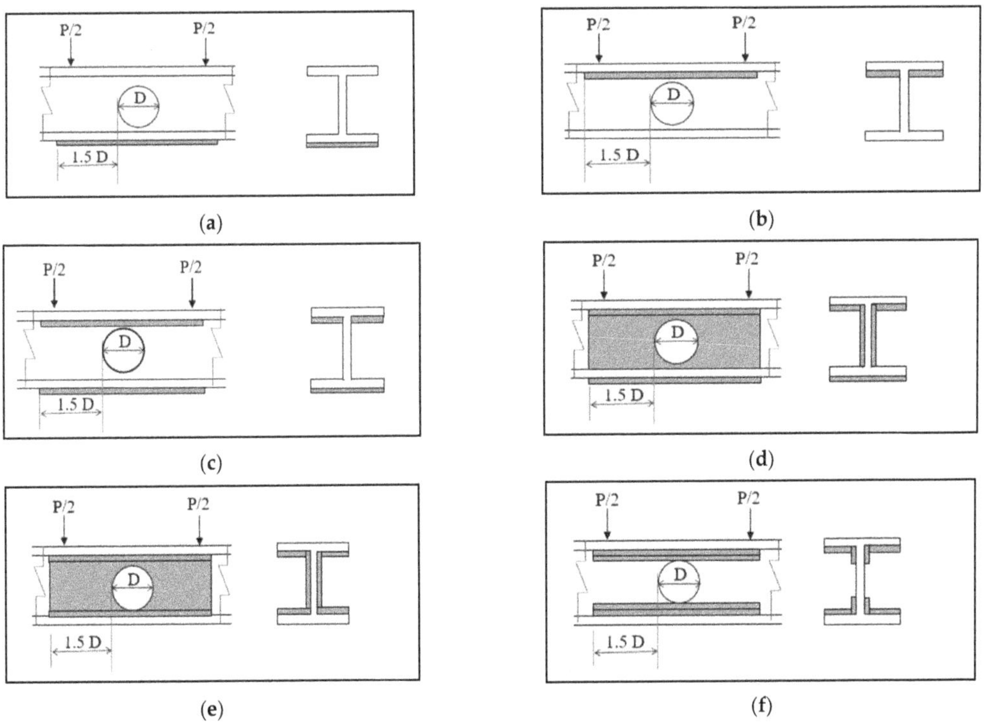

Figure 19. CFRP-strengthening of SBWOs: (**a**) Scheme-I; (**b**) Scheme-II; (**c**) Scheme-III; (**d**) Scheme-IV; (**e**) Scheme-V; and (**f**) Scheme-VI.

The first strengthening strategy, 'Scheme-I' was based on the application of CFRP sheets on the bottom of the beam's bottom flange, as shown in Figure 19a. The total length of the CFRP sheet was taken as four times the size of the opening. Therefore, the CFRP sheets were extended beyond the edge of the opening, by 1.5 times the opening size, from both sides, as shown in Figure 19. The selection of this length was based on a previous study [24] on the optimal length of CFRP to strengthen steel beams with openings. The analysis was carried out starting with one CFRP sheet layer at the bottom flange and more layers were added successively to study their effects on the structural performance of the beam. Beams strengthened with a lower number of layers of CFRP sheets failed due to the CFRP rupture at lower load levels. As the number of layers of CFRP sheets increased, the mode of failure changed from CFRP rupture to plastic hinge formation at the top T-region, as shown in Figure 20. CFRP application at the bottom flange strengthened the bottom T-region and improved its stiffness. However, the un-strengthened top T-region, being subjected to higher flexural compressive stresses, failed in the observed manner.

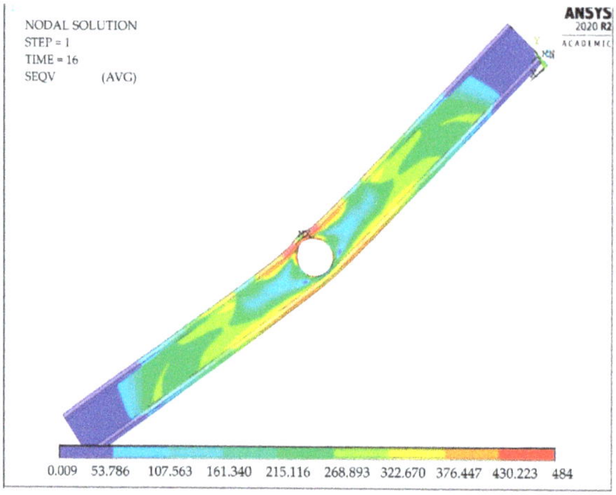

Figure 20. von Mises stress distribution at ultimate stage of strengthened D1H* beam using Scheme-I.

It was also observed that a plateau was reached and any additional layer beyond four layers had a negligible increase in the ultimate load of the beam. Therefore, Scheme-I was ineffective in strengthening SBWOs, since a plastic hinge was formed in the top T-region before the target load was reached for any additional layers of CFRP sheets.

The second strategy 'Scheme-II' was studied with the intent to address the limitation of the first scheme. Therefore, CFRP sheets were applied only to the bottom of the top flange, as shown in Figure 19b. The application of FRPs to the beam's top flange offers several advantages, such as increasing the beam's load capacity and stiffness, enhancing the top flange's resistance to local buckling and providing tensile strength during reversed loading situations, like those encountered by beams in bridges [21]. The CFRP sheet layers were successively increased and the model was analyzed. Scheme-II was also found to be ineffective. The reason was that the bottom T-region around the opening yielded and the plastic hinge was formed, as shown in Figure 21. Moreover, the application of additional layers of CFRP sheets also did not play a role in achieving the target load, due to the prior bottom T-region failure.

In the third strengthening configuration, 'Scheme-III', CFRP sheets were applied to both the top and bottom flanges, as shown in Figure 19c. This strengthening scenario was adopted after the first two schemes proved ineffective in restoring the strength of the undamaged beam. The beam performed better than the previous cases, in terms of improving both the strength and stiffness, resulting in lower stress levels at both the top and bottom flanges at failure. However, localized stress concentrations were observed at the top and bottom T-regions (Figure 22) adjacent to the openings, prior to reaching the desired load level. This indicated that the strengthening of the flanges alone was not sufficient and needed to be extended to the web as well. Therefore, the fourth strengthening configuration, 'Scheme-IV', was considered next.

In 'Scheme-IV', CFRP sheets were applied to the bottom of both flanges, as well as on both sides of the web, as shown in Figure 19d. An equal number of layers of CFRP sheets were added successively. With six layers of CFRP sheets below the top and bottom flanges and on both sides of the web, this scheme was able to strengthen D1H* successfully, which sustained a load of 338 kN prior to failure (CFRP rupture). The stress distribution on the steel beam at failure for 'Scheme-IV' is shown in Figure 23. It can be seen from the figure that the excessive stress was shifted away from the critical region near the opening toward the end of the CFRP sheet. This resulted in a higher load capacity and a more even stress distribution at the mid-span.

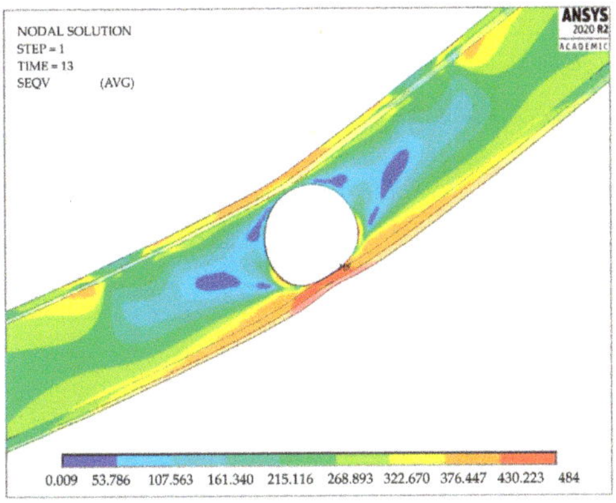

Figure 21. von Mises stress distribution at ultimate stage of strengthened D1H* beam using Scheme-II.

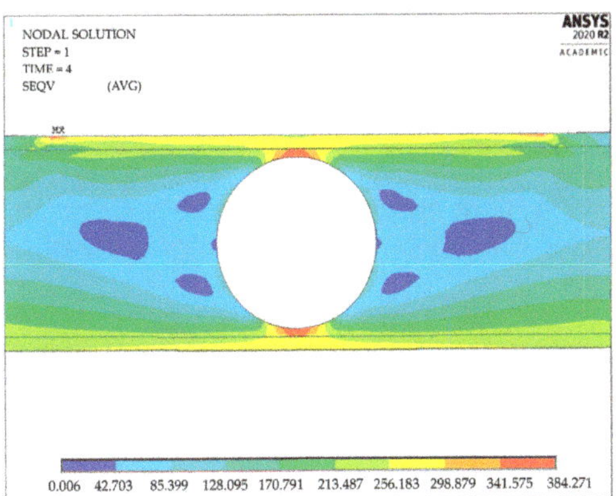

Figure 22. von Mises stress distribution of strengthened D1H* beam using Scheme-III.

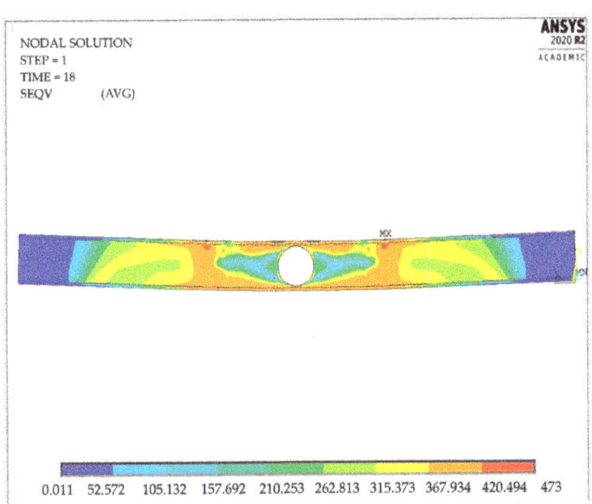

Figure 23. von Mises stress distribution at ultimate stage of strengthened D1H* beam using Scheme-IV.

As an alternative to 'Scheme-IV', another strengthening configuration, 'Scheme-V', was also studied. CFRP sheets were applied to the bottom of the top flange, to the top of the bottom flange, and on both sides of the web, as shown in Figure 19e. Like 'Scheme-IV', this strategy was also able to strengthen D1H* successfully with six CFRP sheet layers, which sustained a load of 332 kN prior to failure. The stress distribution on the steel beam at failure for 'Scheme-V' is shown in Figure 24.

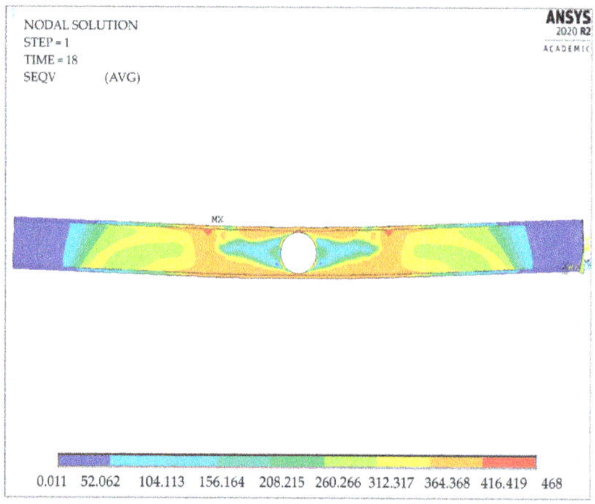

Figure 24. von Mises stress distribution at ultimate stage of strengthened D1H* beam using Scheme-V.

For the application of strengthening scheme-IV, it can be noted that the CFRP sheet must be cut into various pieces. To avoid such cuts, in the scheme-V strengthening configuration, continuous CFRP sheets were used on each side of the web and were extended to the total width of the flange as one piece. The difference in the failure load observed was negligible between the two schemes (338 kN and 332 kN). Additionally, the stress distribution in the CFRP sheets alone, as used in the 'Scheme-V' strengthening configuration, is shown in Figure 25, which indicates a lower stress level in the CFRP sheets towards the middle portion of the web. The reason behind such an observation was because the opening was in the bending-dominant low-shear region and the middle portion of the web in such a region would be subjected to a lower stress level. Thus, to economize 'Scheme-V', a final strengthening configuration of 'Scheme-VI' was proposed.

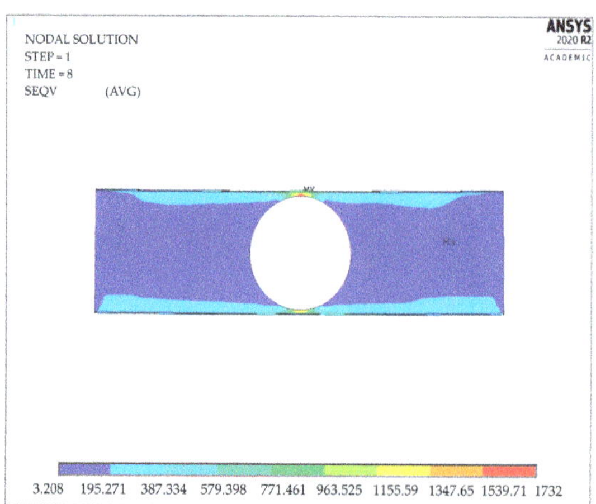

Figure 25. von Mises stress distribution in CFRP sheets (D1H* strengthened using scheme-V).

In 'Scheme-VI', CFRP sheets were applied on the bottom of the top flange, the top of the bottom flange, and on both sides of the web, covering only the top and bottom T-regions of the beam, as shown in Figure 19f. An equal number of layers of CFRP sheets was added successively. As anticipated, this scheme was also able to strengthen D1H* successfully with six CFRP sheet layers, which sustained a load of 331 kN prior to failure. The stress distribution of the strengthened beam at the ultimate stage for 'Scheme-VI' is shown in Figure 26, which closely resembles that of 'Scheme-V' (depicted in Figure 24), showing minimal differences.

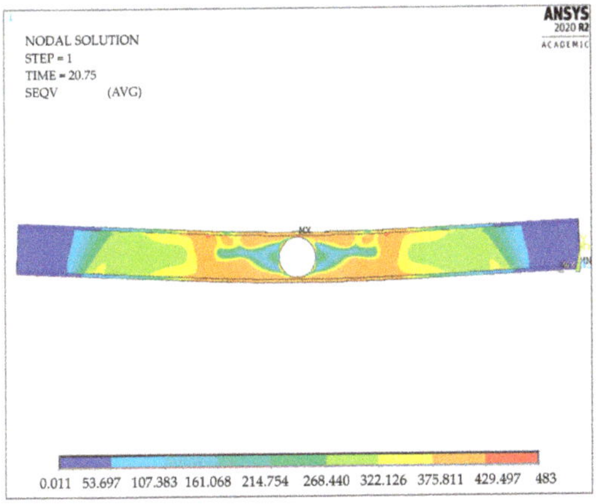

Figure 26. von Mises stress distribution at ultimate stage of strengthened D1H* beam using Scheme-VI.

In conclusion, three strengthening configurations, Scheme-IV, V, and VI, were able to strengthen D1H* to the desired level of strength. In all these strategies, a minimum of six CFRP sheet layers were needed. The load versus mid-span deflection diagram for D1H*, with these strengthening configurations, using six CFRP layers is shown in Figure 27. When compared with the control (CV-00) and damaged D1H* beam models, the strengthened beams had considerable enhancement in the stiffness in the inelastic regime after the yield point is reached. On the other hand, all strengthened models had an almost similar behavior with a slightly higher inelastic stiffness improvement, as the beam strengthened with scheme-IV. Based on the above observations, scheme-VI was found to be more efficient due to the lower material usage and greater cost-effectiveness.

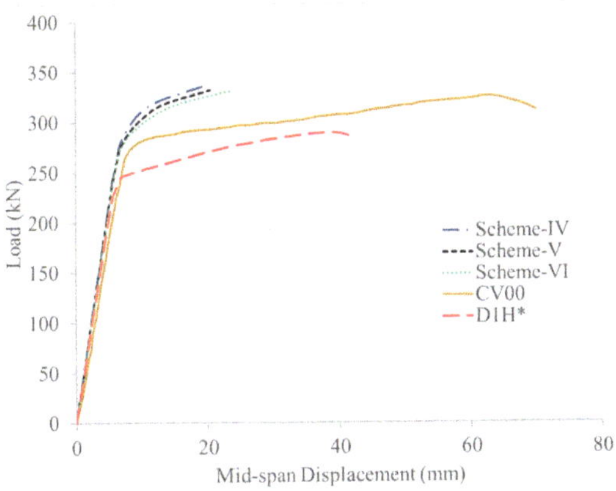

Figure 27. Load versus mid-span deflection diagram of D1H* beam with CFRP strengthening.

6. Conclusions

In the present study, finite element analysis was used to numerically investigate the efficacy of CFRP sheets on strengthening steel I-beams, damaged due to corrosion or cut-out web openings. FEA models were developed and validated against existing test specimens from the literature. Based on these validated models, a parametric investigation was carried out to examine the influence of various corrosion levels and cut-out web openings. Subsequently, CFRP sheets were applied to the bottom flanges of the corrosion-damaged steel I beams and the optimum number of CFRP layers required in restoring their capacities was investigated. Furthermore, six FRP strengthening configurations were proposed to restore the loss of the beam's strength, due to the damage induced by cut-out openings. Based on the parametric study results, the following conclusions were drawn:

- Due to the thinning of the bottom flange of the beam, both the yield and the ultimate load capacities of the beam were reduced. The reduction in the ultimate load followed a nearly perfect linear trend. For every 2.5% thickness loss of the bottom flange along 80% of the beam's span, the ultimate load was observed to reduce by roughly 1.5%.
- An adequate number of layers of CFRP sheets could restore both the yield and the ultimate load capacities of the beam subjected to corrosion damage to the level of the undamaged beam, while also increasing the overall beam stiffness. CFRP sheets, as a result, could be used as an alternative to CFRP plates or other conventional flexural strengthening techniques.
- The overall stiffness of the beams was observed to increase with the application of CFRP sheets. However, when a lower number of layers was used, the sheets ruptured before the desired level of load was reached, due to the overstressing of the thin sheets. Therefore, it is desirable to use a higher number of sheet layers for strengthening beams subjected to corrosion damage to delay sheet rupture, which could cause a sudden load drop in the beam.
- If the total area of the openings (sum of area of all openings) was kept constant, it was observed that beams with a greater number of circular web openings of smaller sizes had higher load capacities than those with a smaller number of larger web openings. Additionally, it is advisable to provide web openings in the bending predominant region of the beam's span.
- Three of the proposed strengthening configurations, which involved the application of CFRP sheets only to the bottom, the top, or both the bottom and the top flanges (Schemes-I, II, and III, respectively), were found to be ineffective. However, the remaining three strengthening configurations of Schemes-IV, V, and VI, which involved the application of CFRP sheets to both the flanges and the web, with a minimum of six layers of CFRP sheets, could restore the strength of the steel beam with circular web openings to the level of the one without openings. An additional number of CFRP sheet layers in these schemes would result in improvements in both the load-carrying capacity and stiffness of the beams.

The simplified FEA models developed in this study can be utilized for future research to highlight the influence of crucial parameters such as the mechanical properties of CFRPs and the types of fibers (i.e., Aramid, glass, and basalt) on the flexural and shear strengthening of damaged steel I-beams, particularly in steel bridges, where beams are subjected to fatigue loadings. Bearing that in mind, the findings of this research must be experimentally verified to ensure their practical applicability before being implemented in real-world scenarios.

Author Contributions: Conceptualization, A.P.; methodology, A.P. and P.K.S.; software, P.K.S. and M.A.; validation, P.K.S. and M.A.; investigation, P.K.S.; data curation, P.K.S.; writing—original draft preparation, P.K.S.; writing—review and editing, A.P. and M.A.; supervision, A.P.; project administration, A.P. All authors have read and agreed to the published version of the manuscript.

Funding: This research received no external funding.

Data Availability Statement: Data is contained within this article.

Conflicts of Interest: Author Mohannad Alhusban is employed by the company Crawford, Murphy & Tilly, Inc. The remaining authors declare that the research was conducted in the absence of any commercial or financial relationships that could be construed as a potential conflict of interest.

References

1. Tavakkolizadeh, M.; Saadatmanesh, H. Fatigue strength of steel girders strengthened with carbon fiber reinforced polymer patch. *J. Struct. Eng.* **2003**, *129*, 186–196. [CrossRef]
2. Ghafoori, E.; Motavalli, M. Normal, high and ultra-high modulus carbon fiber-reinforced polymer laminates for bonded and un-bonded strengthening of steel beams. *Mater. Des.* **2015**, *67*, 232–243. [CrossRef]
3. Kim, Y.J.; Harries, K.A. Fatigue behavior of damaged steel beams repaired with CFRP strips. *Eng. Struct.* **2011**, *33*, 1491–1502. [CrossRef]
4. Tafsirojjaman, T.; Fawzia, S.; Thambiratnam, D. Investigation on the behaviour of CFRP strengthened CHS members under monotonic loading through finite element modelling. *Structures* **2020**, *28*, 297–308. [CrossRef]
5. Tafsirojjaman, T.; Dogar, A.U.R.; Liu, Y.; Manalo, A.; Thambiratnam, D.P. Performance and design of Steel Structures Reinforced with FRP Composites: A state-of-the-art review. *Eng. Fail. Anal.* **2022**, *138*, 106371. [CrossRef]
6. Siwowski, T.W.; Siwowska, P. Experimental study on CFRP-strengthened steel beams. *Compos. Part B Eng.* **2018**, *149*, 12–21. [CrossRef]
7. Peiris, A.; Harik, I. Steel beam strengthening with UHM CFRP Strip Panels. *Eng. Struct.* **2021**, *226*, 111395. [CrossRef]
8. Bastani, A.; Das, S.; Kenno, S. Rehabilitation of thin walled steel beams using CFRP Fabric. *Thin-Walled Struct.* **2019**, *143*, 106215. [CrossRef]
9. Batuwitage, C.; Fawzia, S.; Thambiratnam, D.; Al-Mahaidi, R. Durability of CFRP strengthened steel plate double-strap joints in accelerated corrosion environments. *Compos. Struct.* **2017**, *160*, 1287–1298. [CrossRef]
10. Xu, S.; Li, H.; Wang, Y.; Wang, Y.; Wang, Y. Influence of corrosion on the bond behavior in CFRP-Steel Single Lap Joints. *Constr. Build. Mater.* **2020**, *236*, 117607. [CrossRef]
11. Li, H.; Xu, S.; Wang, Y.; Song, C. Deterioration of the bond behavior between CFRP and corroded steel in wet-dry cycling environment. *Compos. Struct.* **2021**, *272*, 114202. [CrossRef]
12. Morkhade, S.G.; Gupta, L.M. An experimental and parametric study on steel beams with web openings. *Int. J. Adv. Struct. Eng.* **2015**, *7*, 249–260. [CrossRef]
13. Morkhade, S.G.; Gupta, L.M. Experimental investigation for failure analysis of steel beams with web openings. *Steel Compos. Struct.* **2017**, *23*, 647–656. [CrossRef]
14. Morkhade, S.G.; Gupta, L.M. Ultimate load behaviour of steel beams with web openings. *Aust. J. Struct. Eng.* **2019**, *20*, 124–133. [CrossRef]
15. Morkhade, S.G.; Gupta, L.M. Critical study of steel beams with web openings. *Aust. J. Struct. Eng.* **2023**, *24*, 24–35. [CrossRef]
16. Abbas, J.L. Behaviour of steel I beams with web openings. *Civ. Eng. J.* **2023**, *9*, 596–617. [CrossRef]
17. Manoharan, A.C.; Tripathi, R.K. Analysis of steel beams with circular opening. *Int. J. Civ. Eng. Technol.* **2017**, *8*, 411–422.
18. Morkhade, S.G.; Kshirsagar, M.; Dange, R.; Patil, A. Analytical study of effect of web opening on flexural behaviour of hybrid beams. *Asian J. Civ. Eng.* **2019**, *20*, 537–547. [CrossRef]
19. Morkhade, S.G.; Lokhande, R.S.; Gund, U.D.; Divate, A.B.; Deosarkar, S.S.; Chavan, M.U. Structural behaviour of castellated steel beams with reinforced web openings. *Asian J. Civ. Eng.* **2020**, *21*, 1067–1078. [CrossRef]
20. Morkhade, S.G.; Jagtap, K.R.; Ghorpade, P.S.; Ahiwale, D.D.; Najm, H.M. Buckling performance evaluation of steel cellular beams strengthened with flange cover plate. *Asian J. Civ. Eng.* **2022**, *23*, 1277–1290. [CrossRef]
21. Altaee, M.J.; Cunningham, L.S.; Gillie, M. Experimental investigation of CFRP-strengthened steel beams with web openings. *J. Constr. Steel Res.* **2017**, *138*, 750–760. [CrossRef]
22. ANSYS®. *Academic Research Mechanical Release 20.2*; Help System; ANSYS, Inc.: Canonsburg, PA, USA, 2020; Available online: http://www.ansys.com/ (accessed on 28 August 2023).
23. Lawson, R.M.; Hicks, S.J. *Design of Composite Beams with Large Web Openings: In Accordance with Eurocodes and the UK National Annexes*; Steel Construction Institute: Ascot, UK, 2011; ISBN 9781859421970.
24. Altaee, M.; Cunningham, L.; Gillie, M. Novel technique for strengthening steel beams with web penetrations. In Proceedings of the MACE PGR Conference, Manchester, UK, 22 March 2016.

buildings

MDPI

Article

Tests and Seismic Response Analysis of Guided-Rail-Type Anti-Tensile Rubber Bearing

Longfei Zhang [1,2,*], Xiang Lan [1,2], Kechuan Wu [1,2] and Wenzheng Yu [1,2]

[1] College of Architecture and Civil Engineering, Kunming University, Kunming 650214, China; lx523947015@kmu.edu.cn (X.L.); wukc@kmu.edu.cn (K.W.); yuwenzheng@kmu.edu.cn (W.Y.)

[2] Institute of Engineering Earthquake Resistance and Disaster Reduction, Kunming University, Kunming 650214, China

* Correspondence: zhanglongfei08052003@kmu.edu.cn

Abstract: When subjected to seismic activity, tall isolated buildings with a high aspect ratio are susceptible to overturning as a result of the failure of rubber isolation bearings under tension. In order to address this issue, a guided-rail tension device (GR) has been developed to enhance the tensile strength of rubber bearings. Furthermore, a novel guided-rail isolation rubber bearing (GR&RB) has been proposed as a potential solution. Quasi-static tests have been conducted to investigate the mechanical properties of the GR, as well as the GR&LNR600 and LNR600. Additionally, numerical finite element analysis has been employed to study the seismic response of the GR&RB under El Centro seismic action in a high-rise building with an aspect ratio of approximately 4. The experimental results suggest that the inclusion of GRs has a limited effect on the horizontal mechanical attributes of rubber isolation bearings. Nevertheless, it simultaneously enhances their tensile strength, demonstrating a significant augmentation of 4.23 times. Moreover, the mechanical behavior of the GR aligns with the Hook model. The numerical analysis suggests that the utilization of GR&RB can mitigate the tensile stress levels of rubber isolators. Furthermore, it is recommended to augment the elastic tensile stiffness of the GR while reducing the open value to enhance the tensile efficiency, with the potential to attain up to 75% efficiency in tensile performance.

Keywords: seismic isolation; tensile rubber isolation bearings; guided-rail tension device; experimental study; finite element analysis

1. Introduction

Base isolation is a widely utilized and effective technique to mitigate vibrations in mid-rise and low-rise buildings, demonstrating favorable seismic performance [1–4].

As research in isolation technology progresses, new avenues of investigation have emerged. In an effort to enhance the seismic resilience of buildings situated in mountainous regions, Zhang et al. [5] introduced an isolated step-terrace structure and conducted shaking table experiments on a 1:10 scale model of this innovative system. The outcomes revealed that the seismic performance of the isolated step-terrace structure significantly surpassed that of conventional step-terrace configurations. Rayegani et al. [6,7] explored the phenomenon of seismic pounding resulting from the excessive deformation of the isolation layer during near-fault earthquakes. Their study introduced the use of magnetorheological dampers within the isolation layer, which proved effective in mitigating displacement and preventing pounding occurrences. Furthermore, they assessed the seismic collapse probability and life cycle cost associated with the implementation of smart hybrid isolation structures in the presence of pounding effects. Zelleke et al. [8] introduced a reliability-based multi-hazard optimization framework tailored to base-isolated buildings subjected to diverse multi-hazard scenarios. Tena-Colunga et al. [9] conducted a study on four independent slender steel structure buildings in sturdy soil, finding that the seismic responses of slender base-isolated structures can be significantly compromised during strong earthquakes.

With the increasing endorsement of base isolation technology by relevant authorities in China, its application has expanded to encompass high-rise buildings, such as the Kunming Tianhu Jingxiu, a notable high-rise shear wall renovation project [10]. To prevent the potential damage to rubber bearings resulting from tension, which could lead to the overturning of isolated structures, it is imperative that isolators do not experience tensile stress. Even in cases where tensile stress does occur, it should be maintained below 1.0 MPa in accordance with Chinese seismic regulations [11,12]. However, in the context of high-rise buildings with high aspect ratios, the manifestation of the overturning effect under seismic action is conspicuous, often resulting in tensile stress in isolators. Numerous studies have indicated the limited tensile resistance of rubber isolators [13,14]. When subjected to tension, the rubber bearing may develop internal cavities due to negative pressure [15], despite exhibiting no external signs of damage, consequently leading to a reduction in vertical compressive stiffness to approximately half of the isolator's initial stiffness. Moreover, when the tensile stress reaches 1.5–3.0 MPa, the tensile stiffness of the rubber isolator undergoes a sharp decline, signifying complete damage to the rubber bearing [16]. Consequently, the issue of tensile failure in rubber isolators has consistently impeded the widespread application of seismic isolation technology in high-rise buildings [17].

Many scholars have conducted extensive research to address the challenge of tensile failure in rubber isolators. Kelly et al. [18] applied a ball-and-socket joint linkage system to resist the tensile forces in the isolation system. Nagarajaiah et al. [19] proposed a combination of springs and rubber isolators to enhance the tensile performance and recentering capability of rubber isolators. Kasalanti et al. [20] employed the prestress technique to prevent the tensile failure of isolators. Qi et al. [21] recommended the addition of vertical steel bars at the edges of isolators to enhance their tensile safety. Yan et al. [22] developed three types of three-dimensional base isolators by serially connecting horizontal isolation sub-devices with vertical isolation sub-devices to counteract overturning. Su et al. [23] proposed the parallel connection of three laminated rubber isolators to convert tension into compression. Wang et al. [24] introduced tension-laminated rubber bearings (TLRB) with a specified tensile capacity and provided calculation formulas for their horizontal stiffness in compression and tension states, as well as a formula to calculate the limit shear strain of the bearings. Ge et al. [25] established design criteria for the tensile performance of rubber isolators through tensile–shear tests and developed a gate-shaped device for tension. Pauletta et al. [26] investigated the impact of the vulcanization time and the rubber formulation process on the tensile characteristics of rubber bearings by conducting tensile experiments on prototype rubber bearings. The study put forth recommendations aimed at enhancing the tensile properties of rubber bearings. Zhang et al. [27] delved into the tensile behavior of thick rubber bearings (TRBs) via numerical simulations and analytical examinations. Their work presented an analytical framework tailored to represent the tensile behavior of the bearings under cyclic loading conditions. However, many of these devices or methods that enhance the tensile capacity of rubber isolators may concurrently diminish their horizontal isolation performance, resulting in a reduction in the horizontal isolation efficiency of the base isolation structure and constraining their application in engineering.

It should be noted that Vaiana et al. [28] categorized rubber bearings into two primary groups, steel-reinforced elastomeric bearings (SREBs) and fiber-reinforced elastomeric bearings (FREBs), distinguished by their reinforcement methods. SREBs incorporate steel shims for reinforcement, while FREBs utilize fibers, offering cost-effectiveness advantages in production and installation. By default, the rubber bearings discussed in this paper are specifically steel-reinforced elastomeric bearings (SREBs), denoted as RB. In light of the aforementioned challenges, this paper proposes a guided-rail tension device (GR) to augment the tensile strength of rubber isolators. Additionally, a novel guided-rail isolation rubber bearing (GR&RB) is introduced for high-rise buildings with high aspect ratios. The horizontal and vertical mechanical properties of the GR are examined through pseudo-static tests, and the dynamic response of the GR&RB under earthquake action is investigated using numerical methods in high-rise isolated structures with a high aspect ratio.

2. Structure of GR&RB

In general, the RB configuration typically consists of an upper attachment plate, a rubber bearing, and a lower attachment plate, with the rubber bearing situated between the upper and lower attachment plates and typically connected to them using bolts. As depicted in Figure 1a, under tension, the interaction between the upper and lower attachment plates and the internal rubber layer of the rubber bearing can generate a negative pressure void. This phenomenon weakens the lateral constraint of the steel plate on the internal rubber layer, resulting in a significant reduction in the vertical compressive stiffness of the rubber bearings and consequently impacting the overall safety of the isolated structure.

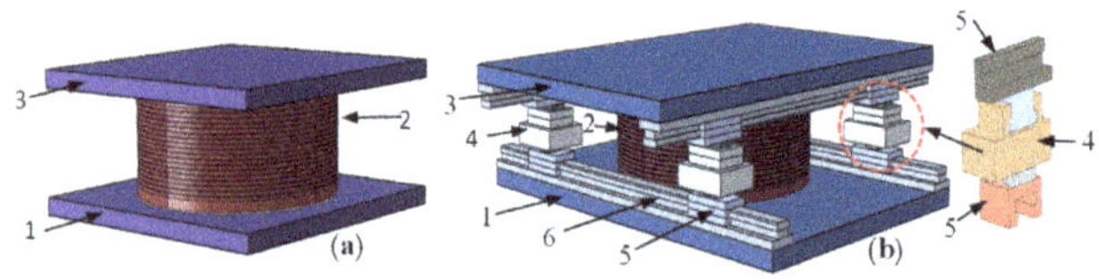

Note: 1—Lower attachment plate, 2—Rubber bearing, 3—Upper attachment plate, 4—Tension device, 5—T-shaped fasteners, 6—Guide rail.

Figure 1. Structures of isolators: (**a**) RB; (**b**) GR&RB.

In an effort to enhance the tensile performance, guided-rail tension devices (GRs) are integrated with RBs to form a guided-rail isolation rubber bearing (GR&RB), as illustrated in Figure 1b. The GR&RB system comprises a lower attachment plate, a rubber bearing, four guide rails, four tension devices, T-shaped fasteners, and an upper attachment plate. The rubber bearing and tension devices are positioned between the upper and lower attachment plates and are connected to them using bolts. Each tension device is equipped with T-shaped fasteners on both ends, which are designed to interact with the guide rails. The guide rails are attached to the upper and lower connecting plates using bolts. When the upper and lower attachment plates undergo horizontal relative deformation, the T-shaped fasteners can freely slide along the guide rails, facilitating the coordination of the tension devices with the horizontal deformation of the rubber bearings. Under tension in the upper and lower attachment plates, the tension devices, through their interaction with the guide rails, share the tensile load with the rubber bearing, thereby reducing the tensile effect on the rubber bearing and preventing the generation of negative pressure voids between the internal rubber layer and the steel plate. Conversely, when the rubber bearing is under compression, the tension devices do not bear any pressure, as the entirety of the pressure is supported by the rubber bearing.

3. Experimental Tests

The uniaxial tensile limit characteristics of a single GR were examined, as well as the properties of the GR&LNR600 (a specific GR&RB configuration where the rubber bearing is LNR600) and LNR600. The findings from these tests were subsequently utilized in the dynamic analysis of a high-rise isolated structure. It is worth noting that in the uniaxial ultimate tensile test of a single GR, the length of the GR was limited to 600 mm, which was shorter than the GR length used in the GR&LNR600 configuration.

3.1. Test Specimens

All GRs utilized in the experiment were constructed from carbon structural steel (referred to as No. 45 in GB). The specific parameters of the LNR600 employed in the GR&LNR600 configuration are detailed in Table 1, while the primary dimensions of the GR&LNR600 system are visually represented in Figure 2.

Table 1. Details of LNR600.

Outer diameter	D (mm)	600
Inner diameter	d (mm)	100
Total height	H (mm)	263
Rubber layer thickness	t_r (mm)	6.22
Number of rubber layers	n_r	18
Total thickness of rubber layers	T_r (mm)	112
Steel shim thickness	ts (mm)	3
Number of steel shims	n_s	17
Total thickness of steel shims	T_s (mm)	51
Primary shape factor	S_1	23.44
Secondary shape factor	S_2	5.35

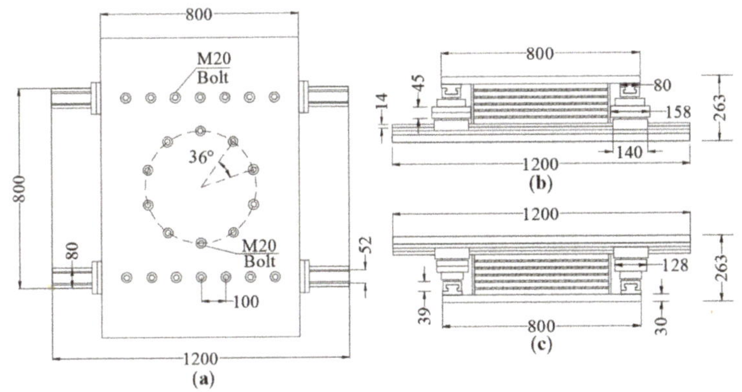

Figure 2. Main dimensions of GR&LNR600 (**a**) vertical view; (**b**) front view; (**c**) left view. Unit: mm.

3.2. Test Equipment

The mechanical performance tests of the LNR600 and GR&LNR600 were conducted on a shear testing device (see Figure 3a), and the uniaxial ultimate tensile test for a single GR was performed on an electro-hydraulic servo testing device (see Figure 3b). Both sets of testing equipment contained a self-balanced system.

Figure 3. Test equipment: (**a**) shear testing device; (**b**) electro-hydraulic servo testing device.

3.3. Test Procedure

The uniaxial tensile tests of the single GR, LNR600, and GR&LNR600 were carried out under displacement-controlled loading, employing a loading rate of 1 mm/s. Notably, the single GR was subjected to failure under tension, whereas the tensile displacements of the LNR600 and GR&LNR600 were limited to 5.2 mm to ensure specimen elasticity.

To examine the impact of the angle between the horizontal shear direction and the guide rail on the mechanical properties of the guided-rail isolation rubber bearings, loading angles of 45° and 0° were considered in the horizontal compression–shear and tensile–shear cyclic tests of the GR&LNR600, as presented in Table 2. The LNR600 utilized in tests T1 to T6 represented the same natural rubber bearing, with Figure 4 illustrating the loading directions for these tests. The shear tests of the LNR600 and GR&LNR600 were conducted in accordance with the industry standard within the Chinese construction sector [29]. Furthermore, the horizontal equivalent stiffness (K_b) of rubber isolation bearings was derived from a quasi-static test consisting of 3 cycles at a loading frequency of 0.02 Hz and a 100% shear strain amplitude.

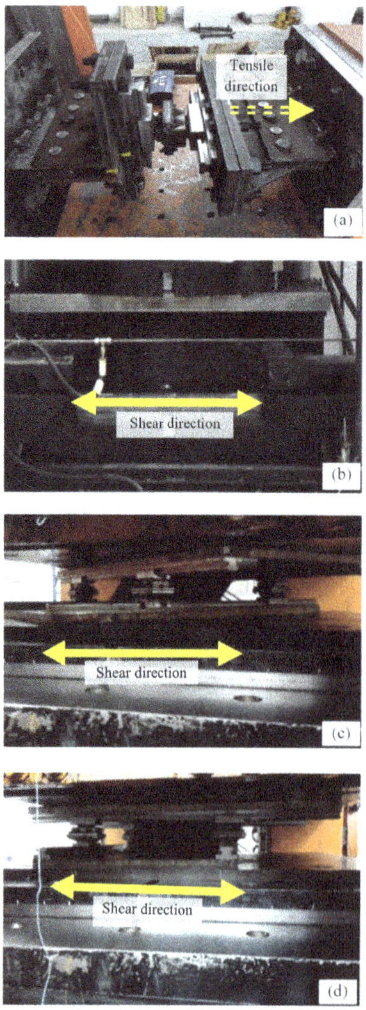

Figure 4. Loading direction: (**a**) single GR; (**b**) LNR600; (**c**) GR&LNR600-45°; (**d**) GR&LNR600-0°.

Table 2. Shear test scheme.

Test Item	Test No.	Subject	Horizontal Loading Angle	Vertical Load
Compression–shear	T1	LNR600	/	Pressure loads of 4239 kN
	T2	GR&LNR600	45°	
	T3	GR&LNR600	0°	
Tensile–shear	T4	LNR600	/	Tensile loads of 282.6 kN
	T5	GR&LNR600	45°	
	T6	GR&LNR600	0°	

3.4. Test Results

3.4.1. Tensile Properties

During the initial phase of the single GR tensile failure test, no visible deformation was observed in the guide rail. As the loading displacement increased, the GR experienced sudden failure, leading to the separation of the central connecting component. Simultaneously, the damaged section on the left slid down onto the test platform, and the middle section of the guide rail underwent bending under tension, as depicted in Figure 5. The tensile force–displacement curve of the GR uniaxial tension is illustrated in Figure 6.

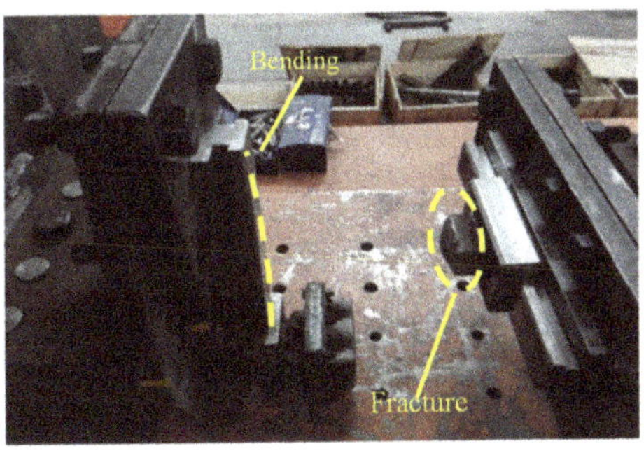

Figure 5. Single GR tensile failure.

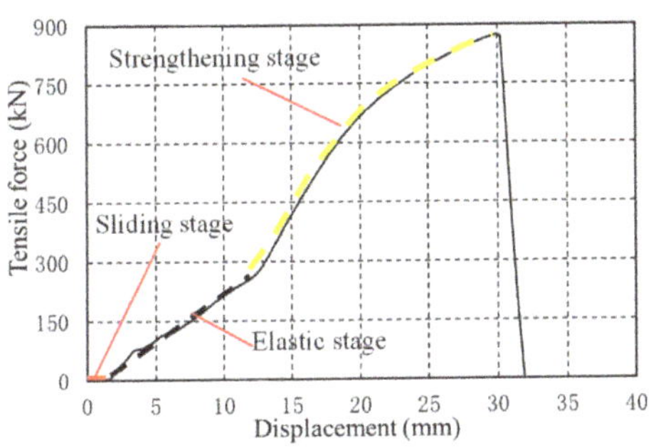

Figure 6. Tensile force–displacement curve of GR uniaxial tension.

The analysis of the tensile force–displacement curve reveals three distinct stages in the tensile process of the GR: the slip stage, elastic stage, and strengthening stage. The slip stage is attributed to the clearance between GR components, which is linked to the processing accuracy, with a gap value of 1.4 mm. Upon reaching tensile displacement of 30.1 mm, the GR specimen experiences sudden failure, leading to a sharp drop in load. Assuming the ultimate tensile bearing capacity of the GR corresponds to the end of the elastic stage, the design bearing capacity of a single GR is estimated to be 266 kN, with elastic tensile stiffness of 27.6 kN/mm and a gap value of 1.4 mm. The tensile behavior of the GR can be described as the Hook model, as depicted in Figure 7, and can be expressed as

$$f = \begin{cases} k(d-l) & (if \quad d-l>0) \\ 0 & (otherwise) \end{cases} \tag{1}$$

where k is the elastic tensile stiffness of the GR, d is the axial deformation value of the GR, and l is the gap value ($l \geq 0$).

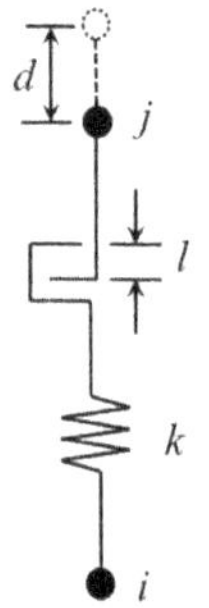

Figure 7. Diagram of Hook model.

Figure 8 presents the load–displacement curves of the GR&LNR600 and LNR600 in the uniaxial tensile tests. The tensile bearing capacity of the GR&LNR600 reached 1195 kN, approximately 4.23 times that of the LNR600, indicating a significant enhancement in the tensile performance of the conventional rubber isolation bearings due to the GRs.

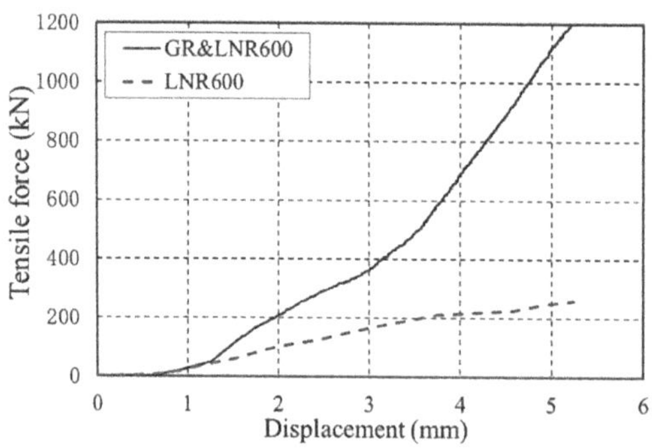

Figure 8. Tensile force–displacement curves of GR&LNR600 and LNR600.

3.4.2. Horizontal Compression–Shear and Tensile–Shear Performance

For ease of reference, the tensile bearings with angles of 45° and 0° between the horizontal loading direction and the guide rail are denoted as GR&LNR600-45° and GR&LNR600-45°, respectively. According to JG/T 118-2018 [29], the horizontal equivalent stiffness K_b is defined as the slope of the hysteretic curve between −100% and +100% shear deformation with reference to the third cycle of the test. Figure 9 illustrates the hysteresis curves of the horizontal compressive–shear and tensile–shear tests for specimens T1 to T6. The measured horizontal equivalent stiffness values of each specimen are summarized in Table 3.

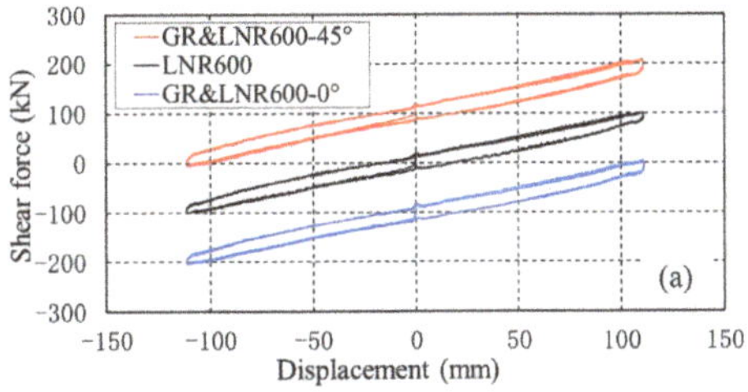

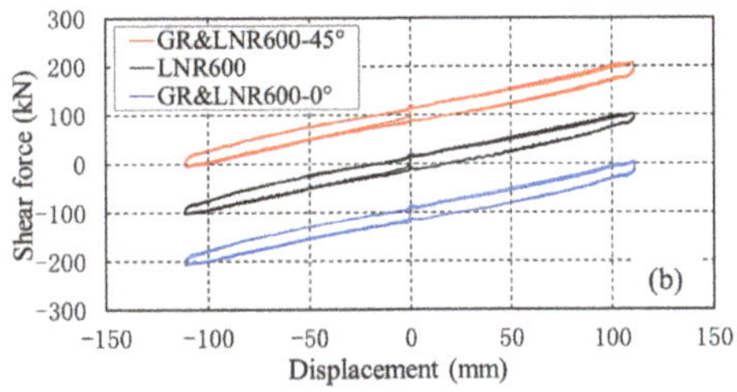

Figure 9. Hysteresis curves of horizontal shear tests: (**a**) compression–shear; (**b**) tensile–shear.

Table 3. Measured horizontal equivalent stiffness (K_b) in kN/mm.

Type	Compression–Shear	Tensile–Shear	Deviation
LNR600	0.891	0.927	4.1%
GR&LNR600-45°	0.927	0.960	3.6%
GR&LNR600-0°	0.926	0.957	3.4%

An analysis of Table 3 reveals that the horizontal equivalent stiffness of the specimens consistently demonstrates greater values under tensile–shear loading in comparison to compression–shear loading, albeit not exceeding 4.1%. Furthermore, the angle between the horizontal shear direction and the guide rail exhibits a certain enhancing effect on the horizontal equivalent stiffness, but within 4%. Consequently, it is concluded that the influence of GRs on the horizontal performance of rubber isolation bearings can be deemed negligible. In other words, the horizontal mechanical properties of the GRT&RB are akin to those of RBs of the same scale.

4. Numerical Analysis of Seismic Response

4.1. Building for Numerical Analysis

The building under investigation is a reinforced concrete high-rise shear wall structure characterized by an aspect ratio of 3.92. Its superstructure comprises 17 storeys, with a total height of 51 m. Specifics regarding the concrete properties are as follows: the compressive strength of the concrete for the first and second storeys is 26.8 MPa, featuring a wall thickness of 300 mm for the first level and 250 mm for the second storey. For the 3rd to 17th storeys, the compressive strength of the concrete is recorded as fc = 20.1 MPa, with a wall thickness ranging from 200 mm to 250 mm. Furthermore, the floor slab thickness for the first to fourth storeys measures 150 mm, while, for the 5th to 17th storeys, it is 120 mm.

The isolation system employed in this study encompasses a total of 24 rubber isolation bearings, inclusive of guided-rail tension-type rubber bearings and traditional rubber bearings. Additionally, a conventional isolation model (referred to as M_0) solely featuring traditional rubber bearings was established. To explore the impact of the elastic tensile stiffness and aperture on the tensile performance of GRs, nine comparative models with varying elastic tensile stiffness and aperture were devised. For ease of reference, the subscript in the model number denotes the specific elastic tensile stiffness and clearance. For instance, $M_{k(1.5)l(1)}$ signifies that the elastic tensile stiffness of the GR is 1.5 times the test value, while the gap value is equivalent to the test value. Subsequent model numbers follow this sequence. The layout and dimensions of the structural plane and isolation layer are depicted in Figure 10.

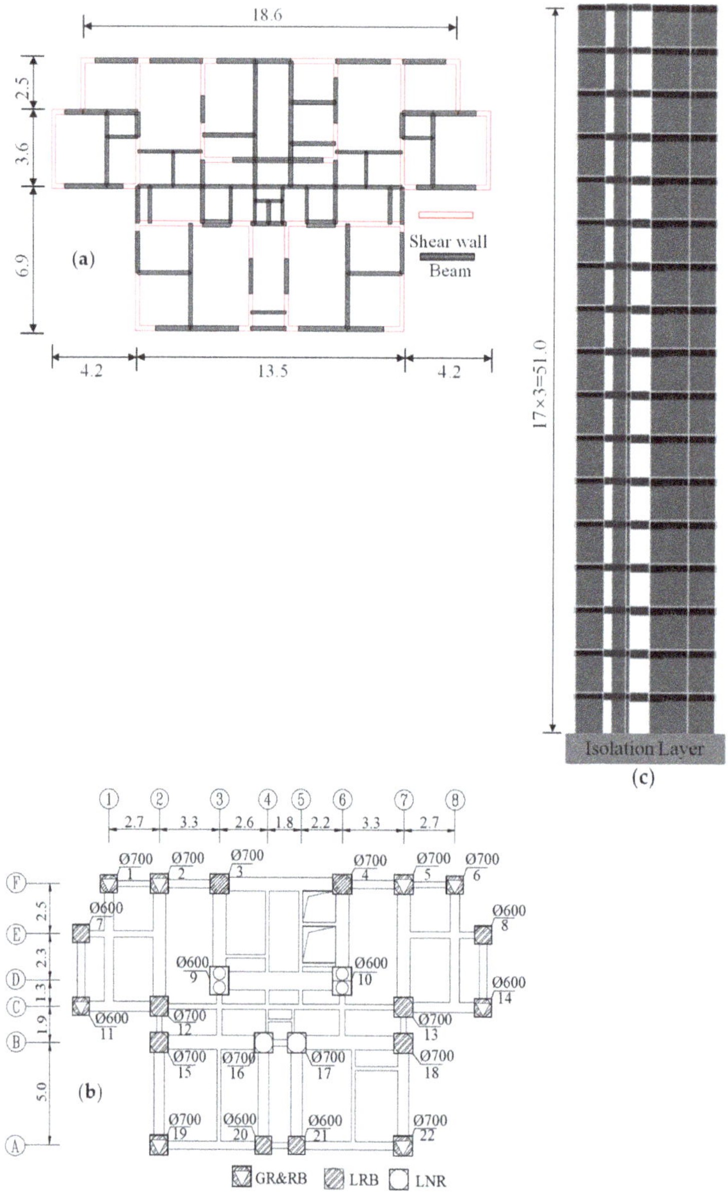

Figure 10. Model of high-rise shear wall building: (**a**) plan view; (**b**) layout of isolation layer; (**c**) elevation. Unit: m.

The mechanical properties of the GRs in each model are shown in Table 4.

Table 4. Mechanical properties of GRs.

Model No.	M_0	$M_{k(1)l(1)}$	$M_{k(2)l(1)}$	$M_{k(3)l(1)}$	$M_{k(1)l(0.5)}$	$M_{k(2)l(0.5)}$	$M_{k(3)l(0.5)}$	$M_{k(1)l(0)}$	$M_{k(2)l(0)}$	$M_{k(3)l(0)}$
Elastic tensile stiffness k (kN/mm)	/	27.6	55.2	82.8	27.6	55.2	82.8	27.6	55.2	82.8
Open l (mm)	/	1.4	1.4	1.4	0.7	0.7	0.7	0	0	0

Numerous empirical investigations have substantiated the nonlinearity exhibited by RBs in both horizontal and vertical orientations, commonly delineated by hysteresis loops. The analysis was executed utilizing the 2019 iteration of the ETABS software. The behavior of the RBs was simulated through the implementation of a hysteretic isolator, which amalgamated the attributes of axial elasticity and interconnected horizontal plasticity. The hysteretic isolator, illustrated in Figure 11, incorporates a plasticity model derived from the hysteresis characteristics introduced by Wen [30], Park, Wen, and Ang [31] and validated for base isolation evaluations by Nagarajaiah, Reinhorn, and Constantinou [32].

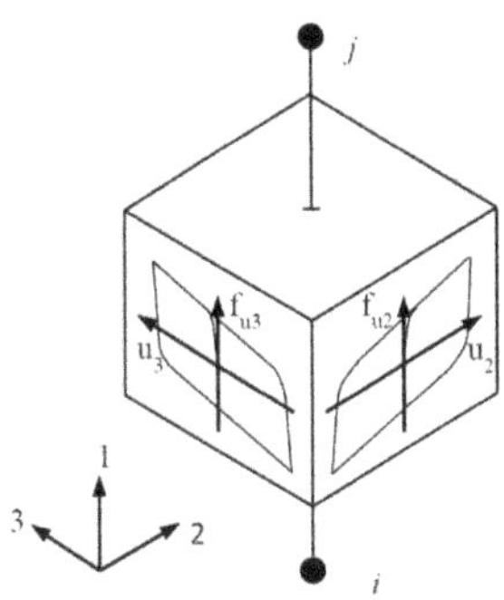

Figure 11. Hysteretic isolator horizontal property.

Although RBs exhibit nonlinear hysteretic behavior in the vertical direction, in this study, the tensile and compressive stiffnesses were modeled as linear to rapidly obtain relatively accurate results. Additionally, references [14,33] suggest that the vertical tensile stiffness of rubber bearings typically ranges from one fifth to one tenth of the compressive stiffness. This study simplifies both the compressive and tensile stiffness of RBs as linear to provide a relatively accurate portrayal of their vertical mechanical properties through the parallel combination of a hysteretic isolator and a gap. Furthermore, the vertical mechanical behavior of the GR&RB is characterized by the inclusion of a Hook element in parallel with the former. The models depicting the vertical mechanical behaviors of the RB and GR&RB are delineated in Figure 12.

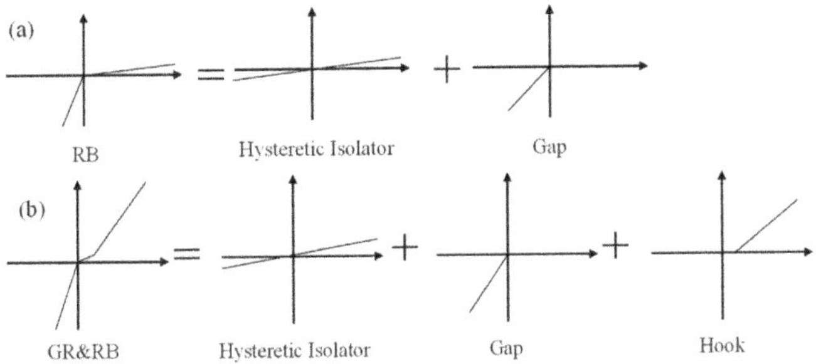

Figure 12. Vertical mechanical performance model: (**a**) RB; (**b**) GR&RB.

To enhance the precision and computational effectiveness, the horizontal and vertical mechanical hysteresis characteristics of RBs can be effectively replicated using the Vaiana–Rosati model (VRM) developed by Vaiana and Rosati [34,35].

Table 5 delineates the mechanical properties of the conventional rubber isolation bearings utilized in the numerical analysis conducted in this study.

Table 5. Mechanical properties of traditional rubber isolation bearings.

Type	Vertical Compressive Stiffness (kN/mm)	Vertical Tensile Stiffness (kN/mm)	Horizontal Equivalent Stiffness (kN/mm)	Initial Shear Stiffness (kN/mm)	Post-Yielding Stiffness (kN/mm)	Yield Strength (kN)
LNR600	1800	180	0.96	/	/	/
LRB600	2100	210	1.58	13.11	1.01	63
LNR700	2350	235	1.17	/	/	/
LRB700	2600	260	1.87	15.19	1.17	90

In consideration of the effective duration and the consistent soil type, the three-directional El Centro earthquake records, widely employed in seismic engineering research, were chosen for the nonlinear dynamic time history analysis. Given the focus of the present study on investigating the tensile performance and influencing factors of guided-rail-type tensile rubber bearings, the selection of a representative group of ground motions for excitation was sufficient. Figure 13 provides a depiction of the waveform of the El Centro earthquake.

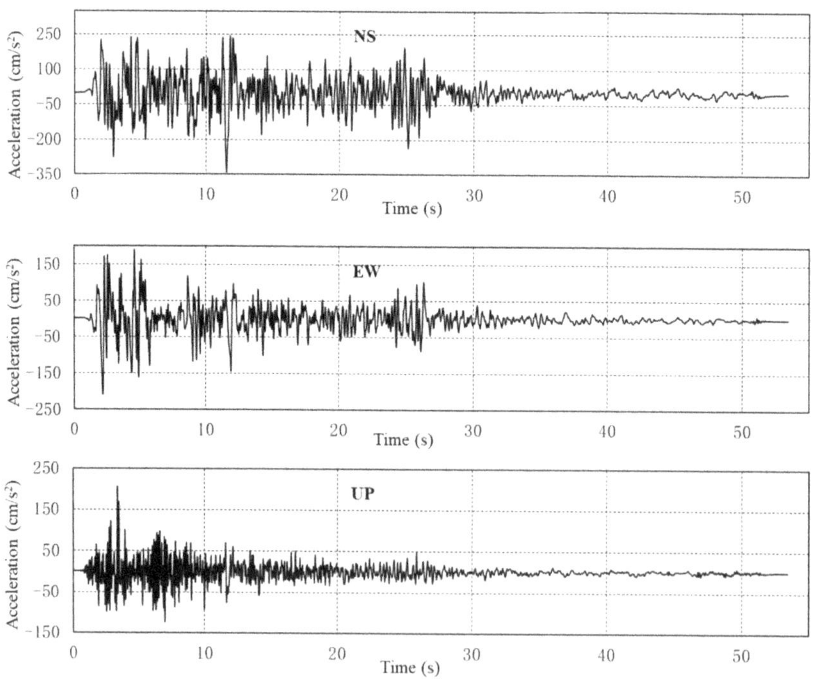

Figure 13. El Centro ground acceleration records.

The nonlinear time domain analysis method employed in ETABS incorporates material linearity and link nonlinearity. Furthermore, the analysis utilizes nonlinear direct integration through the Hilber–Hughes–Taylor alpha method [36].

4.2. Tensile Performance of Isolators

Figure 14 illustrates the vertical stress distribution of the rubber bearings under the El Centro seismic action for eight numerical models. The application of the GR&RB is observed to reduce the tensile stress level of the rubber isolators without altering the loading state of the rubber bearings, as evidenced in Figure 14. A comparison of the tensile effects depicted in Figure 14a,b suggests that the tensile effect is amplified with an increase in the elastic tensile stiffness of the GR, and it also escalates with a decrease in the open value of the GR.

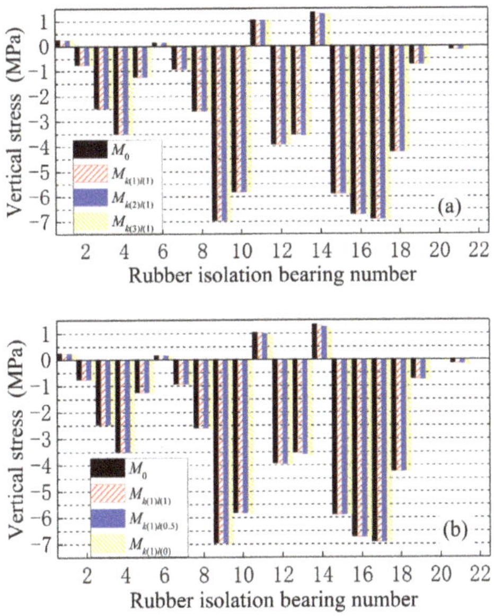

Figure 14. Vertical stress distribution: (**a**) keeping the open value unchanged while changing the elastic tensile stiffness, (**b**) keeping the elastic tensile stiffness unchanged while changing the open value.

Moreover, Figure 15 presents the vertical force–time history curves of the No. 14 rubber isolation bearing.

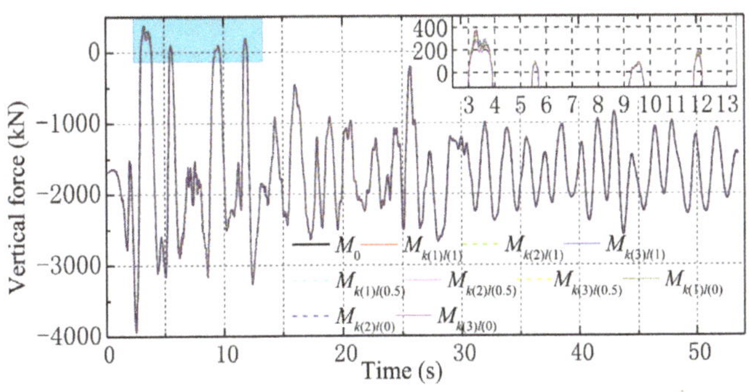

Figure 15. Vertical force–time history curves of No. 14 rubber isolation bearing.

As depicted in Figure 15, the activation of the GR occurs exclusively when the rubber isolation bearing is subjected to tension, functioning in conjunction with the rubber bearing to counteract the tensile forces. Nevertheless, the existence of an opening leads to a delayed tensile response in the GR. It is noteworthy that the greater the magnitude of the opening, the more prominent the phenomenon of delayed tensile force becomes, leading to the reduced tensile resistance effect of the GR.

The tensile stress ratio can be used to characterize the tensile efficiency of isolated structures:

$$\eta = \frac{\sigma}{\sigma_0} \tag{2}$$

where η is the tensile efficiency, σ is the maximum tensile stress of the rubber bearing for the comparative models, and σ_0 is the maximum tensile stress of the rubber bearing for M_0. Figure 16 illustrates the impact of the varying elastic tensile stiffness and open values of GRs on the tensile efficiency.

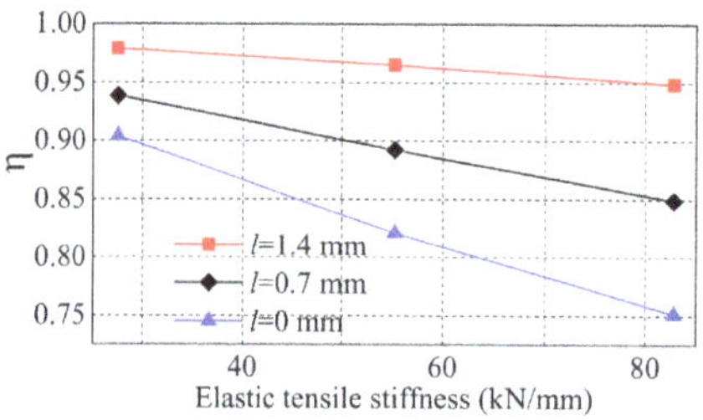

Figure 16. Tensile efficiency.

Based on the findings presented in Figure 16, the application of the GR&RB results in a reduction in tensile stress in the rubber isolation bearings by 75–98%. Optimal outcomes are achieved by enhancing the elastic tensile stiffness and concurrently reducing the open value of the GR, leading to a potential value of 75% tensile efficiency.

5. Conclusions

In order to mitigate the risk of overturning due to tension in rubber isolation bearings, a novel form of tensile rubber isolation bearing, denoted as the GR&RB, has been introduced. The mechanical characteristics of GRs, as well as the GR&LNR600 and LNR 600, have been thoroughly examined through quasi-static testing. Additionally, the impact of tensile efficiency factors such as the elastic tensile stiffness and the open value of the GR have been explored via finite element analysis. The key findings and implications derived from this investigation can be summarized as follows.

(1) The presence of clearance results in a slip stage during the initial phase of single GR stretching, thereby enabling the characterization of the GR's tensile behavior as Hook. Moreover, during finite element simulation, the vertical mechanical behavior of the GR&RB can be effectively represented through the parallel superposition of the hysteretic isolator, gap, and Hook.

(2) Within the elastic tensile range of rubber bearings, the tensile strength of the GR&RB is approximately 4.3 times greater than that of RBs of equivalent scale, underscoring the remarkable tensile capacity of the GR&RB.

(3) The horizontal mechanical performance of the GR&RB exhibits almost complete independence with respect to the angle of the guide rail and the loading direction, thereby equating the horizontal mechanical performance of the GR&RB to that of RBs of similar dimensions.

(4) The implementation of the GR&RB leads to a reduction in the level of tensile stress experienced by the rubber isolators. This reduction signifies that the utilization of the GR&RB diminishes the likelihood of failure related to overturning in seismic isolation structures by enhancing the tensile performance of the bearings. Furthermore, both an increase in the elastic tensile stiffness and a decrease in the open value of the GR can elevate the tensile strength of the GR&RB, with the optimal effect achieved through

an increase in the elastic tensile stiffness while simultaneously decreasing the open value of the GR.

Author Contributions: Investigation, L.Z., W.Y. and X.L.; methodology, L.Z. and W.Y.; formal analysis, L.Z. and K.W.; project administration, L.Z.; writing—original draft preparation, L.Z. and X.L.; writing—review and editing, L.Z. and K.W.; supervision, K.W.; funding acquisition, L.Z. All authors have read and agreed to the published version of the manuscript.

Funding: This research was funded by the general projects with joint special funds for basic research of Local Undergraduate Universities in Yunnan Province (Grant No. 202101BA070001-080 and 202101BA070001-176) and the Talent Introduction Project of Kunming University (Grant No. YJL20026).

Data Availability Statement: The data presented in this research are available upon request from the corresponding author.

Acknowledgments: The work presented in this paper has been supported by the Frontier Research Team of Kunming University 2023. The authors express their gratitude for the invaluable experimental support provided by the Kunming Earthquake Prevention and Disaster Reduction Technology Test Base, as well as for the provision of the experimental rubber isolation bearings by Yunnan GOWE Machinery & Technology Co., Ltd.

Conflicts of Interest: The authors declare no conflicts of interest.

References

1. Skinner, R.I.; Robinson, W.H.; Mcverry, G.H. *An Introduction to Seismic Isolation*; John Wiley & Sons, Inc.: New York, NY, USA, 1993.
2. Jin, J.M.; Tan, P.; Zhou, F.L.; Ma, Y.H.; Shen, C.Y. Shaking table test study on mid-story isolation structure. *Adv. Mater. Res.* **2012**, *446–449*, 378–381. [CrossRef]
3. Zheng, G.; Xu, H. Research on shaking table test of base and story isolation structures. *Appl. Mech. Mater.* **2014**, *580–583*, 1776–1781. [CrossRef]
4. Wu, Y.; Lu, J.; Qi, A. Shaking table test and numerical analysis of mid-story isolation eccentric structure with tower–podium. *Adv. Mech. Eng.* **2019**, *11*, 1687814018819562. [CrossRef]
5. Zhang, L.F.; Tao, Z.; Pan, W.; Wu, Z.J.; Lan, X. Shaking table test on isolated step-terrace frame structure in mountainous area. *J. Build. Struct.* **2020**, *41*, 24–32.
6. Rayegani, A.; Nouri, G. Application of smart dampers for prevention of seismic pounding in isolated structures subjected to near-fault earthquakes. *J. Earthq. Eng.* **2022**, *26*, 4069–4084. [CrossRef]
7. Rayegani, A.; Nouri, G. Seismic collapse probability and life cycle cost assessment of isolated structures subjected to pounding with smart hybrid isolation system using a modified fuzzy based controller. *Structures* **2022**, *44*, 30–41. [CrossRef]
8. Zelleke, D.H.; Saha, S.K.; Matsagar, V.A. Reliability-based multi-hazard design optimization of base-isolated buildings. *Eng. Struct.* **2024**, *301*, 117242. [CrossRef]
9. Tena-Colunga, A.; Parra-García, D.F. Seismic resilient performance of code-designed, base-isolated slender steel buildings in firm soils. *Structures* **2024**, *59*, 105754. [CrossRef]
10. Luo, Q.; Tan, Y.; Guo, M.; Pan, K.; Shang, Q.; Gao, G. Seismic isolation structural design for a 100-meter-high residential building of Tianhu Jingxiu Shantytown Renovation project in Kunming. *Build. Struct.* **2016**, *46*, 33–38.
11. *GB 50011-2010*; Code for Seismic Design of Buildings. China Architecture & Building Press: Beijing, China, 2016.
12. *GB/T 51408-2021*; Standard for Seismic Isolation Design of Buildings. China Architecture & Building Press: Beijing, China, 2021.
13. Zhou, F. *Engineering Structure Vibration Control*; Seismological Press: Beijing, China, 1997.
14. Architectural Institute of Japan. *Recommendation for the Design of Base Isolated Buildings*; Liu, W., Translator; Seismological Press: Beijing, China, 2006.
15. Kumar, M.; Whittaker, A.S.; Constantinou, M.C. Experimental investigation of cavitation in elastomeric seismic isolation bearings. *Eng. Struct.* **2015**, *101*, 290–305. [CrossRef]
16. Yang, Q.R.; Liu, W.G.; He, W.F.; Feng, D.M. Tensile stiffness and deformation model of rubber isolators in tension and tension-shear states. *J. Eng. Mech.* **2010**, *136*, 429–437. [CrossRef]
17. Cheng, H.; Liu, W.; Wang, S. Analysis of the tension of rubber bearings in the design of isolated high-rise buildings. *Earthq. Eng. Eng. Dyn.* **2007**, *27*, 161–166.
18. Kelly, J.M.; Griffith, M.C.; Aiken, I.D. *A Displacement Control and Uplift Restraint Device for Base Isolated Structures*; The Pacific Earthquake Engineering Research Center: Berkeley, CA, USA, 1987.
19. Nagarajaiah, S.; Reinhorn, A.M.; Constantinou, M.C. Experimental-study of sliding isolated structures with uplift restraint. *J. Struct. Eng.* **1992**, *118*, 1666–1682. [CrossRef]
20. Kasalanati, A.; Constantinou, M.C. Testing and modeling of prestressed isolators. *J. Struct. Eng.* **2005**, *131*, 857–866. [CrossRef]

21. Qi, A.; Lin, Y. Research on the approaches to incease the height-width ratio with supplemental steel bar. *Earthq. Eng. Eng. Vib.* **2005**, *46*, 33–38.
22. Yan, X.; Zhang, Y.; Wang, H. Experimental study on mechanical properties of three kind three-dimensional base isolation and overturn resistance devices. *J. Vib. Shock* **2009**, *28*, 49–53.
23. Su, J.; Heisha, W.L.H.; Zhou, F. Study on an innovation tensile-resistant device for laminated rubber bearing. *Ind. Constr.* **2010**, *40*, 43–46.
24. Wang, D.; Lu, X. Experimental research on mechanical properties of anti-tension isolation bearing. *J. Build. Struct.* **2015**, *36*, 124–132.
25. Ge, J.; Zhang, L.; Zhang, G.; Liu, X.; Ma, B.; Guan, Z. Study on design of tensile base performance of the Chengdu museum isolation structure. *J. Build. Struct.* **2016**, *37*, 16–23.
26. Pauletta, M.; Pinzano, F.; Frappa, G.; Russo, G. Tensile tests for the improvement of adhesion between rubber and steel layers in elastomeric isolators. *Appl. Sci.* **2020**, *10*, 8063. [CrossRef]
27. Zhang, Z.; Vassiliou, M.F.; Zhou, Y. Numerical and analytical investigations on tensile behavior of thick rubber bearings. *Eng. Struct.* **2024**, *301*, 117313. [CrossRef]
28. Vaiana, N.; Sessa, S.; Marmo, F.; Rosati, L. An accurate and computationally efficient uniaxial phenomenological model for steel and fiber reinforced elastomeric bearings. *Compos. Struct.* **2019**, *211*, 196–212. [CrossRef]
29. *JGT 118-2018*; Rubber Isolation Bearings for Buildings. Standards Press of China: Beijing, China, 2018.
30. Wen, Y.K. Method for random vibration of hysteretic systems. *J. Eng. Mech.* **1976**, *102*, 249–263. [CrossRef]
31. Park, Y.J.; Wen, Y.K.; Ang, A.H.S. *Earthquake-Resistant Design of Building Structures*; CRC Press: Boca Raton, FL, USA, 1986.
32. Nagarajaiah, S.; Reinhorn, A.M.; Constantinou, M.C. A simplified method for base isolation system design. *Earthq. Eng. Struct. Dyn.* **1991**, *20*, 371–391.
33. Wu, R.; Wang, S.; Liu, W.; Du, D. Research on influence of difference between pulling and pushing stiffness of RB&LRB upon the isolation effectiveness. *Earthq. Resist. Eng. Retrofit.* **2008**, *30*, 24–28.
34. Vaiana, N.; Rosati, L. Classification and unified phenomenological modeling of complex uniaxial rate-independent hysteretic responses. *Mech. Syst. Signal Process.* **2023**, *182*, 109539. [CrossRef]
35. Vaiana, N.; Rosati, L. Analytical and differential reformulations of the Vaiana–Rosati model for complex rate-independent mechanical hysteresis phenomena. *Mech. Syst. Signal Process.* **2023**, *199*, 110448. [CrossRef]
36. Wilson, E. *Three-dimensional Static and Dynamic Analysis of Structures*; Computers & Structures Inc.: Berkeley, CA, USA, 2002.

buildings

MDPI

Article

Numerical Study on Mechanical Behaviors of New Type of Steel Shear-Connection Horizontal Joint in Prefabricated Shear Wall Structure

Xiaohui Wu [1], Yanfeng Wang [1], Shaofeng Ji [1], Mengze Liu [2] and Dayang Wang [2,*]

[1] Power Grid Planning Research Center of Guangdong Power Grid Co., Ltd., Guangzhou 510220, China
[2] School of Civil Engineering, Guangzhou University, Guangzhou 510006, China
* Correspondence: wadaya2015@gzhu.edu.cn

Abstract: The shear connection joint is an important component of the prefabricated shear wall structure system, which plays a dual role in ensuring reasonable force transmission and structural integrity. A new type of steel shear-connection horizontal joint was proposed in this study. In order to verify the effectiveness of the proposed new joint, the established numerical model and its constitutive relationship were first verified based on the existing experimental results. Then, the influence effect on the mechanical behaviors of this new type of joint was further investigated with 20 computational cases. The corresponding influencing laws were established, and optimal parameter ranges were suggested. Finally, a simplified constitutive model—namely, a bilinear constitutive model based on the M-θ relationship—of the new steel shear-connection joint is further advanced and deduced. The results show that the proposed new joint can provide stable shear capacity and superior energy dissipation capacity. The interlocking slot in the new steel shear-connection joint is suggested to be designed with a slot length of 20~30 mm, a slot number of two or three, and a slot thickness of 20~30 mm so as to guarantee superior mechanical behaviors. The advanced bilinear constitutive model can effectively capture the mechanical characteristics of the new joints, in which the maximum error is only 7.67% between theory and simulation.

Keywords: steel shear-connection joint; mechanical behavior; prefabricated shear wall; parameter effect; constitutive model

1. Introduction

The shear wall structure, with its superior seismic and wind resistance performance, has become an essential lateral and load-bearing component for the construction of high and super high-rise buildings [1,2]. In high-rise residential buildings, there is a high demand for prefabricated shear walls [3]. During the construction process, most of the components of the building structure are prefabricated in the factory, and after prefabrication, they are transported to the construction site for on-site assembly. Compared with cast-in-place shear wall structures, it has many advantages such as energy conservation, shortened construction period, improved construction quality, and green environmental protection. It has a relatively broad development space in the future construction industry [4–7].

Many studies have developed a variety of prefabricated shear wall systems and conducted experimental and numerical analyses to explore their seismic behaviors [8–11]. As for the horizontal connections of precast shear wall structures, Soudki et al. [12] investigated an experimental study of horizontal connections for precast wall panels subjected to reversed cyclic shear deformations. Experimental results were used to determine the cyclic behavior of the connections and to identify the contribution of the connection components. Zhao et al. [13] proposed a type of fully assembled horizontal joint using bolt–steel sleeve connectors for prefabricated shear wall structures. The results indicated that the precast shear wall shows a higher horizontal bearing capacity, but slightly lower ductility and energy dissipation than the normal shear wall. Chen et al. [14] proposed a new dry joint, which is a bolted C-shaped steel plate for the horizontal joint of precast shear walls. Four identical sub-assembly precast shear wall panels bolted with four distinct configurations of steel plates (plain, X-stiffeners, horizontal slots, and vertical slots) were tested, which showed that the bolted steel plate with horizontal slots could improve the ductility and energy dissipation of the joint. Wei and Li [15] investigated the mechanical behavior of prefabricated RC shear walls with horizontal joints welded by steel plates by experiment. As for the horizontal connections of precast shear wall structures, Perez et al. [16] studied a fiber based analytical model of unbonded post-tensioned precast concrete walls with vertical joint connectors. The accuracy of the trilinear idealized lateral load behavior was verified based on the derived closed form expressions based on the results of the analytical parameter study. Chu et al. [17] investigated the shear behavior of two-way hollow core precast panel (TWHCPP) shear wall with vertical connections by numerical and experimental methods. The results indicated that the TWHCPP shear wall specimens with vertical connections exhibit monolithic load-bearing mechanisms before the peak point. And the load-bearing mechanism of the TWHCPP shear wall transformed to multiple vertical concrete straps working cooperatively with transverse reinforcements. Huang et al. [18] explored the seismic performance of fully assembled composite walls with vertical joints using different dry connections. Li et al. [19] investigated the dynamic responses of one monolithic reinforced concrete (RC) joint and three PC joints with different wet connection configurations, which showed that the interface damage between PC beam and joint led to the reduced integrity of the PC joints. Zhu et al. [20] developed two innovative precast concrete-encased concrete-filled steel tube (CECFST) column-to-column dry connections (PCDCs), which showed that the PCDCs had excellent tensile resistance and ductility. Li et al. [21] developed a new bolt–plate connection joint assembled by the high strength bolt with embedded steel plate in the precast shear wall (PSW) and investigated the seismic performance of the proposed bolt–plate connection joint under low cyclic loading. Fu et al. [22] studied the seismic behavior of a new connection mode of prefabricated steel reinforced concrete (SRC) shear wall, which show that the failure mode of the new type prefabricated SRC shear wall is the same as that of the integrally cast SRC shear wall, and the seismic behavior and bearing capacity are slightly lower. Fu et al. [23] studied the seismic performance of precast concrete composite shear walls with multiple boundary elements.

It can be found from the above literature review that although the use of bolted connections in prefabricated shear walls may lead to increased local stress damage, the introduction of steel materials has effectively improved the overall mechanical behavior of the assembled components. However, some challenges have been observed in some applications of the previous connection joints, such as tedious lifting procedure on site and crowded reinforcement bars in the joints. Hence, a steel connection joint with high construction efficiency and good seismic performance is seemingly worth investigating for the prefabricated shear wall, which is an effective way of improving the overall seismic performance of structures. A new type of steel shear-connection horizontal joint (SSCH joint) for the prefabricated shear wall structures is proposed in this study. The main innovation of this study focuses on the analysis and verification of the mechanical properties of the new SSCH joint, and it then proposes the design parameters and simplified calculation model for this new SSCH joint. To achieve this goal, numerical analysis with twenty cases for this new joint is then presented under low-cycle reciprocating loading, with calibrated models and parametric variations of research parameters are studied in detail, including length, number and thickness of the interlocking slot and axial compression ratio. The discussions are focused on the bearing capacity, energy dissipation, and ductility to explore the parameter influence laws and design value ranges of the newly proposed joint. And based on these computational data, simplified constitutive relationship model is further derived and validated.

2. Design of the New Steel Shear-Connection Joint

A new type of steel shear-connection horizontal joint was proposed for prefabricated reinforced concrete shear walls, as shown in Figure 1. The SSCH joint consists of pre-embedded shaped steel components, on-site assembled steel components, reserved high-strength bolts, and split bolt sleeves. First, the embedded steel components are embedded at the four end positions of the reinforced concrete shear wall and cast them into shear wall modules in factory. Second, the shear wall module with the pre-embedded steel components is transported to the construction site. The on-site assembled steel components are then inserted into the pre-embedded steel components and connected by high-strength bolts between each other to ensure the collaborative work performance of the SSCH joint. Finally, the gaps and holes in the SSCH joints are filled with expanded concrete to form a shear wall as a whole. In addition, in order to further improve the overall shear performance of the reinforced concrete shear wall using the SSCH joint, the upper and lower boundaries of each shear wall module are connected by interlocking slots, with free contact interfaces between the slots. The interlocking slot is designed as a square shape to achieve higher buckle effect, which is prefabricated and processed Using CNC (computer numerical control) machine tools in the factory after being designed. When subjected to in-plane force, the buckle effect of the interlocking slots between the pre-embedded and on-site assembled steel components tightly connects the upper and lower shear wall modules together. When subjected out-of-plane force, due to the on-site assembled steel components connected to the pre-embedded steel components through high-strength bolts, it can also ensure that the upper and lower shear wall modules work together with the SSCH joints without out of plane instability. Obviously, the number, length, and thickness of the interlocking slot are the most important parameters for achieving the reliability of the SSCH joint connections, which will be discussed in detail in the section of parameter analysis.

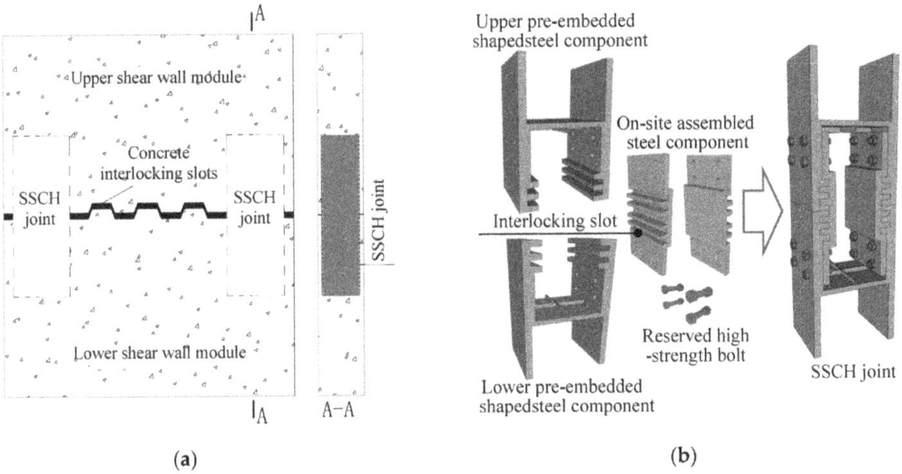

Figure 1. Design of the prefabricated shear wall with the new SSCH joint: (**a**) sketch of the prefabricated shear wall with the SSCH joint; (**b**) details of the SSCH joint.

3. Numerical Model and Validation

3.1. Constitutive Relationship of Concrete and Steel Materials

3.1.1. Constitutive Relationship of Steel Material

The steel adopts ABAQUS's built-in bilinear dynamic strengthening constitutive model BKIN, which follows the Mises stress yield criterion and strengthening criterion and takes into account the Bauschinger effect of the steel, which can accurately simulate the mechanical performance of the steel under reciprocating loads. The numerical model of steel is Q345, with a yield strength of 345 MPa, an elastic modulus of 2×10^5 MPa, and Poisson's ratio of 0.3.

3.1.2. Constitutive Relationship of Concrete Material

ABAQUS has three built-in concrete constitutive models: plastic damage constitutive model, smeared cracking model, and brittle cracking model. Among the three constitutive relationship models, the plastic damage constitutive model takes into account the differences in material tensile and compressive properties and reflects them through different damage factors [24], which is used to simulate the stress characteristics of concrete under cyclic or dynamic loads It is easier to converge compared to other models and is not affected by cracks. The plastic damage constitutive model is then adopted in this study.

There are five material parameters for the plastic damage constitutive model that need to be defined firstly: expansion angle, eccentricity, yield strength ratio f_{b0}/f_{c0}, the second stress invariant k in the meridian plane of tension and compression, and the viscosity parameter. The expansion angle and eccentricity define the shape of the flow potential energy surface. The yield strength ratio and the second stress invariant k determine the shape of the yield surface in the plane stress and off plane. The viscosity parameter is closely related to whether the analysis results can converge, and the default value is 0. The larger the viscosity coefficient, the easier the material stiffness and convergence of the analysis. The smaller the viscosity parameter, the closer the calculation results are to the real situation, but the more difficult it is to converge the analysis results [10]. After comprehensive consideration of factors such as analysis accuracy, computational efficiency, and convergence, the final parameter values determined in this study are shown in Table 1.

Table 1. Parameter values of the concrete plastic damage model.

Elasticity parameter	Young's modulus			3×10^{10} N/m^2	
	Poisson's ratio			0.2	
Plastic parameter	Expansion angle	Eccentricity	f_{b0}/f_{c0}	k	Viscosity parameter
	30	0.1	1.16	0.6667	0.0005

The constitutive definition of concrete consists of two parts: elasticity and plasticity. The elasticity part is determined by the elastic modulus and initial yield strength of concrete. The stress–strain relationship of the plasticity part is relatively complex, which adopts Zhenhai Guo's model in this study [25]. As shown in Figure 2a, the stress–strain equation can be expressed as follows under uniaxial compression:

$$y = \begin{cases} \frac{E_0 \varepsilon_c}{f_c} x & (x \leq 0.211) \\ \alpha_a x + (3 - 2\alpha_a)x^2 + (\alpha_a - 2)x^3 & (0.211 < x \leq 1) \\ \frac{x}{\alpha_d (x-1)^2 + x} & (x > 1) \end{cases} \quad (1)$$

where x is the strain, y is the stress, ε_c is the peak strain, $\varepsilon_c = (700 + 172\sqrt{f_c}) \times 10^{-6}$, E_0 is an elastic modulus of concrete, f_c is concrete axial compressive strength, and α_a and α_d are parameters determined by f_c and $f_c^{0.785}$, namely, $\alpha_a = 2.4 - 0.0125 f_c$, $\alpha_d = 0.157 f_c^{0.785} - 0.905$.

The stress–strain equation can be expressed as follows under uniaxial tension, as shown in Figure 2b:

$$y = \begin{cases} 1.2x - 0.2x^6 & (x \leq 1) \\ \frac{x}{\alpha_t (x-1)^{1.7} + x} & (x > 1) \end{cases}, \quad (2)$$

where $\alpha_t = 0.312 f_t^2$ and f_t is concrete axial tensile strength.

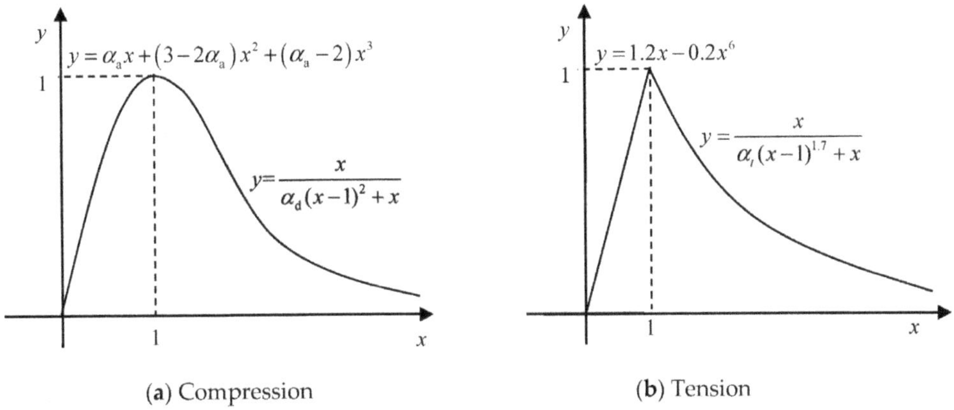

Figure 2. Stress–strain curves of concrete plastic damage constitutive model.

Some phenomena, such as plastic cumulative and stiffness degradation, will be generated for concrete material under tension and compression reciprocating loading, which has great difference with one-way loading. For example, the force–displacement hysteresis curve may be very full, with no pinching phenomenon under reciprocating loading, if the damage factor is not considered, which will then cause over an estimation of energy dissipation capacity of concrete members. Actually, there is a certain pinch phenomenon of hysteresis curves for concrete members in experimental loading [11]. Hence, the concept of damage factor is introduced to describe the complex behavior of concrete under reciprocating loading. Energy equivalence method and graphic method [26] are the commonly used methods employed to calculate the damage factor, and the latter is adopted in this paper, as shown in Figure 3.

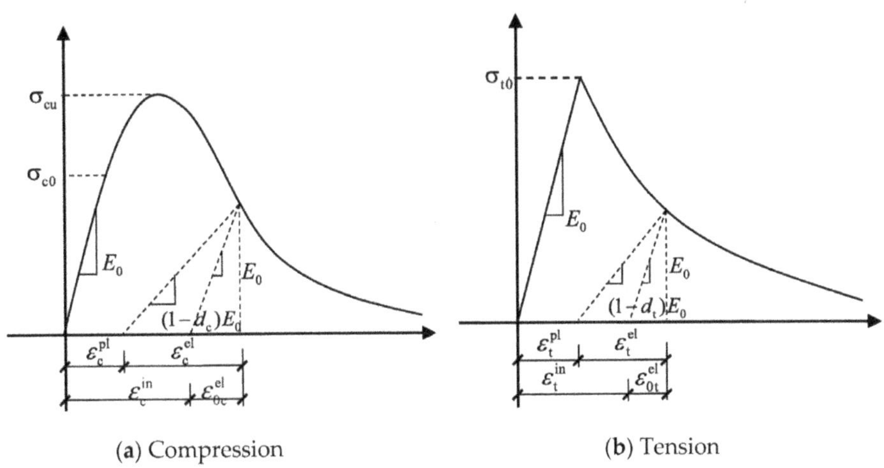

Figure 3. Stress–strain curve of concrete damage model.

As shown in Figure 3a, the initial relationship between stress and strain is linear when the condition of uniaxial compression. But after the stress arrives to its elastic maximum compressive stress σ_{c0}, the relationship will assume the status of a rising curve and then start to decrease in compression when the maximum compressive stress σ_{cu} is reached.

Finally, unloading begins at stress point σ_{cn}. The compressive plastic strain of concrete can be expressed as follows [26]:

$$\begin{cases} \tilde{\varepsilon}_c^{pl} = \tilde{\varepsilon}_c^{in} - \left(\varepsilon_c^{el} - \varepsilon_{oc}^{el}\right) = \tilde{\varepsilon}_c^{in} - \left[\frac{\sigma_{cn}}{(1-d_c)E_0} - \frac{\sigma_{cn}}{E_0}\right] = \tilde{\varepsilon}_c^{in} - \frac{d_c}{1-d_c}\frac{\sigma_{cn}}{E_0} \\ \tilde{\varepsilon}_c^{in} = \varepsilon_{cn} - \varepsilon_{oc}^{el} \\ \varepsilon_{oc}^{el} = \frac{\sigma_{cn}}{E_0} \end{cases} . \quad (3)$$

Hence, the compressive damage factor can be obtained:

$$d_c = 1 - \frac{\sigma_{cn} {E_0}^{-1}}{\tilde{\varepsilon}_c^{pl}(1/b_c - 1) + \sigma_{cn} {E_0}^{-1}}, \quad (4)$$

where $b_c = \tilde{\varepsilon}_c^{pl}/\tilde{\varepsilon}_c^{in}$, d_c is compressive damage factor and $0 \le d_c \le 1$. The concrete will be fully damaged when $d_c = 1$, partly damaged when $0 \le d_c \le 1$, and not damaged when $d_c = 0$. σ_{cn} and ε_{cn} are the compressive stress and strain, respectively, of the unloading point. $\tilde{\varepsilon}_c^{pl}$ is compressive plastic strain, $\tilde{\varepsilon}_c^{in}$ is compressive nonlinear strain, ε_c^{el} is compressive linear strain considering damage effect, and ε_{oc}^{el} is compressive linear strain without damage effect.

The initial relationship between stress and strain is linear when the condition of uniaxial tension. After the stress arrive to its elastic maximum tension stress σ_{t0}, the relationship assumes a curve status and then starts to soften in tension. Finally, unloading begins at stress point σ_{tn}. The tension plastic strain of concrete can be expressed as follows:

$$\begin{cases} \tilde{\varepsilon}_t^{pl} = \tilde{\varepsilon}_t^{ck} - \left(\varepsilon_t^{el} - \varepsilon_{ot}^{el}\right) = \tilde{\varepsilon}_t^{ck} - \left[\frac{\sigma_{tn}}{(1-d_t)E_0} - \frac{\sigma_{tn}}{E_0}\right] = \tilde{\varepsilon}_t^{ck} - \frac{d_t}{1-d_t}\frac{\sigma_{tn}}{E_0} \\ \tilde{\varepsilon}_t^{ck} = \varepsilon_{tn} - \varepsilon_{ot}^{el} \\ \varepsilon_{ot}^{el} = \frac{\sigma_{tn}}{E_0} \end{cases} . \quad (5)$$

Hence, the tension damage factor can be obtained:

$$d_t = 1 - \frac{\sigma_{tn} {E_0}^{-1}}{\tilde{\varepsilon}_t^{pl}(1/b_t - 1) + \sigma_{tn} {E_0}^{-1}}, \quad (6)$$

where $b_t = \varepsilon_t^{pl}/\varepsilon_t^{ck}$, d_t is tension damage factor and $0 \le d_c \le 1$. The concrete will be fully damaged in tension when $d_c = 1$, partly damaged when $0 \le d_c \le 1$, and not damaged when $d_c = 0$. σ_{tn} and ε_{tn} are tension stress and strain, respectively, of the unloading point. ε_t^{pl} is tension plastic strain, ε_t^{ck} is tension nonlinear strain, ε_t^{el} is tension linear strain considering damage effect, and ε_{ot}^{el} is tension linear strain without damage effect. In Equations (4) and (6), the values of b_c and b_t adopt the suggestion of Birtel and Mark [27], and the two parameters can generally be defined as $b_c = 0.7$ and $b_t = 0.1$. The values of constitutive relationship considering the plastic damage effect of concrete and corresponding damage factors can then be calculated for simulation purpose.

3.2. Validation of a Numerical Model

A numerical model using a computer program ABAQUS 6 software is implemented for the nonlinear dynamic analysis of a reinforced concrete shear wall using finite modeling, material constitutive, and computational procedures described in the previous sections. The numerical model has been studied by Qian et al. [28] using mechanical performance test. Figure 4 shows the specific geometry and rebar details of the shear wall specimen. Table 2 shows the yield strength and ultimate strength of the rebars. The concrete strength grade used in this shear wall is C40, and the axial compression ratio is 0.55. The loading method is first to use force loading, repeat each level of load once, and then use displacement loading with each level of load cycling three times. The maximum loading displacement is 40 mm. The height of the loading point is 2960 mm from the bottom section of the wall. When the bearing capacity of the specimen decreases to 85% of the peak bearing capacity, or when the specimen cannot continue to bear the load, the test stops.

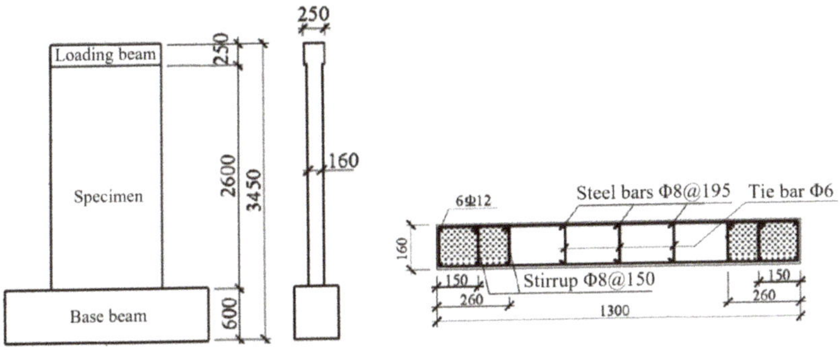

Figure 4. Specific geometry and rebar details of the shear wall specimen [28].

Table 2. Yield and ultimate strength of steel.

Steel Type	Yield Strength/MPa	Ultimate Strength/MPa
B12	389	540
B8	330	484
B6	400	546

Based on the sample parameters provided above, the numerical model of the shear wall established in this section is shown in Figure 5, and the model details is shown in Table 3. Solid modeling is adopted in this section. The element type used in concrete is the three-dimensional eight-node reduced integral element C3D8R. The element type used for the reinforcement is the three-dimensional two-node truss element T3D2. The interface between steel bars and concrete adopts a binding relationship. The loading beam, base beam, and the specimen are bound and consolidated together through internal steel mesh and concrete. The boundary condition is defined as bottom consolidation, which constrains the displacement and rotation in three directions. The loading method is consistent with the referenced literature, namely, force loading followed by displacement loading, and with a maximum loading displacement of 40 mm.

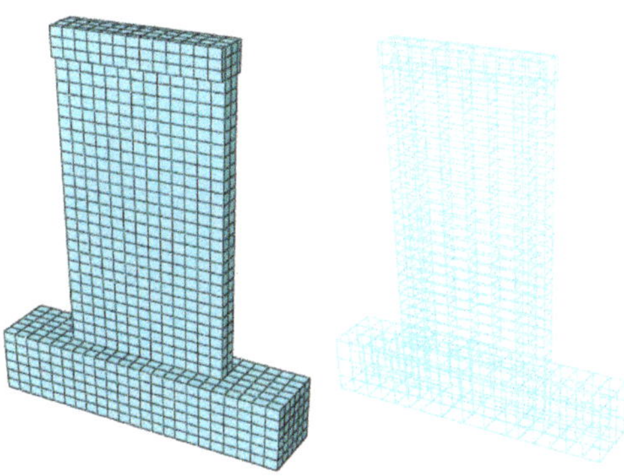

(**a**) Shear wall body (**b**) Rebar mesh

Figure 5. Numerical model of the shear wall.

Table 3. Numerical model parameters.

Component Category	Material	Element Type	Mesh Size	Element Number	Node Number	Constitutive Model
Wall body	Concrete	C3D8R	10 mm	1660	25,772	Plastic damage constitutive
Rebar mesh	Steel	T3D2	10 mm	2506	24,716	Bilinear dynamic strengthening constitutive

Result comparisons of the load–displacement skeleton curves between the referenced literature and this study are given in Figure 6. It can be found that there is a good fit between the simulation and experiment results. The positive loading forces obtained from experiment and numerical calculation are 561 kN and 532 kN, respectively, with an error of 5.2%. The negative loading forces obtained via experiment and numerical calculation are −445 kN and −458 kN, respectively, with an error of 2.9%. The small difference between the referenced literature and this study may be due to differences in the material models used and in the structure and approximations made in the numerical techniques. Hence, it can be concluded that the numerical model used in this study is reasonable and feasible, which guarantees the reliability of following simulations.

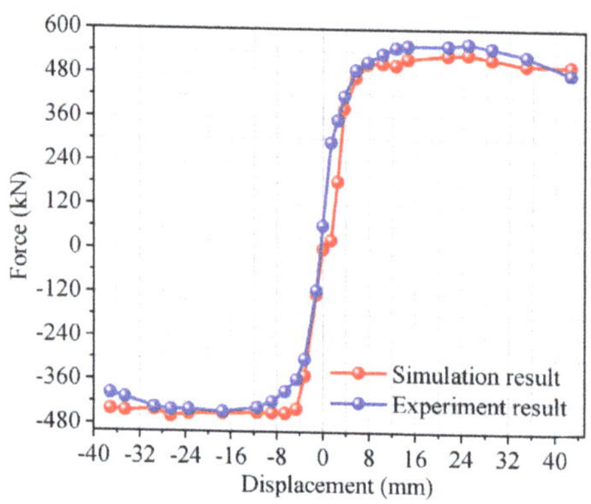

Figure 6. Load−displacement skeleton curves.

3.3. Model Establishment of the SSCH Joint

ABAQUS software is adopted to build a three-dimensional finite element model of the SSCH joint, as shown in Figure 7. The SSCH joint has dimensions of 200 mm × 240 mm × 680 mm, with the thickness of the steel plate of 20 mm. Q345 is used by the SSCH joint with the yield strength of 345 MPa. The holes in the SSCH joint are filled with C40 concrete. The steel plate and concrete of the SSCH joint are modeled using a three-dimensional solid element C3D8R, which is a linear reduced integral element with eight nodes and six facets. Plastic damage constitutive was adopted by the concrete material. Bilinear dynamic strengthening constitutive was adopted by the steel of the SSCH joint [29]. Finite element mesh generation is a very important step in the entire finite element numerical simulation analysis, and the number of grids directly affects the calculation accuracy and cost. In order to find the most reasonable grid division, comparative model calculations are conducted for 10 different grid divisions, as shown in Table 4.

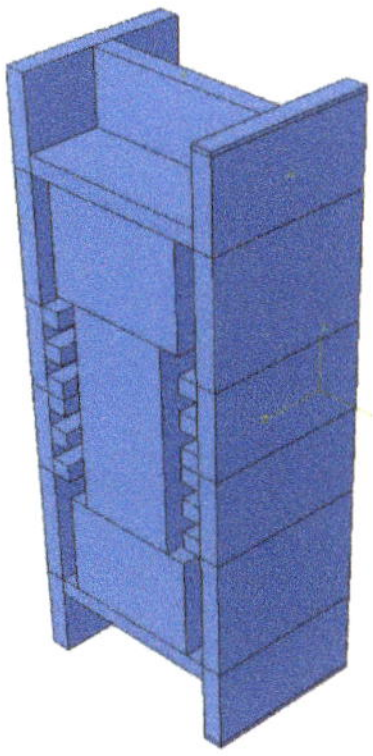

Figure 7. Numerical model of the SSCH joint.

Table 4. Model design for different grid divisions.

Model Number	Mesh Size	Node Number	Element Number
1	6 mm	185,646	172,436
2	8 mm	72,568	65,130
3	10 mm	44,262	38,694
4	12 mm	23,340	19,964
5	16 mm	8100	6435
6	20 mm	4800	3676

The bottom stress and maximum force of the different models are compared, and the results are shown in Figure 8. It can be found that the stress and maximum load at the bottom of the SSCH joints show a trend of first rapidly increasing, then slowly increasing, and finally tending to remain unchanged. The inflection point of the curve appears in Model 6, where the mesh size of the model is 10 mm. Therefore, the subsequent calculation models all use 10 mm grid size to ensure the rationality of the calculation results.

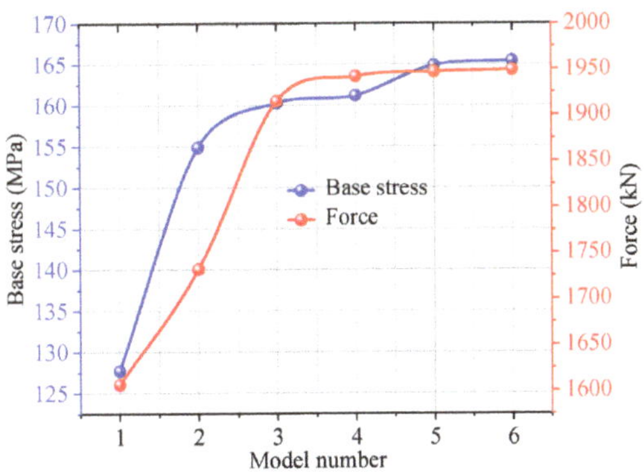

Figure 8. Comparison bottom stress and maximum force of the different models.

Another important issue in establishing the SSCH joint model is the contact problem, as there are a large number of contact interfaces, such as the interfaces between the pre-embedded steel components and the on-site assembled steel components, between the steel components and concrete of the SSCH joint. The concrete inside SSCH joint are filled by the micro expansion concrete filled. After the concrete solidifies, it will generate strong pressure on the steel plate groove and increase the friction force between each other. In addition, there are high-strength bolts connecting the steel plate and concrete through it. When the gap between the bolt hole and the bolt is not considered, there is almost no relative displacement between the filled micro expansion concrete and the steel plate. Therefore, the contact interface between steel plate and concrete is defined by binding relationship in simulation [29].

The interfaces between the pre-embedded steel components and the on-site assembled steel components are assembled and transmitted through the interlocking slots. This type of contact interface can transmit pressure to each other under pressure. But when pressure is not present, the contact interface will separate from each other. Obviously, this contact interface cannot be defined as a binding relationship in numerical models. This contact relationship is simulated using a hard contact with friction coefficient in this study. According to the Code for Design of Steel Structures [30], the friction coefficient between the steel interfaces is usually within the range of 0.3–0.5. Therefore, the contact interfaces between the pre-embedded steel components and the on-site assembled steel components is defined as hard contact, with a friction coefficient of 0.5.

4. Parameter Effect of the SSCH Joint

4.1. Calculation Cases

It can be found that the mechanical performance of the SSCH joint is mainly achieved through the buckle connection among the interlocking slots. The design parameters of the interlocking slots will directly affect the transmission of mechanical properties and mechanical behavior of prefabricated shear walls. Therefore, this section will focus on discussing the effect of slot parameters on the mechanical behavior of the SSCH joints, in order to achieve optimal parameter design. The calculation cases are shown in Table 5, which are explained specifically as follows:

LE10~LE40: Parameter effect of the length of the interlocking slot is discussed on the mechanical behavior of the SSR joint in these four cases, as shown in Figure 9a.

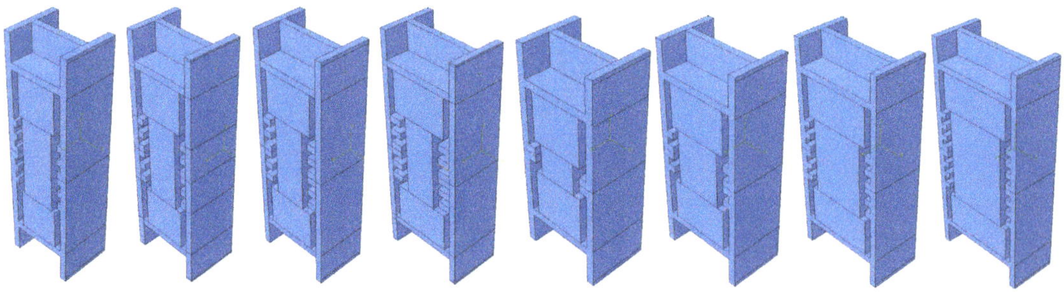

(**a**) LE10~LE40 (**b**) BNU1~ BNU4

Figure 9. Schematic diagram of the calculation models.

BNU1~BNU4, CNU1~CNU4: Parameter effect of the number of the interlocking slot is discussed on the mechanical behavior of the SSR joint in these eight cases. As for the cases of BNU1~BNU4, the length of the interlocking slot is defined as 20 mm, as shown in Figure 9b. As for the cases of CNU1~CNU4, the length of the interlocking slot is defined as 30 mm.

TH10~TH40: Parameter effect of the thickness of the interlocking slot is discussed on the mechanical behavior of the SSR joint in these four cases.

ACR3~ACR6: Parameter effect of the axial compression ratio of the interlocking slot is discussed on the mechanical behavior of the SSR joint in these four cases.

Table 5. Design of the calculation cases.

Case Name	Interlocking Slot			Axial Compression Ratio	Mesh Size	Contact Interface	Friction Coefficient
	Length	Number	Thickness				
LE10	10 mm	3	30 mm	0.5	10 mm	Hard contact	0.5
LE20	20 mm	3	30 mm	0.5	10 mm	Hard contact	0.5
LE30	30 mm	3	30 mm	0.5	10 mm	Hard contact	0.5
LE40	40 mm	3	30 mm	0.5	10 mm	Hard contact	0.5
BNU1	20 mm	1	20 mm	0.5	10 mm	Hard contact	0.5
BNU2	20 mm	2	20 mm	0.5	10 mm	Hard contact	0.5
BNU3	20 mm	3	20 mm	0.5	10 mm	Hard contact	0.5
BNU4	20 mm	4	20 mm	0.5	10 mm	Hard contact	0.5
CNU1	30 mm	1	20 mm	0.5	10 mm	Hard contact	0.5
CNU2	30 mm	2	20 mm	0.5	10 mm	Hard contact	0.5
CNU3	30 mm	3	20 mm	0.5	10 mm	Hard contact	0.5
CNU4	30 mm	4	20 mm	0.5	10 mm	Hard contact	0.5
TH10	30 mm	3	10 mm	0.5	10 mm	Hard contact	0.5
TH20	30 mm	3	20 mm	0.5	10 mm	Hard contact	0.5
TH30	30 mm	3	30 mm	0.5	10 mm	Hard contact	0.5
TH40	30 mm	3	40 mm	0.5	10 mm	Hard contact	0.5
ACR3	30 mm	3	20 mm	0.3	10 mm	Hard contact	0.5
ACR4	30 mm	3	20 mm	0.4	10 mm	Hard contact	0.5
ACR5	30 mm	3	20 mm	0.5	10 mm	Hard contact	0.5
ACR6	30 mm	3	20 mm	0.6	10 mm	Hard contact	0.5

4.2. Effect of the Length of the Interlocking Slot

Figure 10 shows a comparison of the force–displacement relationship and the stiffness degradation of the SSCH joints in different lengths of the interlocking slot. Table 6 gives the key mechanical properties of the SSCH joints in different lengths of the interlocking slot. It can be observed that the yield displacement of SSCH joints with different slot lengths is around 6.7 mm, and the ultimate displacement is around 16 mm, without significant difference. However, as the slot length increases, the bearing capacity and the initial stiffness of the SSCH joint gradually increases with the increasing length of the interlocking slot. For example, the yield forces of the cases LE10, LE20, LE30, and LE40 are 1009.4 kN, 1227.8 kN, 1489.6 kN, and 1632.4 kN, respectively. Correspondingly, the peak force are 1243.5 kN, 1428 kN, 1657.5 kN, and 1867.5 kN, respectively. In addition, the energy consumption capacity of the SSCH joints also increases with the increase in slot length. Compared to the SSCH joint with the slot length of 10 mm, the average energy dissipation per cycle of the joint with the slot length of 20 mm increased by 15.27%, and the total energy increased by 16.7%. The average energy dissipation per cycle of the joint with the slot length of 30 mm increased by 16.47%, and the total energy increased by 18.82%. The average energy dissipation per cycle of the joint with the slot length of 40 mm increased by 22.82%, and the total energy increased by 21.78%.

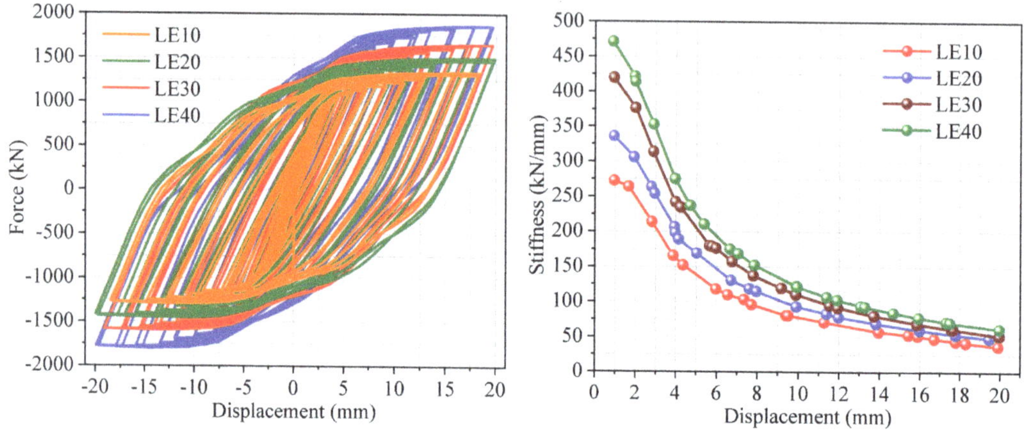

(**a**) Force − displacement relationship (**b**) Stiffness degradation

Figure 10. Curves depicting the force − displacement relationship and stiffness degradation for the cases LE10~LE40.

Table 6. Key mechanical properties of the SSCH joint in different lengths of the interlocking slot.

Cases	Yield Point			Peak Point			Total Energy Dissipation		Ductility Coefficient
	Δ_y (mm)	P_y (kN)	*AP*	Δ_u (mm)	P_u (kN)	*AP*	Energy (kN·m)	*AP*	
LE10	3.63	1009.40	——	15.89	1243.50	——	473.00	——	4.37
LE20	3.73	1227.80	21.64%	16.02	1428.00	14.84%	552.00	16.70%	4.30
LE30	3.76	1489.60	47.57%	16.56	1657.50	33.29%	562.00	18.82%	4.41
LE40	3.83	1632.40	61.72%	16.87	1867.50	50.18%	576.00	21.78%	4.41

Note: *AP* = Value of (LE20/30/40 − LE10)/Value of LE10 × 100%. Meanings of the *AP* in following tables are the same.

The stiffness degradation of the SSCH joints with different slot lengths shows a consistent variation pattern; that is, the decrease is significant in the initial stage of loading, but the decrease trend slows down in the later stage of loading. For the SSCH joint with different slot lengths, although their stiffness gradually increases with the increase in slot length, the amplification in stiffness decreases with the increase in slot length. This phenomenon indicates that an increase in the slot length is beneficial for improving the stiffness of the SSCH joint, but it is only effective within a certain range of slot length. The same phenomenon can also be found in the bearing capacity and energy dissipation of the SSCH joint. Hence, the recommended range for the length of the interlocking slot is 20 mm~30 mm.

Contours of Mises stress with changing of the loading displacement for the Case LE10 and LE30 are shown in Figure 11. It can be found that the Mises stress has an almost consistent development pattern for the Cases. In the initial stage of loading, stress concentration areas appear at the bottom of the steel component and at both ends of the interlocking slot. After that, the stress concentration area slowly expands and begins to spread from the bottom of the steel component to the upper part. In the condition of the same loading displacement, it can be found that the stress diffusion is faster, and the stress concentration range is larger when the length of the interlocking slot is smaller, indicating that increase in the length of the slot is beneficial for improving the bearing capacity of the SSCH joint.

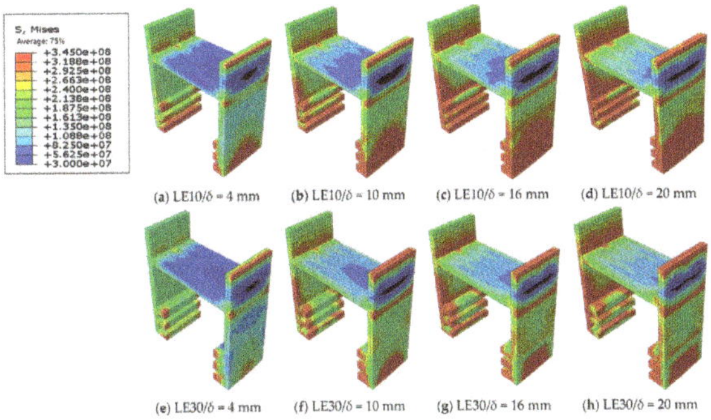

Figure 11. Contours of Mises stress with changing of the loading displacement for different slot length.

4.3. Effect of the Number of the Interlocking Slot

Figures 12 and 13 show the force–displacement relationship and the stiffness degradation of the SSCH joint in different numbers of the interlocking slot. Table 7 gives the key mechanical properties of the SSCH joint in different numbers of the interlocking slot. Under the same length of tooth slots, it can be found that the mechanical behaviors of the SSCH joints, such as load-bearing capacity, stiffness degradation, and ductility, show a consistent trend with the increase in the slot number. For example, the yield and peak forces of the SSCH joint with one interlocking slot and 20 mm slot length are 1026 kN and 1347 kN. Correspondingly, the yield and peak forces with four interlocking slots and 20 mm slot length are 1064.4 kN and 1447.5 kN. The latter only increased by 3.74% and 7.46% compared to the former, respectively. Similarly, when the slot length is 30 mm, the yield and peak forces of the SSCH joints with four interlocking slots only increase by 6.21% and 2.48% compared to joint with one interlocking slot. In addition, under the same condition of the slot length, although the ductility coefficient of the SSCH joints decreases to a certain extent with the increase in the slot number, the degree of decrease is also very small. Under the condition of 20 mm slot length, the ductility coefficient of the SSCH joints with four slots decreases by only 1.86% compared to one slot. Under the condition of 30 mm slot length, the ductility coefficient decreases by 8.48%.

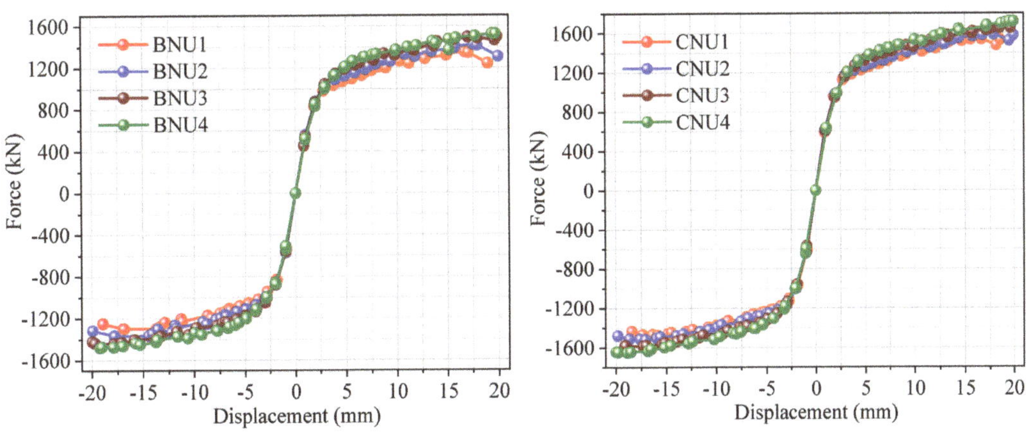

(**a**) Cases BNU1~BNU4

(**b**) Cases CNU1~CNU4

Figure 12. Force − displacement curves of the SSCH joint in different numbers of the interlocking slot.

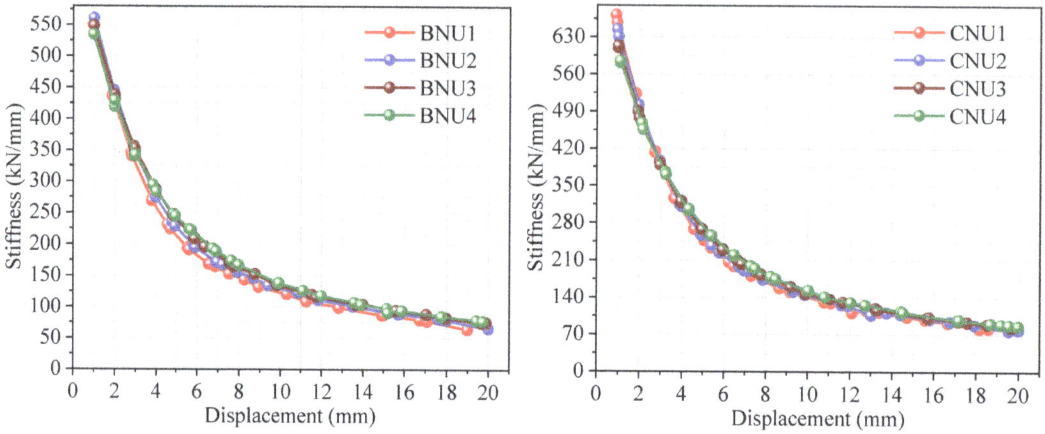

(**a**) Cases BNU1~BNU4 (**b**) Cases CNU1~CNU4

Figure 13. Stiffness degradation curves of the SSCH joint in different numbers of the interlocking slots.

Table 7. Key mechanical properties of the SSCH joint in different numbers of the interlocking slots.

Cases	Yield Point			Peak Point			Total Energy Dissipation		Ductility Coefficient
	Δ_y (mm)	P_y (kN)	*AP*	Δ_u (mm)	P_u (kN)	*AP*	Energy (kN·m)	*AP*	
BNU1	3.69	1026.0	——	15.52	1347.0	——	442.48	——	4.20
BNU2	3.81	1038.0	1.17%	15.89	1389.0	3.12%	490.28	10.80%	4.17
BNU3	3.73	1052.4	2.57%	16.02	1428.0	6.01%	552.13	24.78%	4.30
BNU4	3.84	1064.4	3.74%	16.45	1447.5	7.46%	575.28	30.01%	4.28
CNU1	3.34	1218.0	——	16.02	1633.5	——	479.83	——	4.79
CNU2	3.53	1254.0	2.96%	16.15	1647.0	0.83%	531.92	10.86%	4.58
CNU3	3.76	1276.8	4.83%	16.56	1657.5	1.47%	561.91	17.11%	4.41
CNU4	3.83	1293.6	6.21%	16.78	1674.0	2.48%	586.97	22.33%	4.38

Although the mechanical parameters, such as joint bearing capacity, stiffness degradation, and ductility coefficient, do not change significantly with the increase in the slot number, the energy dissipation capacity of the SSCH joints has significantly improved. For example, the total energy dissipation of the cases BNU1, BNU2, BNU3, and BNU4 are 442.48 kN·m, 490.28 kN·m, 552.13 kN·m, and 575.28 kN·m, respectively, and the corresponding growth rates are 10.80%, 24.78%, and 30.01% relative to the joint with one slot. Correspondingly, the total energy dissipation of the cases CNU1, CNU2, CNU3, CNU4 are 479.83 kN·m, 531.92 kN·m, 561.91 kN·m, and 586.97 kN·m, and the corresponding growth rates are 10.86%, 17.11%, and 22.33%. It can be seen that although the energy consumption capacity of the SSCH joints is also increasing with the increase in the slot number, the growth rate is continuously decreasing. As is well-known, the more slots there are, the more complex the processing, transportation, and assembly of the joints. Therefore, considering factors of the complexity of joint construction, the economy of material use, the convenience of on-site assembly, and so on, this study suggests that the number of interlocking slots for the SSCH joint should be two or three.

4.4. Effect of the Thickness of the Interlocking Slot

Figure 14 shows the comparisons of hysteretic curves, skeleton curves, stiffness degradation, and energy dissipation of the SSCH joints for the cases TH10~TH40. Table 8 gives the key mechanical properties of the SSCH joint in different thicknesses of the interlocking slot. Contours of Mises stress with changing of the loading displacement for the CaseTH30 and TH40 are shown in Figure 15. It can be observed that the yield displacement, yield force, ultimate displacement, and ultimate force of the SSCH joints all significantly increase with the increase in slot thickness. Compared to the yield force and ultimate force of the SSCH joints with 10 mm slot thickness, the SSCH joints with 20 mm slot thickness are increased by 18.54% and 17.92%, the SSCH joints with 30 mm slot thickness are increased by 47.53% and 42.04%, and the SSCH joints with 40 mm slot thickness are increased by 62.29% and 56.66%, respectively. This indicates that the thickness of the slot steel plate of the SSCH joint is beneficial for significantly improving the load-bearing capacity.

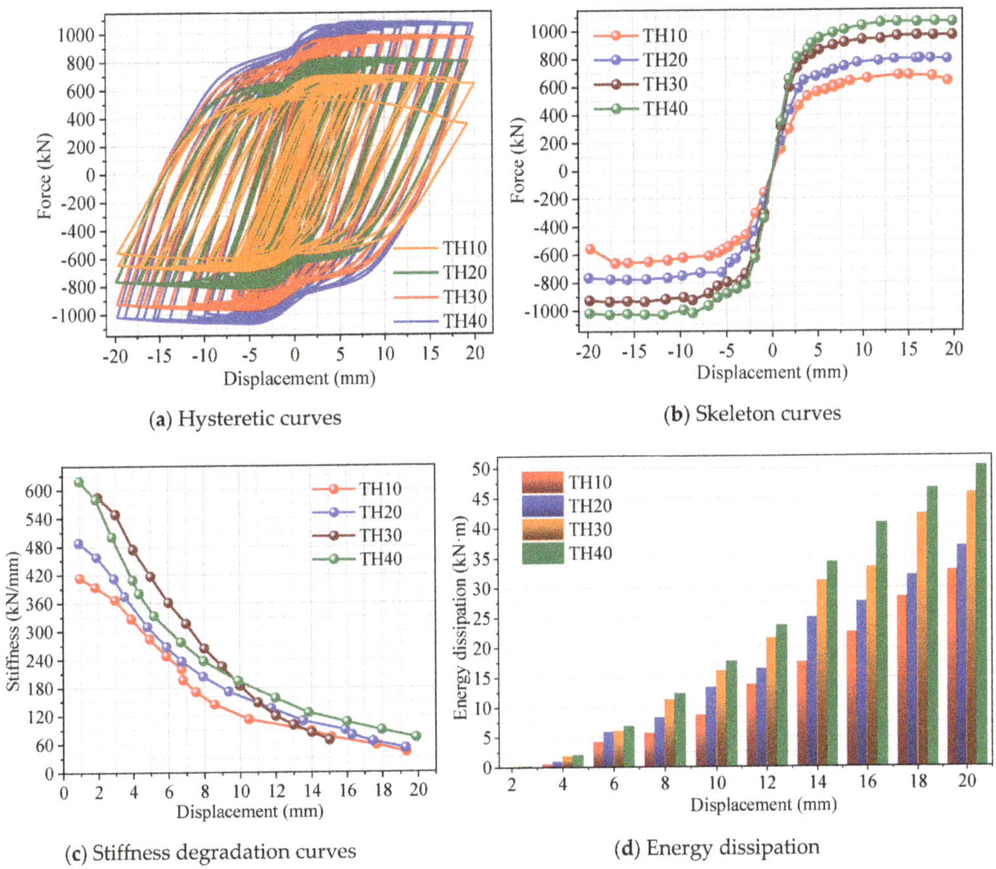

Figure 14. Mechanical behaviors of the SSCH joints for the cases TH10 ~ TH40.

Table 8. Key mechanical properties of the SSCH joint in different thickness of the interlocking slot.

Cases	Yield Point			Peak Point			Total Energy Dissipation		Ductility Coefficient
	Δ_y (mm)	P_y (kN)	*AP*	Δ_u (mm)	P_u (kN)	*AP*	Energy (kN·m)	*AP*	
TH10	3.73	506.66	——	15.11	680.11	——	296.38	——	4.05
TH20	3.73	600.62	18.54%	17.46	801.96	17.92%	401.25	35.38%	4.68
TH30	3.74	747.48	47.53%	17.96	966.02	42.04%	465.97	57.22%	4.80
TH40	3.74	822.23	62.29%	17.98	1065.48	56.66%	512.56	72.94%	4.80

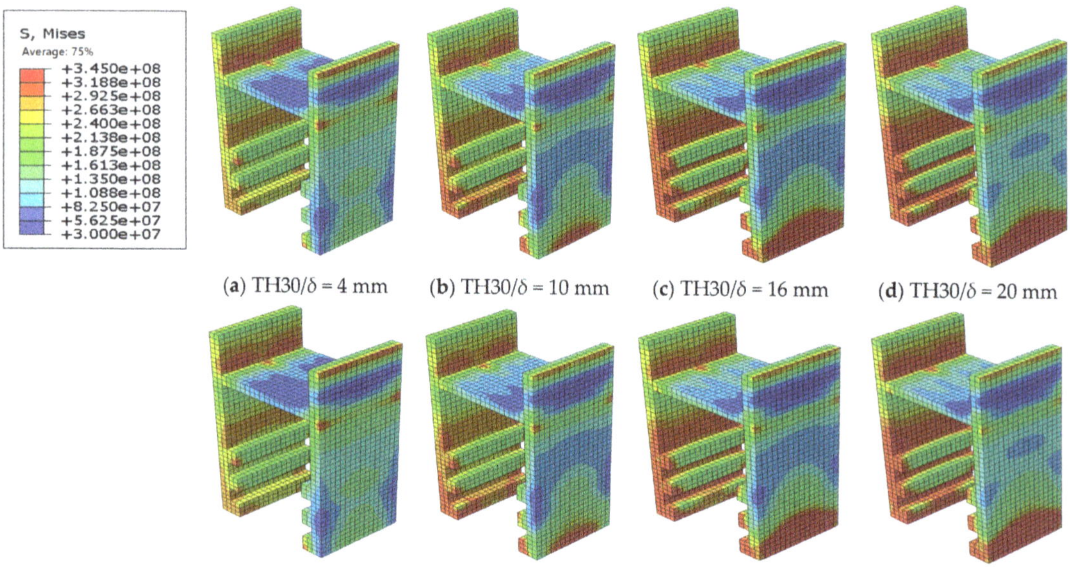

(**a**) TH30/δ = 4 mm (**b**) TH30/δ = 10 mm (**c**) TH30/δ = 16 mm (**d**) TH30/δ = 20 mm

(**e**) TH40/δ = 4 mm (**f**) TH40/δ = 10 mm (**g**) TH40/δ = 16 mm (**h**) TH40/δ = 20 mm

Figure 15. Contours of Mises stress with changing of the loading displacement for different slot thickness.

It can be observed that there is also a similar trend in the energy consumption capacity of the SSCH joints. Namely, the energy consumption capacity of the SSCH joints gradually increases with the increase in the thickness of the slot steel plate. Compared to the SSCH joint with 10 mm slot thickness, the total energy dissipations of the SSCH joints with a slot thickness of 20 mm, 30 mm, and 40 mm are increased by 35.38%, 57.22%, and 72.94%, respectively. In addition, the stiffness degradation of the SSCH joints also follows a similar pattern. The thicker the slot steel plate, the greater the overall stiffness of the joint, and the slower the degradation effect. Although an increase in the thickness of the slot steel plate will bring superior characteristics, such as bearing capacity, energy consumption, and stiffness degradation, it will also lead to an increase in the joint height and width, which may further cause certain constraints on joint design, cost control, on-site construction, etc. So, the recommended range for the thickness of slot steel plate of the SSCH joint is 20–30 mm. In addition, it can be found that the Mises stress of the SSCH joints gradually increases with increase in loading displacement, and the stress concentration area continuously diffuses from the bottom to the upper part. The stress distribution and diffusion range of the SSCH joint with different slot thicknesses exhibit similar variation patterns.

4.5. Effect of the Axial Compression Ratio

Figure 16 shows the comparisons of the skeleton curves and energy dissipation of the SSCH joints for the cases ACR3~ACR6. Table 9 gives the key mechanical properties of the SSCH joint in different axial compression ratios. It can be observed that the skeleton curves of the SSCH joints under different axial compression ratios exhibit almost identical variation patterns. Under the same loading displacement, the bearing capacity, ductility, and energy dissipation of the SSCH joints also does not differ significantly. For example, the yield force of the cases ACR3~ACR6 are 1039 kN, 1051 kN, 1064 kN, and 1078 kN, respectively, in which the difference between the maximum and minimum values is only 3.75%. Correspondingly, the ductility of the cases ACR3~ACR6 are 4.38, 4.36, 4.41, 4.44, respectively, in which the difference between the maximum and minimum values is only 1.25%. It can then be concluded that the axial compression ratio has a certain effect on the mechanical behaviors of the SSCH joints, but this effect is very small and can be basically ignored.

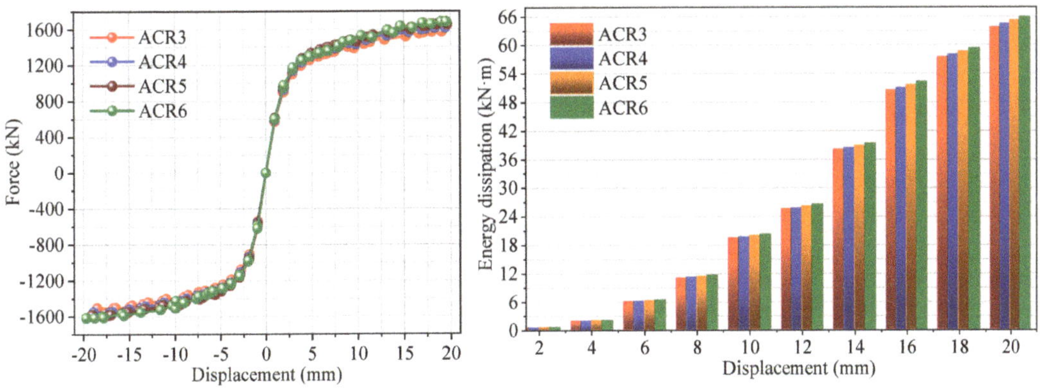

(**a**) Skeleton curves (**b**) Energy dissipation

Figure 16. Mechanical behaviors of the SSCH joints for the cases ACR3 ~ ACR6.

Table 9. Key mechanical properties of the SSCH joint in different axial compression ratios.

Cases	Yield Point			Peak Point			Total Energy Dissipation		Ductility Coefficient
	Δ_y (mm)	P_y (kN)	*AP*	Δ_u (mm)	P_u (kN)	*AP*	Energy (kN·m)	*AP*	
ACR3	3.62	1039.00	——	15.87	1591.50	——	543.55	——	4.38
ACR4	3.69	1051.00	1.15%	16.12	1630.50	2.45%	552.21	1.59%	4.36
ACR5	3.76	1064.00	2.41%	16.56	1657.50	4.15%	561.91	3.38%	4.41
ACR6	3.83	1078.00	3.75%	16.98	1687.50	6.03%	567.85	4.47%	4.44

5. Simplified Constitutive Model of the SSCH Joint

5.1. Theory Deduction

According to the deformation characteristics and the degree of transferable bending moment of the connection joints under external forces, joint connections can be divided into rigid connection joints, semi-rigid connection joints, and hinge connection joints. Semi-rigid nodes have greater advantages, which can effectively reflect the bending moment and rotational effects of joints, have simpler construction forms and higher construction efficiency, and also comply with the real situation of common joint connections [31]. The

M-θ relationship is also used to characterize the restoring force model of semi-rigid connections, that is, the relationship between the bending moment at the joint connection and the relative rotation angle. Strictly speaking, various types of M-θ relationships are all nonlinear during the entire loading stage, and the diversity of connection forms in practical engineering directly leads to the complexity of the M-θ relationships, which makes it difficult to theoretically derive universally applicable expressions. This section will derive and establish the restoring force constitutive relationship of the SSCH joints based on the bilinear constitutive model based on the M-θ relationship.

The key to establishing a bilinear M-θ constitutive model lies in the calculation of the parameters are yield and ultimate points, namely, yield and ultimate bending moments and yield and ultimate rotation angles, as shown in Figure 17. Once the four parameters are determined, a bilinear constitutive model can then be established. To obtain the bending moment M_y at the yield point, it is necessary to first determine the position of the neutral axis of the cross-section. The neutral axis divides the cross-section into two equal areas above the neutral axis, and each element within the cross-section has a compressive stress equaling to the yield stress σ_y, as shown in Figure 16, in which $A_1 = A_2 = A/2$. For rectangular components, the neutral axis is the same for ultimate moment and elastic moment.

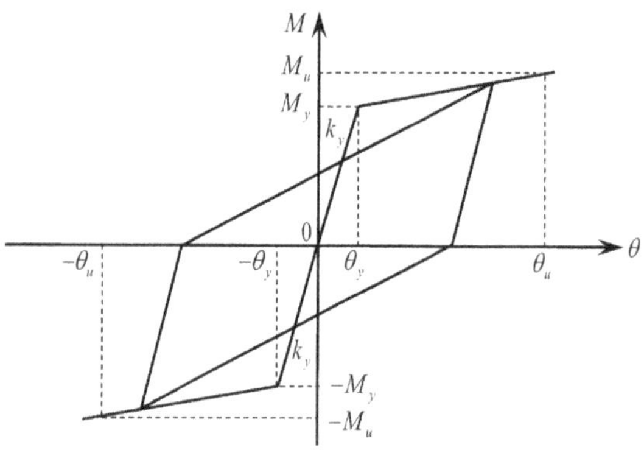

Figure 17. Bilinear constitutive model with the M-θ relationship.

By taking the moment of the force in Figure 18a towards the neutral axis, the ultimate moment M_u of the rectangular cross-section component can be obtained as follows [31]:

$$M_u = \frac{\sigma_y A(y_1 + y_2)}{2}, \tag{7}$$

where y_1 and y_2 are the distances from the neutral axis to the centroids c_1 and c_2 of areas A_1 and A_2, respectively.

The ratio of the ultimate moment of a component to its yield moment is called the cross-sectional shape function, usually represented by the shape factor f:

$$f = \frac{M_u}{M_y} = \frac{Z}{S}. \tag{8}$$

The ultimate modulus Z of rectangular cross-section components is as follows:

$$Z = \frac{bh}{2}\left(\frac{h}{4} + \frac{h}{4}\right) - \frac{bh^2}{4}. \tag{9}$$

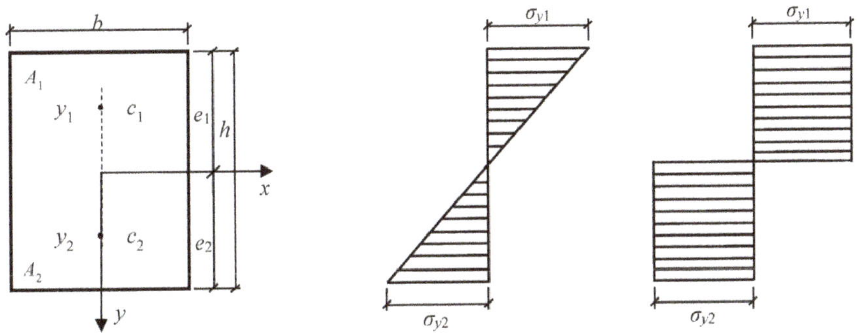

(**a**) Section parameter definition (**b**) Stress at yield state (**c**) Stress at ultimate state

Figure 18. Stress distribution of rectangular cross-section component.

As shown in Figure 18a, the distance from the neutral axis to the edge of the elastic core is represented by e. The stress distributions of rectangular cross-section component in the yield state and fully plastic state are shown in Figures 18b and 18c, respectively. The expression for the yield moment can be obtained as follows:

$$M_y = \sigma_y b(\frac{h}{2} - e)(\frac{h}{2} + e) + \sigma_y b(\frac{2\sigma^2}{3}). \tag{10}$$

The SSCH joint is modeled as an equivalent cantilever column, and it is assumed that the SSCH joint can mainly be subject to bending failure under the combined action of axial pressure and horizontal force. When using a rod model with plastic hinges at the cantilever column end for nonlinear analysis, the equivalent plastic hinge lengths at the column end in the yield state and ultimate state are l_{py} and l_{pu}, respectively. If the curvature of the cross-section within this length range is the same and is determined as the yield curvature φ_y and ultimate curvature φ_u of the cross-section. The curvature distribution of the SSCH joints at the yield and ultimate stages is shown in Figure 19, in which H is the length from the horizontal force loading point to the column end, and φ_p is the plastic curvature ($\varphi_p = \varphi_u - \varphi_y$).

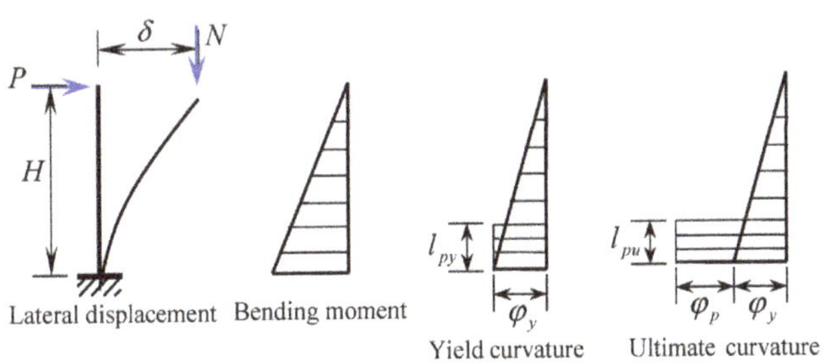

Figure 19. Lateral displacement and curvature of equivalent cantilever columns for the SSCH joint.

The yield angle θ_y and ultimate angle θ_u of the plastic hinge zone can be obtained from Figure 19 according to references [32,33]:

$$\begin{cases} \theta_y = \varphi_y l_{py} \\ \theta_u = \varphi_u l_{pu} \end{cases}, \tag{11}$$

in which, $\begin{cases} \varphi_y = \frac{\varepsilon_{c0}}{x_{ny}} \\ \varphi_u = \frac{\varepsilon_{cc}}{x_{nu}} \end{cases}$, $\begin{cases} l_{py} = (5 - 5.47\alpha_y)(0.08h + 0.022 f_s d_{eq}) \\ l_{pu} = 0.37\alpha_u(0.08h + 0.022 f_s d_{eq}) \end{cases}$, where ε_{c0} and ε_{cc} are the strains of the concrete at the edge of the compression zone of the cross-section when it reaches yield, and ultimately, respectively, $\varepsilon_{c0} = 0.002$, $\varepsilon_{cc} = 0.0033$. x_{ny} and x_{nu} are depths of the compression zone relative to ε_{c0} and ε_{cc}. α_y and α_u are ratios of the compressive area of concrete to the full cross-sectional area in the yield and ultimate states, f_s is steel strength, and d_{eq} is the equivalent diameter of the steel in the SSCH joint when it is equivalent to the longitudinal load-bearing steel bar.

Therefore, for the equivalent cantilever column of the SSCH joint, the expressions for the vertex yield displacement and ultimate displacement are as follows [34]:

$$\begin{cases} \delta_y = \varphi_y \left[\frac{(H-l_{py})^3}{3h} + l_{py}(H - l_{py}) \right] \\ \delta_u = \varphi_y \frac{(H-l_{pu})^3}{3h} + \varphi_u l_{pu} \left(H - \frac{l_{pu}}{2} \right) \right] \end{cases} . \quad (12)$$

5.2. Result Discussion

Figure 20 shows the comparisons between the horizontal force–displacement curves calculated based on the simplified constitutive model and the numerical results for some cases above. Table 10 gives the comparative data of the force and displacement at the yield and limit points. It can be found that the hysteretic curves for the simulation and theory have significant differences. The main reason for this is that the proposed M-θ theoretical model adopts the bilinear constitutive law, which causes the inability to capture, effectively, the force–displacement relationship of the SSCH joints in the second and fourth quadrants. However, it can also be found that both the skeleton curves between the simulation result and theory results fit well. The maximum errors for the yield force and the peak force among the six cases given in Table 10 are 7.67% and 4.56%, respectively. The corresponding average error values are 3.11% and 2.67%, respectively. Hence, it can be found that the deduced bilinear constitutive model based on the M-θ relationship can effectively capture the mechanical behavior of the SSCH joints, including stiffness characteristics, yield, and peak mechanical behaviors.

Table 10. Comparative data of parameter indicators at the yield and limit points.

Case	P_y/(kN)			δ_y/(mm)			P_u/(kN)			δ_u/(mm)		
	Simulation	Theory	Error	Simulation	Theory	Error	Simulation	Theory	Error	Simulation	Theory	Error
LE20	1227.80	1265.56	2.98%	3.73	3.58	4.19%	1428.00	1465.23	2.54%	16.02	16.15	0.80%
BNU2	1038.00	1025.98	1.17%	3.81	3.65	4.38%	1389.00	1435.61	3.25%	15.89	16.82	5.53%
TH20	600.62	650.50	7.67%	3.73	3.72	0.27%	801.96	808.28	0.78%	17.46	17.16	1.75%
TH40	822.23	805.36	2.09%	3.74	3.62	3.31%	1065.48	1114.32	4.38%	17.98	18.24	1.43%
ACR4	1051.00	1032.45	1.80%	3.69	3.64	1.37%	1630.50	1638.56	0.49%	16.12	15.72	2.54%
ACR6	1078.00	1047.21	2.94%	3.83	3.59	6.69%	1687.50	1768.21	4.56%	16.98	17.59	3.47%

Note: Error = ABS(Simulation − Theory)/Theory × 100%.

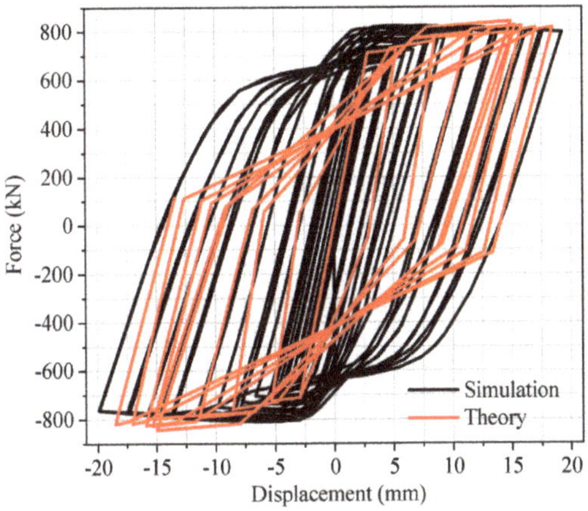

(a) Hysteretic curves of Case TH20

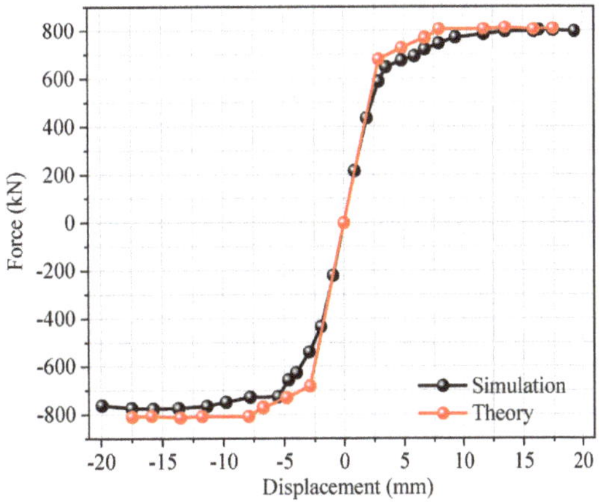

(b) Skeleton curve of Case TH20

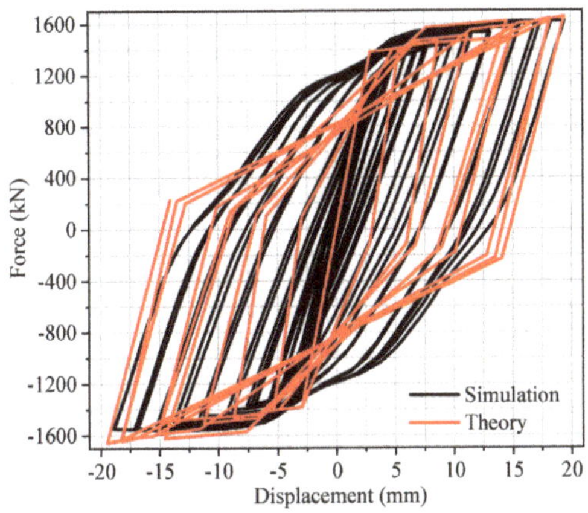

(c) Hysteretic curves of Case ACR4

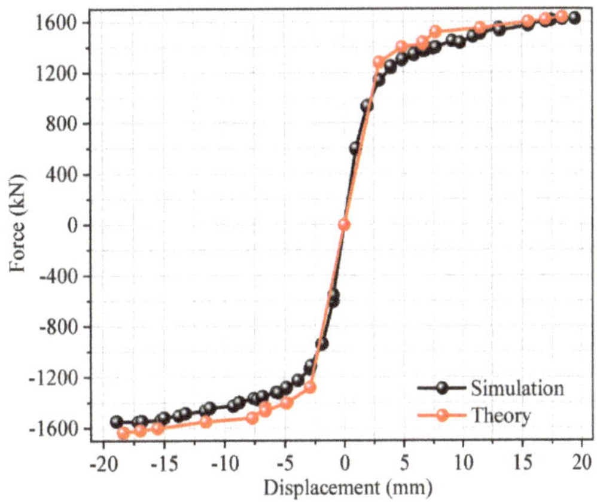

(d) Skeleton curve of Case ACR4

Figure 20. Comparisons of the hysteresis and skeleton curves for different cases.

6. Conclusions

A new type SSCH joint used in prefabricated shear wall structure is proposed in this study. Parameter effects of the SSCH joint, such as the length, number, thickness of the interlocking slot, and axial compression ratio, are investigated in detail based on validated numerical model. A simplified constitutive model of the SSCH joint is further deduced. The main results can be summarized as follows:

(1) The force–displacement hysteresis curve of the SSCH joint is relatively full and no pinching phenomenon is found. The ductility coefficients of the SSCH joint are all beyond 4 for the investigated 20 Cases. It can be concluded that the proposed SSCH joint can provide stable shear capacity and superior energy dissipation capacity, which can serve as a reliable shear connection key for the assembly of prefabricated shear wall structures.

(2) The length and thickness of the interlocking slot have a great impact on the mechanical behavior of the SSCH joint, especially for the slot thickness. Compared to the SSCH joint with 10 mm thickness, the ultimate force and energy dissipation of the SSCH joint with 40 mm thickness are increased by 62.29% and 72.94%, respectively. On the contrary, the slot number and axial compression ratio have less effect on mechanical properties and can be ignored. To guarantee more superior mechanical behaviors, it is

suggested to design the SSCH joint with a slot length of 20~30 mm, a slot number of two or three, and a slot thickness of 20~30 mm.

(3) The bilinear constitutive model based on the M-θ relationship is advanced, in which the moment and rotation angle calculation formulae for two control points, namely, the yield and limit points, on the skeleton line of the restoring force model are deduced. Based on comparison with numerical results, it can be found that the advanced simplified model can effectively capture the mechanical characteristics of the SSCH joints.

Author Contributions: Conceptualization, Data curation, Funding acquisition, X.W.; Investigation, Project administration, Resources, Y.W.; Conceptualization, Visualization, Writing—review & editing, S.J.; Formal analysis, Methodology, Software, Validation, Writing—original draft, M.L.; Conceptualization, Methodology, Supervision, Writing—original draft, Writing—review & editing, D.W. All authors have read and agreed to the published version of the manuscript.

Funding: This project is supported by the National Natural Science Foundation of China (52378496, 52178467), Technology Project of China Southern Power Grid (037700KK52220039) and the Science and Technology Project of Guangzhou City (202235058).

Data Availability Statement: The data presented in this study are available in article.

Acknowledgments: All statements, results, and conclusions are those of the researchers and do not necessarily reflect the views of these foundations. The authors also sincerely thank the anonymous reviewers for their insightful comments and suggestions.

Conflicts of Interest: Authors X.W., Y.W. and S.J. were employed by the company Power Grid Planning Research Center of Guangdong Power Grid Co., Ltd. The remaining authors declare that the research was conducted in the absence of any commercial or financial relationships that could be construed as a potential conflict of interest.

References

1. Singhal, S.; Chourasia, A.; Chellappa, S.; Parashar, J. Precast reinforced concrete shear walls: State of the art review. *Struct. Concr.* **2019**, *20*, 886–898. [CrossRef]
2. Hemamalini, S.; Vidjeapriya, R.; Jaya, K.P. Performance of precast shear wall connections under monotonic and cyclic loading: A state-of-the-art review. *Iran. J. Sci. Technol.-Trans. Civ. Eng.* **2021**, *45*, 1307–1328. [CrossRef]
3. Men, J.J.; Shi, Q.X.; He, Z.J. Optimal design of tall residential building with RC shear wall and with rectangular layout. *Int. J. High-Rise Build.* **2014**, *3*, 285–296.
4. Chu, M.; Liu, J.; Sun, Z. Experimental study on mechanical behaviors of new shear walls built with precast concrete hollow moulds. *Eur. J. Environ. Civ. Eng.* **2019**, *23*, 1424–1443. [CrossRef]
5. Ali, M.M.; Osman, S.A.; Husam, O.A. Numerical study of the cyclic behavior of steel plate shear wall systems (SPSWs) with differently shaped openings. *Steel Compos. Struct.* **2018**, *26*, 361–373.
6. Mao, H.; Qin, G.C.; Lan, T. Study on the seismic performance of box-plate steel structure modular unit. *Adv. Mater. Sci. Eng.* **2019**, *4*, 4561631. [CrossRef]
7. Nguyen, N.H.; Whittaker, A.S. Numerical modelling of steel-plate concrete composite shear walls. *Eng. Struct.* **2017**, *150*, 1–11. [CrossRef]
8. Chu, M.; Wang, B.; Liu, J.; Zhang, P.; Li, X.; An, N. Experimental study on mechanical behavior of precast concrete shear walls with mortise-tenon joints. *J. Build. Struct.* **2021**, *42*, 173–182.
9. Nie, J.-G.; Hu, H.-S.; Fan, J.-S.; Tao, M.-X.; Li, S.-Y.; Liu, F.-J. Experimental study on seismic behavior of high-strength concrete filled double-steel-plate composite walls. *J. Constr. Steel Res.* **2013**, *88*, 206–219. [CrossRef]
10. Wang, D.Y.; Xu, S.C.; Yang, Y.; Mao, J.H.; Zhu, Y.; Guo, D.W.; Nie, Z.L. Study on seismic behaviors of steel–concrete composite shear walls with novel corner designs. *J. Build. Eng.* **2023**, *70*, 106339. [CrossRef]
11. Han, Q.H.; Wang, D.Y.; Zhang, Y.S.; Tao, W.J. Experimental investigation and simplified stiffness degradation model of precast concrete shear wall with steel connectors. *Eng. Struct.* **2020**, *220*, 110943. [CrossRef]
12. Soudki, K.A.; West, J.S.; Rizkalla, S.H.; Blackett, B. Horizontal connections for precast concrete shear wall panels under cyclic shear loading. *PCI J.* **1996**, *41*, 64–80. [CrossRef]
13. Zhao, B.; Wang, Q.; Lu, X. Research on seismic behavior of precast concrete walls with fully assembled horizontal joints. *J. Build. Struct.* **2018**, *39*, 48–55.
14. Cheng, B.; Cai, Y.; Looi, D.T.W. Experiment and numerical study of a new bolted steel plate horizontal joints for precast concrete shear wall structures. *Structures* **2021**, *32*, 760–777. [CrossRef]
15. Wei, H.; Li, Q. Experimental study on seismic behavior of prefabricated RC shear walls with horizontal joints welded by steel plates. *J. Build. Struct.* **2020**, *41*, 77–87.

16. Perez, F.J.; Pessiki, S.; Sause, R. Lateral load behavior of unbonded post-tensioned precast concrete walls with vertical joints. *PCI J.* **2004**, *49*, 48–64. [CrossRef]
17. Chu, M.; Xiong, C.; Liu, J.; Sun, Z. Experimental study on shear behavior of two-way hollow core precast panel shear wall with vertical connection. *Struct. Des. Tall Spec. Build.* **2021**, *30*, e1814. [CrossRef]
18. Huang, W.; Miao, X.W.; Zhao, Y.Y.; Yu, G.; Zhang, J.R.; Fan, Z.H. Experimental study on seismic performance of fully assembled composite walls with vertical joints using different dry connections. *J. Build. Struct.* **2020**, *41* (Suppl. 2), 114–122.
19. Li, H.; Chen, W.; Huang, Z.; Hao, H.; Ngo, T.T.; Pham, T.M.; Yeoh, K.J. Dynamic response of monolithic and precast concrete joint with wet connections under impact loads. *Eng. Struct.* **2020**, *250*, 113434. [CrossRef]
20. Zhu, G.; Ma, Y.X.; Tan, K.H. Experimental and analytical investigation on precast concrete-encased concrete-filled steel tube column-to-column dry connections under axial tension. *Structures* **2022**, *45*, 523–541. [CrossRef]
21. Li, W.; Gao, H.; Xiang, R.; Du, Y. Experimental study of seismic performance of precast shear wall with a new bolt-plate connection joint. *Structures* **2021**, *34*, 3818–3833. [CrossRef]
22. Fu, Y.; Fan, G.; Tao, L.; Yang, Y.; Wang, J. Seismic behavior of prefabricated steel reinforced concrete shear walls with new type connection mode. *Structures* **2022**, *37*, 483–503. [CrossRef]
23. Xue, W.C.; Li, Y.; Cai, L.; Hu, X. Seismic Performance of Precast Concrete Composite Shear Walls with Multiple Boundary Elements. *J. Earthq. Tsunami* **2019**, *13*, 1940006. [CrossRef]
24. Fintel, M. Performance of buildings with shear walls in earthquakes of the last thirty years. *PCI J.* **1995**, *40*, 62–80. [CrossRef]
25. Guo, Z.H. *Principles of Reinforced Concrete Structure*; Science Press: Beijing, China, 1998.
26. Qian, J.R.; Zhao, Z.Z.; Duan, A.; Xia, Z.F.; Wang, M.D. Pesudo-dynamic tests of a 1:10 model of pre-stressed concrete containment vessel for CNP 1000 nuclear power plant. *China Civ. Eng. J.* **2007**, *40*, 7–13.
27. Birtel, V.; Mark, P. Parameterised finite element modelling of RC beam shear failure. In Proceedings of the 19th Annual International ABAQUS Users' Conference, Boston, MA, USA, 23–25 May 2006; pp. 95–108.
28. Qian, J.R.; Zao, J.; Ji, X.D. Experimental study on seismic behavior of steel tube-reinforced concrete composite shear walls with high axial compressive load ratio. *J. Build. Struct.* **2010**, *31*, 5–13.
29. Gao, L.B.; Chen, Q.G. An anisotropic damage constitutive model for concrete and its applications. *Appl. Mech.* **1988**, *65*, 578–583.
30. *GB 50017-2017*; Standard for Design of Steel Structures. China Architecture and Building Press: Beijing, China, 2017.
31. Wang, X.W.; Sun, L. Research on the performance of steel frame semi rigid connections. *J. Wuhan Univ. Technol.* **2002**, *21*, 65–72.
32. Qian, J.R.; Liu, X.M. Test of seismic behavior of FRP-concrete-steel double-skin tubular columns. *China Civ. Eng. J.* **2008**, *41*, 29–36.
33. Seible, F.; Priestley, M.J.N.; Hegemier, G.A. Seismic retrofit of RC columns with continuous carbon fiber jackets. *J. Compos. Constr.* **1997**, *1*, 52–62. [CrossRef]
34. Qian, J.R.; Liu, X.M. A hysteretic model of moment-rotation relationship for plastic hinge zone of FRP-concrete-steel double shin tubular columns. *Eng. Mech.* **2008**, *25*, 48–52.

buildings

MDPI

Article

Nonlinear Dynamic Stability of Cylindrical Reticulated Shells with Initial Damage

Lei Li [1,*], Guangfeng Li [1], Nasim Uddin [2], Limin Tian [3], Zhibing Zhu [1], Chong Bai [3] and Chen Shen [4]

1 School of Civil and Architectural Engineering, Zhengzhou University of Aeronautics, Zhengzhou 450015, China
2 Department of Civil, Construction, and Environmental Engineering, University of Alabama at Birmingham, Birmingham, AL 35294, USA
3 Shaanxi Key Lab of Structure and Earthquake Resistance, School of Civil Engineering, Xi'an University of Architecture and Technology, Xi'an 710055, China
4 China Construction Seventh Engineering Division, Zhengzhou 450015, China
* Correspondence: lilei@zua.edu.cn

Abstract: As large-span structures, reticulated shells are widely used in large-scale public building and act as emergency shelters in the event of sudden disasters. However, spatial reticulated shells are dynamic-sensitive structures; the effect of the initial structural damage on dynamic stability should be considered. In this study, a new nonlinear dynamic model of cylindrical reticulated shells with initial damage is proposed to investigate the effect of initial damage accurately. Firstly, the damage constitutive relations of the building steels are built based on the irreversible thermodynamic theory; furthermore, its fundamental equations are obtained using simulated shell methods. Then, the nonlinear vibration differential equations with damage are obtained and studied with support. Meanwhile, the nonlinear natural vibration frequency with initial damage is derivatized. After that, a bifurcation problem with initial damage is studied by using Flouquet Index, and the dynamic stability state at the equilibrium point is analyzed in depth. It is found that the local dynamic stability of the system is determined via its initial condition, geometric parameters, and initial damage. Moreover, the initial damage dominates over other influence factors due to its strong randomness and uncertainty for the same structure. The damage accumulation results in the transition of the equilibrium point. In addition, the nonlinear natural vibration frequency decreases to zero with the accumulation of the damage reaching 0.618; the local stability of cylindrical reticulated shells fails and they even lose whole stability. This study provides a theoretical foundation for the future investigation of whole stability with initial damage.

Keywords: initial damage; cylindrical reticulated shell; bifurcation; nonlinear stability; nonlinear natural vibration frequency

1. Introduction

Because of the graceful structure style and big space of reticulated shell structures, they meet the requirements of people's lives and have earned more and more attention. With the wide application of reticulated shell structures in engineering, the mechanical properties of the structures attract more and more concern, and many researchers focus on their static stability. There has been ongoing research into static stability for decades [1–4]. Kangsheng Ye, Tiantian Lu, and Si Lu [5] proposed a direct Newton method for the computation of critical points based on the arc-length method of tracing a structural nonlinear solution. Thomas Bulenda and Jan Knippers [6] found that some parameters influence the failure load of domes and barrel vaults and gave suggestions for the imperfect shape, which has to be assumed using a commercial FE-program.

However, in recent years, many structural accidents have occurred in reticulated shells all over the world. Typical accidents include the collapse of the National Economic

Exhibition Hall caused by a snowstorm in Bucharest, Romania, in 1963 and a gymnasium in Hartford, Connecticut, in 1978. Furthermore, Typhoon 9415 in China caused the destruction of the reticulated shell roof of Wenzhou Airport. The Gyeongju Resort Stadium in South Korea collapsed after several days of heavy snow in 2014. These examples make the safety issues of reticulated shells attract increasing attention. In the past decades, the problem of dynamic stability in long-span reticulated shell structures has been investigated by many researchers. Ceshi Sun, Xuekun Zhou, and Shuixing Zhou [7] investigated the nonlinear response frequency of reticulated shell structures. Jihong Ye and Mingfei Lu [8] studied the stability of doubly curved shallow shells under dynamic and dead loads and found interesting phenomena: a subharmonic response, doubling bifurcation, and chaotic behavior. Amabili M., Changjun Cheng, Qiang, and Han et al. [9–14] investigated the nonlinear stability and, in these studies, dynamics such as bifurcation, chaos motion, and the structures were assumed to be composed of perfect components. However, the structural components inevitably become defective in the process of manufacturing, transportation, site construction, and applications in an actual project. These defects are inevitable and should be controlled.

Spatial reticulated shells are dynamic-sensitive structures because of their large span and spatial multiformity [15–17], and the defects are mainly physical and geometric. These defects affect the mechanical performances of the spatial reticulated shells greatly. Qiang Zeng, Xiaonong Guo, and Xu Yang et al. [18] proposed a simple and efficient method of generating stochastic initial geometric imperfections (IGIs) for single-layer reticulated shells considering topology constraints. By applying nonlinear heuristically perturbed virtual interaction forces to the joints, the joint coordinates can be updated using the iterative forward Euler method, aiming to generate realistic stochastic IGIs. Guohua Nie and Zhiwei Li [19] investigated the nonlinear behavior of single-layer squarely reticulated shallow spherical shells with geometrical imperfections under a central concentrated load via the asymptotic iteration method. Sheng He, Zhengrong Jiang, and Jian Cai [20] developed the new simulation method of initial geometric imperfection distribution to study elastoplastic stability analysis for single-layer reticulated shells via ANSYS. Qiongyao Wu, Huajie Wang, and Hongliang Qian [21] studied the mechanical behavior of a bolt-ball joint with an insufficient screwing depth of bolt via the experiment and numerical simulation method and found it had an effect on the stability of a single-layer reticulated shell. Huijuan Liu, Fukun Li, and Hao Yuan [22] proposed the precise bearing capacity equation for a spiral single-layer reticulated shell structure with imperfection. Jingnan Liang, Yugang Li, He Huang, and Feng Fan [23] built a new seismic damage assessment method for single-layer spherical reticulated shells based on structural residual displacement under sequential earthquakes. According to previous studies, a structure with an initial defect has complex dynamic stability performance. The nonlinear dynamic mechanical behavior of cylindrical reticulated shells is highly sensitive to initial imperfections, which are geometric defects and physical defects. Additionally, the initial defects of cylindrical reticulated shells have strong randomness and uncertainty. It is of theoretical and practical significance to investigate the static and dynamic stability of the structure by coupling the geometric defects and physical defects and analyzing the effects of initial damage on these structures. Defects of the structure include initial deflection, residual stresses, and damage. However, most of these results were only concentrated on the effect of initial deflection and residual stresses on the stability of the structure. The previous studies and engineering applications ignored the effects of initial damage, treating the structures as if they had no damage. However, initial damage does exist, and it has strong randomness and uncertainty. Though several empirical models, such as Johnson–Cook [24], Bao–Wierzbicki [25,26], and Lematire [27,28], have been applied to analyze the damage constitutive relations, they are limited due to bad adaptability. To the best of the authors' knowledge, no results have been reported about the effect of initial damage on cylindrical reticulated shells.

In this paper, the nonlinear stability of cylindrical reticulated shells with both geometric defects and physical defects is studied. The damage constitutive relations of the building steels are built and large deflection equations are given for a cylindrical reticulated shell with initial damage. Then, the exact solution of the free vibration equation is obtained and the nonlinear natural vibration frequency with initial damage is derivatized. The dynamic stability state at the equilibrium point is analyzed in depth based on the results of the bifurcation problem of cylindrical reticulated shells with initial damage found by using Flouquet Index. In addition, theoretical support is provided for the nonlinear stable theory of cylindrical reticulated shells with damage.

2. Constitutive Relations of the Structural Steels with Damage

The derivation process of constitutive relations of the structural steels with damage is shown in Figure 1. The detailed process will be discussed below.

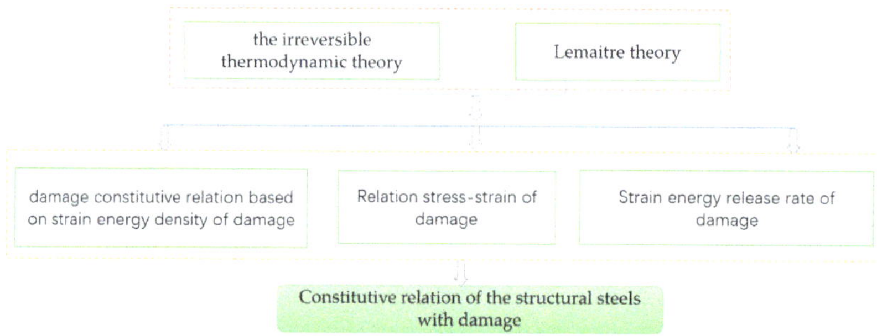

Figure 1. Flowchart to establish the constitutive relations.

By introducing the stress–strain elasticity intrinsic law of the damaged material [29,30], one can have

$$\sigma_{ij} = \frac{\partial \psi}{\partial \varepsilon_{ij}} \tag{1}$$

The damage dissipation power during the process of damage is

$$Y\dot{\omega} \geq 0 \tag{2}$$

Y is the strain energy release rate of damage:

$$Y = -\frac{\partial \psi}{\partial \omega} \tag{3}$$

$\dot{\omega} = \frac{\partial \psi^*}{\partial \omega}$, ψ^* is thermodynamic potential.

Assuming the initial conditions of the material are $\sigma_{ij} = \varepsilon_{ij} = 0$, $Y = 0$, $\omega = 0$, the Helmoltz specific energy $\psi(\varepsilon_{ij}, \omega)$ per unit volume can be expanded with a Taylor series. ε_{ij} is a small strain and can be treated as infinitely small. The degree of damage ω has finite value, so the truncation of series expansion of $\psi(\varepsilon_{ij}, \omega)$ at the second-order term of ε_{ij} and the Nth-order term is

$$\psi(\varepsilon_{ij}, \omega) = \psi_0 + \sum_{i=1}^{N} C^{(n)}\omega^n + \sum_{i=0}^{N} B_{ij}{}^{(n)}\varepsilon_{ij}\omega^n + \frac{1}{2}\sum_{i=0}^{N} A_{ijkl}{}^{(n)}\varepsilon_{ij}\varepsilon_{kl}\omega^n \tag{4}$$

ψ_0 is the free energy of the matrixial at the initial condition and can be treated as zero. $C^{(n)}$, B_{ij}^{n}, and A_{ijkl}^{n} are the tensor value coefficients at the second and fourth orders.

The damage strain energy density, damage stress–strain relationship, and release rate of damage strain energy were obtained through a series of calculations [31]:

$$\psi(\varepsilon_{ij}, \omega) = \mu(\omega)\varepsilon_{ij}\varepsilon_{ij} + \frac{1}{2}\lambda(\omega)(\varepsilon_{kk})^2 \tag{5}$$

$$\sigma_{\mathrm{ij}} = 2\mu(\omega)\varepsilon_{\mathrm{ij}} + \lambda(\omega)\varepsilon_{\mathrm{kk}}\delta_{\mathrm{ij}} \tag{6}$$

$$Y = -\frac{1}{2}\sum_{\mathrm{n}=1}^{\mathrm{N}} \alpha^{\mathrm{n}}\lambda\cdot(\omega)(\varepsilon_{\mathrm{kk}})^2 - \sum_{\mathrm{n}=1}^{\mathrm{N}} \beta^{\mathrm{n}}\mu\cdot(\omega)\varepsilon_{\mathrm{ij}}\varepsilon_{\mathrm{ij}} \tag{7}$$

where λ is the Lame coefficient.

For metal materials (construction steels), the effect of Poisson's ratio can be ignored. Based on the Lematire equivalent strain principle and irreversible thermodynamics theory [27,28], the element damage indicator ω is defined as $\omega = 1 - \frac{\varepsilon_{\mathrm{d}}}{\varepsilon}$, where ε_d is the strain of the structural member with damage and ε is the strain of the structural member without damage. When ω = 1, the structural member is completely damaged; when ω = 0, no damage is observed for the structural member.

Then, the damage constitutive relations of the building structural steels are established mathematically, as shown in Equations (8)–(10):

$$\mu(\omega) = \mu\left(1 - \sum_{\mathrm{n}=1}^{\mathrm{N}} \beta^{\mathrm{n}}\omega^{\mathrm{n}}\right) = \frac{E}{2}\left(1 - \sum_{\mathrm{n}=1}^{\mathrm{N}} \beta^{\mathrm{n}}\omega^{\mathrm{n}}\right) \tag{8}$$

$$\psi(\varepsilon_{\mathrm{ij}}, \omega) = \mu(\omega)\varepsilon_{\mathrm{ij}}\varepsilon_{\mathrm{ij}} \tag{9}$$

$$\sigma_{\mathrm{ij}} = 2\mu(\omega)\varepsilon_{\mathrm{ij}} \tag{10}$$

$$Y = -\sum_{\mathrm{n}=1}^{\mathrm{N}} \beta^{\mathrm{n}}\mu\cdot(\omega)\varepsilon_{\mathrm{ij}}\varepsilon_{\mathrm{ij}} \tag{11}$$

where β^n is the dimensionless non-negative coefficient and ε_{ij} is the strain.

3. Fundamental Equations of Cylindrical Reticulated Shells with Damage

By using simulated shell method, the element effective stiffness of cylindrical reticulated shells with initial damage is built from the damage constitutive Equations (9)–(11) of the building structural steels established above. The element effective stiffness is shown through Equations (12) and (13) [32]:

$$T_{xx} = T_{yy} = \frac{3\sqrt{3}}{4a_1}\left(1 - \omega - \omega^2\right)EA,\ T_{xy} = T_{yx} = \frac{\sqrt{3}}{4a_1}\left(1 - \omega - \omega^2\right)EA \tag{12}$$

$$B_{xx} = B_{yy} = \frac{3\sqrt{3}}{4a_1}\left(1 - \omega - \omega^2\right)EI;\ B_{xy} = B_{yx} = \frac{\sqrt{3}}{4a_1}\left(1 - \omega - \omega^2\right)EI \tag{13}$$

where A is the cross-sectional area of the member, E is the elastic modulus of steel, I is the inertia moment, and $a_1 = \sqrt{3}l/2$, where l is the length of the member.

Similarly, the nonlinear dynamic differential equation of cylindrical reticulated shells considering damage is built by using the simulated shell method. The physical equations concerning the in-plane stability of cylindrical reticulated shells with damage are shown in Equations (14)–(16) [32]:

$$N_x = T_{xx}\varepsilon_x + T_{xy}\varepsilon_y = \frac{3\sqrt{3}}{4a_1}\left(1 - \omega - \omega^2\right)EA(\varepsilon_x + \frac{1}{3}\varepsilon_y) \tag{14}$$

$$N_y = T_{yy}\varepsilon_y + T_{yx}\varepsilon_x = \frac{3\sqrt{3}(1 - \omega - \omega^2)EA}{4a_1}(\varepsilon_y + \frac{1}{3}\varepsilon_x) \tag{15}$$

$$N_{xy} = N_{yx} = T_{xy}\varepsilon_{xy} = \frac{\sqrt{3}(1 - \omega - \omega^2)EA}{4a_1}\varepsilon_{xy} \tag{16}$$

Additionally, the moment in-plane stability of cylindrical reticulated shells with damage can be obtained from Equations (14)–(16), as shown in Equations (17)–(19):

$$M_x = B_{xx}\chi_x + B_{xy}\chi_y = \frac{3\sqrt{3}(1-\omega-\omega^2)EI}{4a_1}(\chi_x + \frac{1}{3}\chi_y) \tag{17}$$

$$M_y = B_{xy}\chi_x + B_{yy}\chi_y = \frac{3\sqrt{3}(1-\omega-\omega^2)EI}{4a_1}(\chi_y + \frac{1}{3}\chi_x) \tag{18}$$

$$M_{xy} = B_{xy}\chi_{xy} = \frac{\sqrt{3}(1-\omega-\omega^2)EI}{4a_1}\chi_{xy} \tag{19}$$

where $\chi_x = -\frac{\partial^2\omega}{\partial x^2}, \chi_y = -\frac{\partial^2\omega}{\partial y^2}, \chi_{xy} = -2\frac{\partial^2\omega}{\partial x\partial y}$.

The geometric equation of cylindrical reticulated shells with damage is obtained from Equations (14)–(16), as shown in Equation (20):

$$\begin{aligned} \varepsilon_x &= \frac{\partial u}{\partial x} + \frac{1}{2}\left(\frac{\partial w}{\partial x}\right)^2 \\ \varepsilon_y &= \frac{\partial v}{\partial y} - \frac{1}{R}w + \frac{1}{2}\left(\frac{\partial w}{\partial y}\right)^2 \\ \varepsilon_{xy} &= \frac{\partial u}{\partial y} + \frac{\partial v}{\partial x} + \frac{\partial w \partial w}{\partial y \partial x} \end{aligned} \tag{20}$$

where u, v, and w correspond to the moving velocity in the x, y, and z direction, respectively. R is the radius of the curvature of the cylindrical reticulated shells in the y direction (Figure 2).

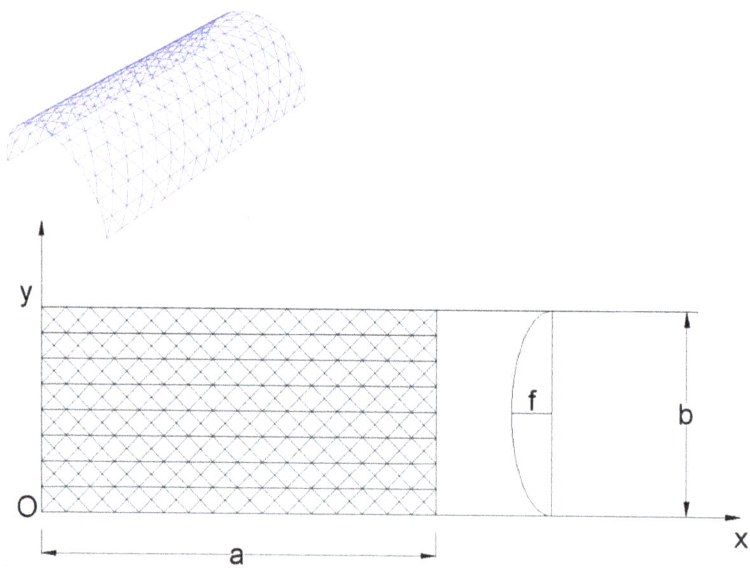

Figure 2. Single-layer cylindrical reticulated shells.

4. The Governing Equations and Boundary Condition of Cylindrical Reticulated Shells

To obtain the nonlinear dynamic differential equations with damage, the relation of displacement and strain with damage, boundary condition, and initial condition should be introduced. In the following sections, these factors are discussed as shown in Figure 3.

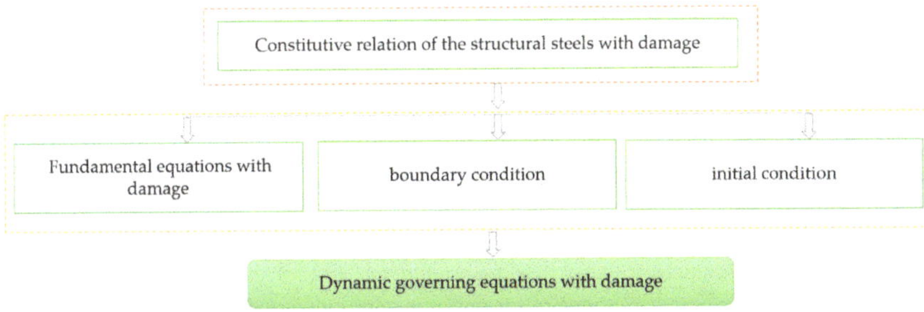

Figure 3. Flowchart of the derivation nonlinear dynamic differential equations with damage.

In this section, based on geometric equations and physical equations, the bending equations and boundary of the cylindrical reticulated shells are deduced. Under transverse loads, the equilibrium equations are shown as follows in Equations (21)–(25) [33]:

$$\frac{\partial N_x}{\partial x}+\frac{\partial N_{xy}}{\partial y}=0 \tag{21}$$

$$\frac{\partial N_y}{\partial y}+\frac{\partial N_{xy}}{\partial x}=0 \tag{22}$$

$$\frac{\partial M_x}{\partial x}+\frac{\partial M_{xy}}{\partial y}-Q_x=0 \tag{23}$$

$$\frac{\partial M_y}{\partial y}+\frac{\partial M_{xy}}{\partial x}-Q_y=0 \tag{24}$$

$$\frac{\partial Q_x}{\partial x}+\frac{\partial Q_y}{\partial y}+N_x\left(\kappa_x+\frac{\partial^2 W}{\partial x^2}\right)+N_y\left(\kappa_y+\frac{\partial^2 W}{\partial y^2}\right)+2N_{xy}\frac{\partial^2 W}{\partial xy}+q-c\frac{\partial w}{\partial t}-\gamma\frac{\partial^2 W}{\partial t^2}=0 \tag{25}$$

where $k_x = 0$ and $k_y = 1/R$ are the curvature in the x and y direction, respectively. q is the uniformly distributed load, c is the damping coefficient, and γ is the mass unit area of the cylindrical reticulated shells.

Then, the stress function is defined as follows:

$$N_x=\frac{\partial^2\phi}{\partial y^2}\ ,N_y=\frac{\partial^2\phi}{\partial x^2},Nxy=-2\frac{\partial^2\phi}{\partial xy} \tag{26}$$

Based on the physical Equations (17)–(19) and equilibrium Equations (21)–(25), the governing equations of cylindrical reticulated shells can be derived, as shown in Equation (27):

$$\begin{gathered}-\frac{3\sqrt{3}EI}{4\mathrm{a}}(1-\omega-\omega^2)\left(\frac{\partial^4 w}{\partial x^4}+2\frac{\partial^4 w}{\partial x^2\partial y^2}+\frac{\partial^4 w}{\partial y^4}\right)+\frac{\partial^2\phi}{\partial y^2}\left(\kappa_x+\frac{\partial^2 w}{\partial x^2}\right)+\frac{\partial^2\phi}{\partial x^2}\left(\kappa_y+\frac{\partial^2 w}{\partial y^2}\right)-\\ 2\frac{\partial^2\phi}{\partial xy}\frac{\partial^2 w}{\partial x^2\partial y^2}+q-c\frac{\partial w}{\partial t}-\gamma\frac{\partial^2 w}{\partial t^2}=0\end{gathered} \tag{27}$$

By defining $L=\frac{\partial^4}{\partial x^4}+2\frac{\partial^4}{\partial x^2\partial y^2}+\frac{\partial^4}{\partial y^4}$ and $L_2(w,\phi)=\frac{\partial^2 w}{\partial x^2}\frac{\partial^2\phi}{\partial y^2}-2\frac{\partial^2 w}{\partial x\partial y}\frac{\partial^2\phi}{\partial x\partial y}+\frac{\partial^2 w}{\partial y^2}\frac{\partial^2\phi}{\partial x^2}$, Equation (27) is simplified as follows:

$$\begin{gathered}-\frac{3\sqrt{3}EI}{4\mathrm{a}}(1-\omega-\omega^2)\left(\frac{\partial^4 w}{\partial x^4}+2\frac{\partial^4 w}{\partial x^2\partial y^2}+\frac{\partial^4 w}{\partial y^4}\right)+\frac{\partial^2\phi}{\partial y^2}\left(\kappa_x+\frac{\partial^2 w}{\partial x^2}\right)+\frac{\partial^2\phi}{\partial x^2}\left(\kappa_y+\frac{\partial^2 w}{\partial y^2}\right)-\\ 2\frac{\partial^2\phi}{\partial xy}\frac{\partial^2 w}{\partial x^2\partial y^2}+q-c\frac{\partial w}{\partial t}-\gamma\frac{\partial^2 w}{\partial t^2}=0\end{gathered} \tag{28}$$

The element effective stiffness Formulas (12) and (13) are deduced as follows [32]:

$$F_{yy} = \frac{T_{yy}}{T_{yy}T_{xx} - T_{xy}^2} = \frac{\sqrt{3}\mathrm{a}}{2EA} \tag{29}$$

$$F_{xx} = \frac{T_{xx}}{T_{yy}T_{xx} - T_{xy}^2} = \frac{\sqrt{3}\mathrm{a}}{2EA} \tag{30}$$

$$F_{xy} = F_{yx} = \frac{T_{yy}}{T_{yy}T_{xx} - T_{xy}^2} = -\frac{\mathrm{a}}{2\sqrt{3}EA} \tag{31}$$

The coordinate deformation equation of cylindrical reticulated shells can be deduced by substituting Equations (29)–(31) into Equations (14)–(16):

$$\varepsilon_x = F_{yy}N_x + F_{xy}N_y,\ \varepsilon_y = F_{xx}N_y + F_{xy}N_x,\ \varepsilon_{xy} = F_{yy}N_{xy} \tag{32}$$

From Equations (31) and (32), the different strain can be obtained as follows:

$$\varepsilon_x = \frac{\sqrt{3}\mathrm{a}}{2EA}\left(N_x - \frac{1}{3}N_y\right),\ \varepsilon_y = \frac{\sqrt{3}\mathrm{a}}{2EA}\left(N_y - \frac{1}{3}N_x\right),\ \varepsilon_{xy} = -\frac{\mathrm{a}}{2\sqrt{3}EA}N_{xy} \tag{33}$$

So, the basic equation and coordinate deformation equation of cylindrical reticulated shells is deduced using Equation (33) and compatibility equations as follows:

$$\begin{aligned} n_1(1-\omega-\omega^2)L_1(w) &= \tfrac{1}{2}L_2(w,\phi) + q - c\tfrac{\partial w}{\partial x} - \tfrac{1}{R}\tfrac{\partial^2 w}{\partial x^2} \\ n_2(1-\omega-\omega^2)L_1(\phi) &= \tfrac{1}{2}L_2(w,w) - \tfrac{1}{R}\tfrac{\partial^2 w}{\partial x^2} \end{aligned} \tag{34}$$

where $n_1 = \frac{\sqrt{3}EI}{4\mathrm{a}}$ and $n_2 = \frac{\sqrt{3}\mathrm{a}}{2EA}$.

The governing equations of cylindrical reticulated shells were used to deduce further equations. Finally, the governing equations and the coordinate deformation equation of cylindrical reticulated shells with initial damage were obtained. In addition, the fixed boundary conditions of the four edges clamped were applied to cylindrical reticulated shells with initial damage in this paper:

$$\text{At } x = 0 \text{ or } m, \overline{e_x} = 0, P_x = 0 \tag{35}$$

$$\text{At } y = 0 \text{ or } n, \overline{e_y} = 0, P_y = 0 \tag{36}$$

where $\overline{e_x}$ and $\overline{e_y}$ are the mean displacement of the four edges of cylindrical reticulated shells in the x and y direction, respectively, and P_x and P_y are thrust along the perimetrical side [5].

5. Natural Vibration Frequency of Cylindrical Reticulated Shells with Initial Damage

For a cylindrical reticulated shell with a size of m (length) × n (width), its curvature along the x and y direction is $\kappa_x = 0$ *and* $\kappa_y = \frac{1}{R}$, and the numerical value is constant. The shell is subjected to general dynamic loads of load intensity q.

Following the governing equations, the coordinate deformation equation, and the boundary conditions of cylindrical reticulated shells with initial damage, the displacement equation of cylindrical reticulated shells with damage can be given from the dynamical Equations (28) and (34). The displacement of cylindrical reticulated shells with initial damage is introduced as

$$w = f(t)\sin(\alpha_0 x)\sin(\beta_0 x) \tag{37}$$

where $\alpha_0 = \frac{\pi}{m}$ and $\beta_0 = \frac{\pi}{n}$.

By combining the Equation (37) and the coordinate deformation Equation (34),

$$n_2(1-\omega-\omega^2)L_1(\phi) = -\frac{1}{R}f(t)\alpha_0{}^2\sin(\alpha_0 x)\sin(\beta_0 x)+ \\ f(t)\alpha_0^2\beta_0{}^2(\cos^2(\alpha_0 x)\cos^2(\beta_0 x)+\sin^2(\alpha_0 x)\sin^2(\beta_0 x)) \tag{38}$$

Additionally, Equation (38) is simplified as

$$n_2(1-\omega-\omega^2)\phi = \frac{f^2(t)}{32}\left[\left(\frac{m}{n}\right)^2\cos(2\alpha_0 x)+\left(\frac{n}{m}\right)^2\cos(2\beta_0 y)\right] \\ +\frac{f(t)}{R}(\cos^2(\beta_0 x)-\sin^2(\alpha_0 x)\sin^2(\beta_0 x))+\frac{1}{R}f(t)\alpha_0{}^2\sin(\alpha_0 x)\sin(\beta_0 x) \tag{39}$$

Furthermore, Equation (39) is simplified as

$$n_2(1-\omega-\omega^2)\phi = \frac{f^2(t)}{32}\left[\left(\frac{m}{n}\right)^2\cos(2\alpha_0 x)+\left(\frac{n}{m}\right)^2\cos(2\beta_0 y)\right]+ \\ \frac{f(t)}{Rm^2\pi^2}\frac{1}{\left(\frac{1}{m^2}+\frac{1}{n^2}\right)^2}(\sin(\alpha_0 x)\sin(\beta_0 x))+n_2\left(\frac{p_x y^2}{2}+\frac{p_y x^2}{2}\right) \tag{40}$$

By using the Galerkin method, Equation (40) is rewritten as

$$\int_1^2\int_0^n\int_0^m\left[\begin{array}{l} n_1(1-\omega-\omega^2)\left(\frac{\partial^4 w}{\partial x^4}+2\frac{\partial^4 w}{\partial x^2\partial y^2}+\frac{\partial^4 w}{\partial y^4}\right) \\ -L_2(w,\phi)+q+c\frac{\partial w}{\partial t}+\gamma\frac{\partial^2 W}{\partial t^2}\end{array}\right]\delta w dxdydt = 0 \tag{41}$$

Considering $\delta f(t) \neq 0$ from Equations (37) and (41),

$$(1-\omega-\omega^2)n_1\frac{\pi^6}{16}f(t)\left(\frac{1}{m^2}+\frac{1}{n^2}\right)^2+\frac{1}{n_2}(1-\omega-\omega^2)\frac{\pi^6}{256}f^3(t)\left(\frac{1}{m^4}+\frac{1}{n^4}\right)+\kappa_y p_y \\ -\frac{1}{n_2}(1-\omega-\omega^2)\frac{2\pi^2}{3}f^2(t)\frac{\kappa_y}{m^2}\frac{1}{m^2m^2}\frac{1}{\left(\frac{1}{m^2}+\frac{1}{n^2}\right)^2}-\frac{1}{n_2}(1-\omega-\omega^2)\frac{\pi^2}{24}f^2(t)\frac{\kappa_y}{n^2} \\ +\frac{1}{n_2}(1-\omega-\omega^2)\frac{\pi^2}{16}f(t)\left(\frac{\kappa_y}{m^2}\right)^2\frac{1}{\left(\frac{1}{m^2}+\frac{1}{n^2}\right)^2}-p_x f(t)\frac{\pi^4}{16m^2}-p_y f(t)\frac{\pi^4}{16n^2}-q+ \\ \frac{\pi^2}{16}c\frac{\partial f(t)}{\partial t}+\frac{\pi^2}{16}\gamma\frac{\partial^2 f(t)}{\partial t^2}=0 \tag{42}$$

The dimensionless variable is defined as follows:

$$\xi(t)=\frac{f(t)}{\beta},\ \lambda=\frac{\mathrm{a}}{\mathrm{b}},\ k'_y=\frac{\mathrm{a}^2}{R\beta},\ P'_x=\frac{n_2 p_x \mathrm{b}^2}{\beta^2},\ P'_y=\frac{n_2 p_y \mathrm{a}^2}{\beta^2} \tag{43}$$

$$Q=\frac{q n_2 \mathrm{a}^2\mathrm{b}^2}{\beta^3},\ n_3=\frac{q\beta^2}{8A},\ c_1=\frac{c\pi^2\mathrm{a}^2\mathrm{b}^2 n_2}{16\beta^2},\ \gamma_1=\frac{n_2\pi^2\mathrm{a}^2\mathrm{b}^2\gamma}{16\beta^2} \tag{44}$$

where $I = \beta^4$.

Natural vibration frequency is defined as $\varpi^* = \tau/t$ and general dynamic loads are $q^* = q_0\cos\Omega\overline{\tau}$, where Ω is exciting frequency.

From Equations (43) and (44), Equation (42) can be nondimensionalized as follows:

$$\gamma\frac{\partial^2\xi}{\partial\overline{\tau}^2}+c\frac{\partial\xi}{\partial\overline{\tau}}+\frac{\pi^6}{256}\left(\frac{1}{\lambda}+\lambda^2\right)\xi^3-\frac{2\pi^2}{3}\kappa_y\frac{1}{\left(\frac{1}{\lambda}+\lambda\right)^2}\xi^2-\frac{\pi^2}{24}\kappa_y\xi^2 \\ +\eta\kappa_y^2\frac{\pi^2}{16}\frac{1}{\left(\lambda+\frac{1}{\lambda}\right)^2}\xi-\frac{\pi^4}{16}\left(p_x^*+p_y^*\right)\xi+\eta\frac{\pi^6}{16}\frac{1}{\left(\lambda+\frac{1}{\lambda}\right)^2}\xi=q_0\cos\Omega\overline{\tau} \tag{45}$$

Based on the boundary conditions in (35) and (36), the relative displacement of the edge of cylindrical reticulated shells with damage, i.e., $x = 0$ and $x = m$, e_x can be calculated as

$$e_x=-\frac{1}{m}\int_0^m\frac{\partial u}{\partial x}dx \tag{46}$$

Equation (20) is substituted for Equation (46):

$$e_x = n_2(p_x - \mu p_y) + \frac{\pi^2}{8}\left(\frac{f}{m}\right)^2 + \frac{1}{Rm^2} f\left(\frac{1}{n^2} - \frac{\mu}{m^2}\right)\frac{1}{\left(\frac{1}{m^2} + \frac{1}{n^2}\right)^2}\frac{2}{\pi}\sin(\beta_0 y) \tag{47}$$

Variable e_x is related to y. The mean displacement of edge $\overline{e_x}$ can be obtained from

$$\overline{e_x} = \frac{1}{n}\int_0^m \frac{\partial u}{\partial x} dx \tag{48}$$

Combining Equations (47) and (48),

$$\overline{e_x} = n_2(p_x - \mu p_y) + \frac{\pi^2}{8}\left(\frac{f}{m}\right)^2 + \frac{1}{Rm^2} f\left(\frac{1}{n^2} - \frac{\mu}{m^2}\right)\frac{1}{\left(\frac{1}{m^2} + \frac{1}{n^2}\right)^2}\frac{4}{\pi^2} \tag{49}$$

By using a similar method, the mean displacement of edge $\overline{e_y}$ is

$$\overline{e_y} = n_2(p_y - \mu p_x) + \frac{\pi^2}{8}\left(\frac{f}{m}\right)^2 - \frac{1}{R} f \frac{4}{\pi^2} + \frac{1}{Rm^2} f\left(\frac{1}{n^2} - \frac{\mu}{m^2}\right)\frac{1}{\left(\frac{1}{m^2} + \frac{1}{n^2}\right)^2}\frac{4}{\pi^2} \tag{50}$$

According to the boundary conditions in (35) and (36), Equations (49) and (50) are

$$n_2(p_x - \mu p_y) = -\frac{\pi^2}{8}\left(\frac{f}{m}\right)^2 - \frac{1}{Rm^2} f\left(\frac{1}{n^2} - \frac{\mu}{m^2}\right)\frac{1}{\left(\frac{1}{m^2} + \frac{1}{n^2}\right)^2}\frac{4}{\pi^2} \tag{51}$$

$$n_2(p_y - \mu p_x) = -\frac{\pi^2}{8}\left(\frac{f}{m}\right)^2 - \frac{1}{R} f \frac{4}{\pi^2} - \frac{1}{Rm^2} f\left(\frac{1}{n^2} - \frac{\mu}{m^2}\right)\frac{1}{\left(\frac{1}{m^2} + \frac{1}{n^2}\right)^2}\frac{4}{\pi^2} \tag{52}$$

Similarly, Equations (51) and (52) are solved:

$$p'_x = -\frac{\pi^2}{8}\left(\frac{1}{\lambda^2} + \mu\right)\frac{1}{1-\mu^2}\xi^2(t) + \mu k'_y \frac{4}{\pi^2(1-\mu^2)}\xi(t) - k'_y \frac{4\lambda^2}{\pi^2(1+\lambda^2)^2}\xi(t) \tag{53}$$

$$p'_y = -\frac{\pi^2}{8}\frac{(\lambda^2+\mu)}{1-\mu^2}\xi^2(t) + \lambda^2 k'_y \frac{4}{\pi^2(1-\mu^2)}\xi(t) - k'_y \frac{4\lambda^2}{\pi^2(1+\lambda^2)^2}\xi(t) \tag{54}$$

Introducing the following equation:

$$Q = Q_0 \cos\Omega t \tag{55}$$

Equations (54) and (55) are substituted into Equation (45):

$$\begin{aligned}
&\gamma\frac{\partial^2\xi}{\partial\overline{\tau}^2} + c\frac{\partial\xi}{\partial\overline{\tau}} + (1-\omega-\omega^2)\left[\eta\frac{\pi^2}{16}\left(\frac{1}{\lambda}+\lambda\right)^2 + \frac{\pi^4-64}{16\pi^2}\kappa_y^2\frac{1}{\left(\lambda+\frac{1}{\lambda}\right)^2} + \kappa_y^2\lambda^2\frac{4}{\pi^2}\frac{1}{1-\mu^2}\right] \\
&-(1-\omega-\omega^2)\left\{\frac{3\pi^2}{8}\kappa_y^2(\lambda^2+\mu)\frac{1}{1-\mu^2} + \frac{5\pi^2}{12}\kappa_y\frac{1}{\left(\frac{1}{\lambda}+\lambda\right)^2} + \frac{\pi^2}{24}\kappa_y\lambda^2\right\}\xi^2 \\
&+(1-\omega-\omega^2)\left\{\frac{\pi^6}{256}\left(\frac{1}{\lambda}+\lambda^2\right) + \frac{\pi^6}{128(1-\mu^2)}\left[2\mu+\left(\frac{1}{\lambda^2}+\lambda^2\right)\right]\right\}\xi^3 - q_0\cos\Omega\overline{\tau} = 0
\end{aligned} \tag{56}$$

Based on the dimensionless variable Formula (43) and (44), Equation (56) can be simplified as

$$\begin{aligned}
&\frac{\partial^2\xi(t)}{\partial t^2} + \varpi^2(1-\omega-\omega^2)\xi(t) + c_2\frac{\partial\xi(t)}{\partial t} - \alpha_1(1-\omega-\omega^2)\xi^2(t) \\
&+\alpha_2(1-\omega-\omega^2)\xi^3(t) = F_1\cos\Omega t
\end{aligned} \tag{57}$$

where

$$\varpi^2 = \frac{1}{\gamma_1}\left[n_3\frac{\pi^2}{16}\left(\frac{1}{\lambda}+\lambda\right)^2 + \lambda^2\frac{\pi^4-64}{16\pi^2(\lambda^2+1)^2}k'_y{}^2 + k'_y{}^2\lambda^2\frac{4}{\pi^2(1-\mu^2)}\right]$$

$$\alpha_1 = \frac{1}{\gamma_1}\left[\frac{3\pi^2 k'_y(\lambda^2+\mu)}{8(1-\mu^2)} + \frac{5\pi^2 k'_y\lambda^2}{12(1+\lambda^2)^2} + \frac{\pi^2 k'_y\lambda^2}{24}\right]$$

$$\alpha_2 = \frac{1}{\gamma_1}\left[\frac{\pi^2(1+\lambda^4)}{256\lambda^2} + \frac{\pi^6(\lambda^4+2\lambda^2\mu+1)}{128(1-\mu^2)\lambda^2}\right]$$

where $c_1 = \frac{c_1}{\gamma_1}$, $F_1 = \frac{Q_0}{\gamma_1}$, μ is Poisson's ratio, and ϖ^2 is the natural vibration frequency of cylindrical reticulated shells.

Defining $\tau = \varpi t, \eta(t) = \frac{\varpi}{\sqrt{\alpha_2}}\xi(t)$, Equation (57) can be simplified:

$$\begin{aligned}&\frac{d^2\eta(\tau)}{d\tau^2} + (1-\omega-\omega^2)\eta(\tau) - \beta_2(1-\omega-\omega^2)\eta^2(\tau)\\&+(1-\omega-\omega^2)\eta^3(\tau) = F\cos\frac{\Omega}{\varpi}\tau - \beta_0\frac{d\eta(\tau)}{d\tau}\end{aligned} \quad (58)$$

where $\beta_0 = \frac{c_2}{\varpi}$, $\beta_2 = \frac{\alpha_1}{\varpi\sqrt{\alpha_2}}$, and $F = \frac{F_1}{\varpi^3}$.

Selecting $w = y_0\sin(\frac{\pi x}{m})\sin(\frac{\pi y}{n})$, $\phi = \phi_1\cos\tau + \phi_2\cos^2\tau$, and $\tau = \varpi t$, based on energy variation principle, the basic Equation (34) is

$$\int_0^m\int_0^n\int_0^{2\pi}\begin{matrix}\left[n_1(1-\omega-\omega^2)L_1(w) - \frac{1}{2}L_2(w,\phi) - \frac{1}{R}\frac{\partial^2\phi}{\partial x^2} - \gamma\varpi^2\frac{\partial^2 w}{\partial\tau^2}\right]\\ \delta y_0\sin(\frac{\pi x}{m})\sin(\frac{\pi y}{n})\cos\tau dx dy d\tau\end{matrix} = 0 \quad (59)$$

Based on the boundary condition and coordinate deformation of Equation (35), introducing $\Omega^2 = \gamma\varpi^2$ and $p_x = p_y = 0$, the nonlinear natural vibration frequency of cylindrical reticulated shells with initial damage is

$$\Omega^2 = \left[\begin{matrix}\frac{64\pi^4 n_1 n^8 n_2 R^2 + 256\pi^4 n_1 m^2 n^6 n_2 R^2 + 384\pi^4 n_1 m^4 n^4 n_2 R^2 + 256\pi^4 n_1 m^6 n^2 n_2 R^2}{64m^4 n^4 n_2 R(n^4+2m^2n^2+m^4)}\\ +\frac{(64\pi^4 n_1 m^8 n_2 R^2 - 3\pi^4 n^8 R^2 - 6\pi^4 m^2 n^6 R^2 - 6\pi^4 m^4 n^4 R^2 - 6\pi^4 m^6 n^2 R^2)y_0^2}{64m^4 n^4 n_2 R(n^4+2m^2n^2+m^4)}\\ -\frac{(3\pi^4 m^8 R^2 + 64m^4 n^8)y_0}{64m^4 n^4 n_2 R(n^4+2m^2n^2+m^4)}\end{matrix}\right](1-\omega^2-\omega) \quad (60)$$

By using the computer simulation, the relation between maximum amplitude Y_0 and nonlinear natural vibration frequency Ω^2 is shown in Figures 4 and 5. The different ratio of the size of m (length) to n (width) effects on nonlinear natural vibration frequency of cylindrical reticulated shells without initial damage is presented in Figure 4. The damage cumulative value effects on relations between maximum amplitude Y_0 and nonlinear natural vibration frequency Ω^2 are visible (in Figure 5). Increasing the ratio of the size of m (length) to n (width) results in a corresponding rise in the nonlinear natural vibration frequency. On the other hand, increasing maximum amplitude Y_0 results in a corresponding rise in nonlinear natural vibration frequency Ω^2. The increasing damage cumulative value (from 0 to 0.618) results in reducing nonlinear natural vibration frequency Ω^2 and maximum amplitude Y_0. Specifically, the nonlinear natural vibration frequency decreases with the accumulation of the damage, and Ω^2 drops to zero when ω = 0.618, meaning the cylindrical reticulated shells collapse.

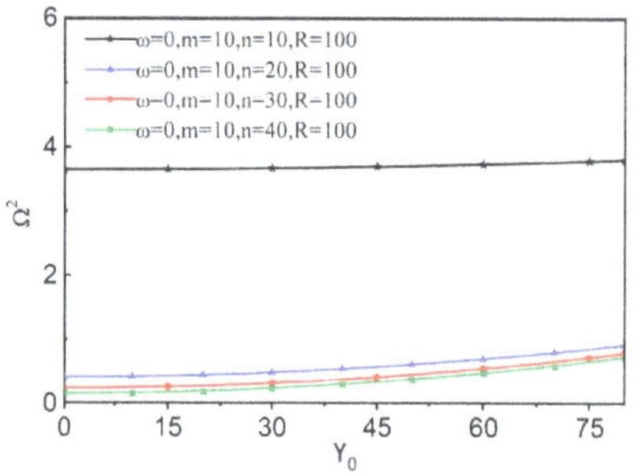

Figure 4. Y_0~Ω^2 relation curves without damage.

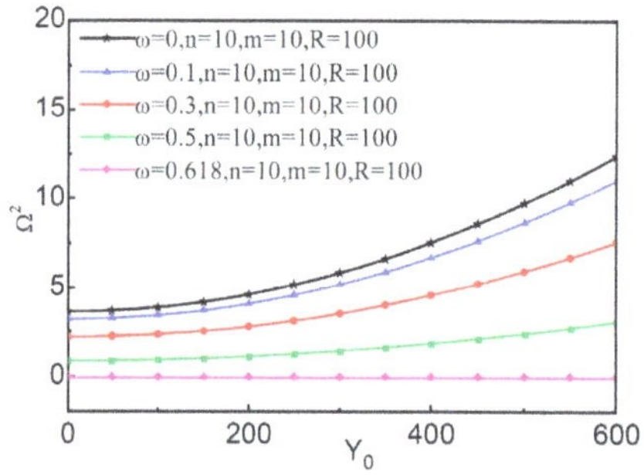

Figure 5. Y_0~Ω^2 relation curves with damage.

6. Equilibrium Bifurcation of Cylindrical Reticulated Shells with Initial Damage

Based on the results above, the equilibrium bifurcation of cylindrical reticulated shells with initial damage is discussed in this section, and the numerical and theoretical results are compared. The technical roadmap of this section is shown in Figure 6.

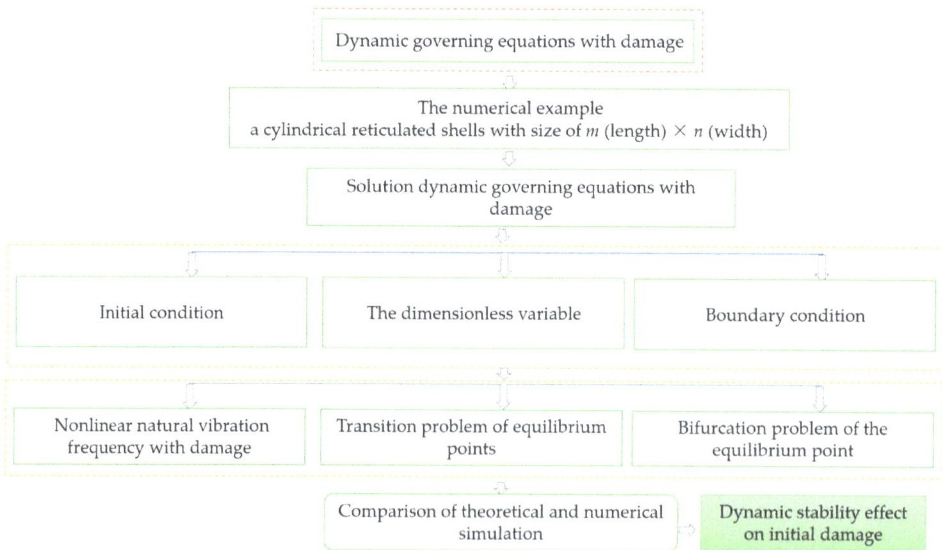

Figure 6. Flowchart of analysis of equilibrium bifurcation of cylindrical reticulated shells with initial damage.

Solving the free vibration equation of cylindrical reticulated shells with initial damage and without external excitation, Equation (58) can be rewritten as

$$\begin{cases} \eta_1'(\tau) = \eta_2(\tau) \\ \eta'_2(\tau) = -(1-\omega-\omega^2)\eta_1(\tau) + \beta'(1-\omega-\omega^2)\eta_1{}^2(\tau) \\ \qquad - (1-\omega-\omega^2)\eta_1{}^3(\tau) - \alpha' \frac{d\eta(\tau)}{d\tau} \end{cases} \tag{61}$$

The Jacobi matrix of Equation (61) is given as follows:

$$\begin{pmatrix} \frac{\partial f_1}{\partial \eta_1} & \frac{\partial f_1}{\partial \eta_2} \\ \frac{\partial f_2}{\partial \eta_1} & \frac{\partial f_2}{\partial \eta_2} \end{pmatrix} = \begin{pmatrix} 0 & 1 \\ -(1-\omega-\omega^2) + 2\beta'(1-\omega-\omega^2)\eta_1 - (1-\omega-\omega^2)\eta_1^2 & -\alpha' \end{pmatrix} \tag{62}$$

By solving $\begin{cases} f_1(\eta_1, \eta_2) = 0 \\ f_2(\eta_1, \eta_2) = 0 \end{cases}$, the system has three equilibrium points (0, 0), and $\left\{ \frac{\beta' \pm \sqrt{(\beta')^2 - 4}}{2}, 0 \right\}$ when $\beta' \geq 2$.

Then, the stability state of cylindrical reticulated shells with damage at the equilibrium point is discussed.

The Jacobi matrix at the equilibrium point (0, 0) is

$$\begin{pmatrix} 0 & 1 \\ -(1-\omega-\omega^2) & -\alpha' \end{pmatrix} \tag{63}$$

Additionally, the characteristic of Equation (63) is

$$\begin{vmatrix} 0-\lambda & 1 \\ -(1-\omega-\omega^2) & -\alpha' - \lambda \end{vmatrix} = 0 \tag{64}$$

The roots of the characteristic equation can be obtained by solving Equation (64):

$$\lambda_{12} = \frac{-\alpha' \pm \sqrt{(\alpha')^2 - 4(1-\omega-\omega^2)}}{2} \tag{65}$$

Specifically, when $0 < \alpha' < 2$, the characteristic roots of Equation (65) are

$$\lambda_{1,2} = \frac{-\alpha' \pm \mathrm{i}\sqrt{(\alpha')^2 - 4(1-\omega-\omega^2)}}{2} \tag{66}$$

The characteristic roots λ are two unequal complex numbers. At this equilibrium point, the stable bifurcation of the complex plane of cylindrical reticulated shells with damage arises, as shown in Figures 7 and 8. When $\alpha' = 1.5$ and $\beta' = 2.5$, and $\omega = 0$, the phase plane of cylindrical reticulated shells without damage is as presented in Figure 7. At the same time, α' and β' have the same values as in Figure 7; their phase plane is shown in Figure 8. The damage value ω of the system is zero when cylindrical reticulated shells have no damage condition. While the damage value ω of the system is 0.1, cylindrical reticulated shells are damaged. Obviously, the bifurcation of the system in Figure 8 is more remarkable than that in Figure 7.

When $\alpha' = 2$ and $\omega = 0$, cylindrical reticulated shells are not damaged, the characteristic roots λ are two equal negative numbers, and the equilibrium point is the critical node. At $0 < \omega < 0.618$, the characteristic roots λ are two unequal negative numbers and the equilibrium point is the stable node. At $\omega = 0.618$, the system is at the critical state of instability. At $\omega > 0.618$, the system has no physical meaning to collapse; it is as shown in Figures 9 and 10 and will be discussed in a future study.

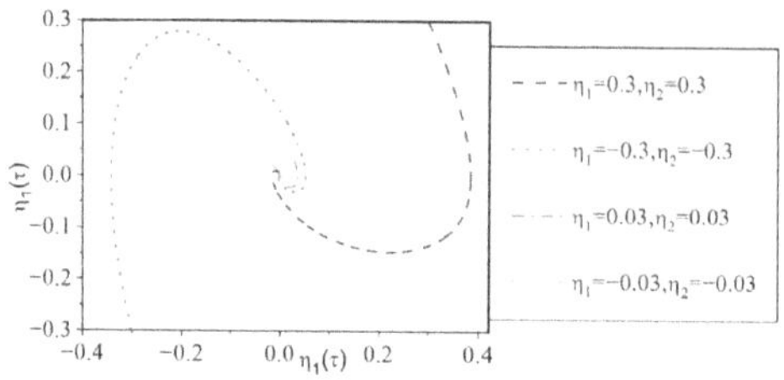

Figure 7. Phase plane $\omega = 0,\ \alpha' = 1.5,\ \beta' = 2.5$.

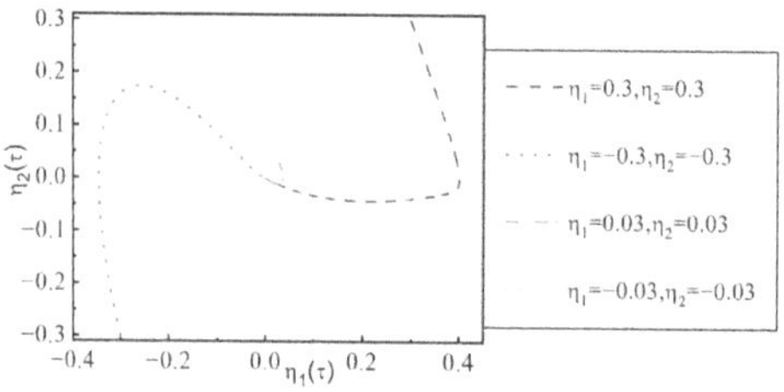

Figure 8. Phase plane $\omega = 0.1,\ \alpha' = 1.5,\ \beta' = 2.5$.

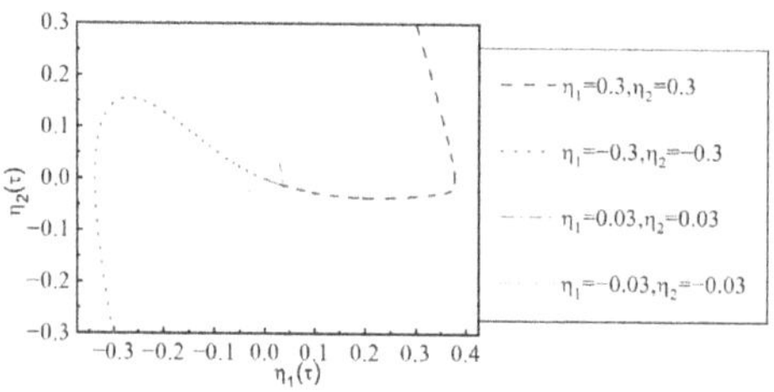

Figure 9. Phase plane $\omega = 0,\ \alpha' = 2, \beta' = 2.5$.

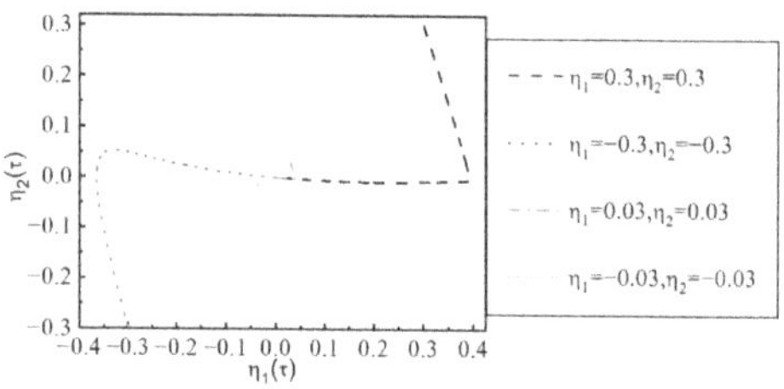

Figure 10. Phase plane $\omega = 0.1,\ \alpha' = 2,\ \beta' = 2.5$.

When $\alpha' \succ 2$, the characteristic roots λ of Equation (65) are

$$\lambda_{1,2} = \frac{-\alpha' \pm \sqrt{(\alpha')^2 - 4(1 - \omega - \omega^2)}}{2} \tag{67}$$

While the characteristic roots λ of cylindrical reticulated shells with initial damage are two unequal complex numbers at equilibrium point, the stable bifurcation of cylindrical reticulated shells with damage arises in the complex plane, and it is as shown in Figures 11 and 12. When $\alpha' = 2.5$ and $\beta' = 2.5$, and the damage value of the system is $\omega = 0$; the phase plane of cylindrical reticulated shells with initial damage is presented in Figure 11. At the same time, α' and β' are the same value as in Figure 11; their phase plane is shown in Figure 12.

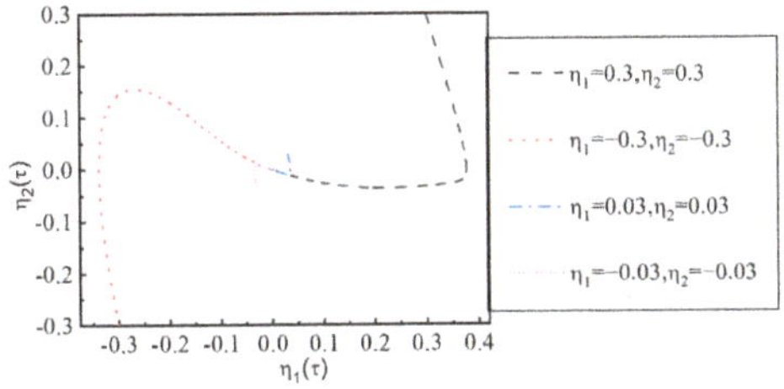

Figure 11. Phase plane $\omega = 0$, $\alpha' = 2.5$, $\beta' = 2.5$.

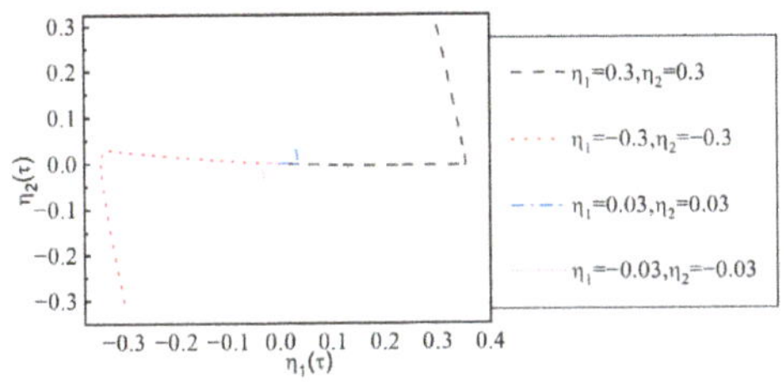

Figure 12. Phase plane $\omega = 0.1$, $\alpha' = 2.5$, $\beta' = 2.5$.

When $\alpha' = 0$, the characteristic roots λ of Equation (65) are

$$\lambda_{1,2} = \pm \mathrm{i}\sqrt{(1-\omega-\omega^2)} \tag{68}$$

When the characteristic roots λ of cylindrical reticulated shells with initial damage are the purely imaginary number at this equilibrium point, the Hopf bifurcation of cylindrical reticulated shells with initial damage arises in the complex plane, as shown in Figures 13 and 14. When $\alpha' = 0$ and $\beta' = 2.5$, and the damage value of the system is $\omega = 0$, the phase plane of cylindrical reticulated shells with damage is presented in Figure 13. At the same time, α' and β' are the same value as in Figure 13; their phase plane is shown in Figure 14.

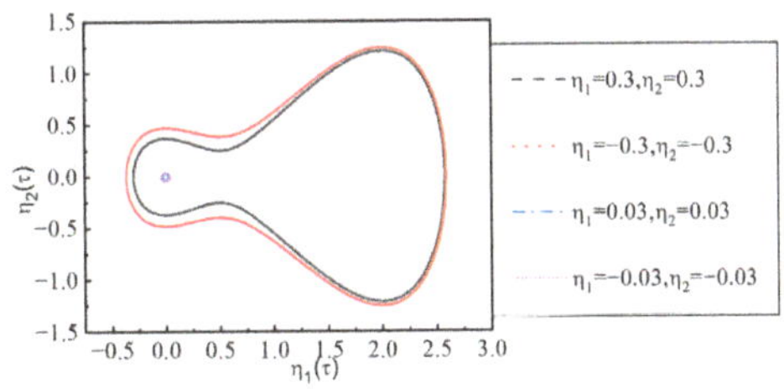

Figure 13. Phase plane $\omega = 0$, $\alpha' = 0$, $\beta' = 2.5$.

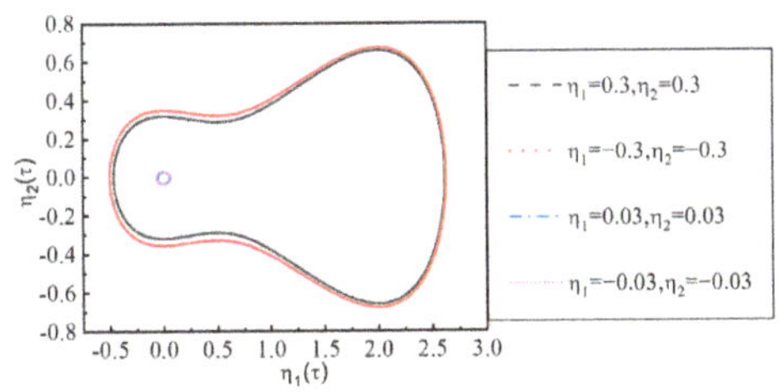

Figure 14. Phase plane $\omega = 0.1$, $\alpha' = 0$, $\beta' = 2.5$.

The phase plane (Figures 7–14) of the system with damage or without damage are drawn through numerical solution. The influence of damage and damper on bifurcation of the system at equilibrium point are seen in Figures 7–14.

Then, the damage accumulation effects on the stability of the equilibrium point is studied in this study via numerical solution, and its simulation results are shown in Figures 15 and 16. The relative position of the equilibrium point $\left\{\frac{\beta'+\sqrt{(\beta')^2-4}}{2},0\right\}$ is shown in Figure 15, and the relative position of the equilibrium point $\left\{\frac{\beta'-\sqrt{(\beta')^2-4}}{2},0\right\}$ is shown in Figure 16. From Figures 15 and 16, when the damage values gradually increase, the equilibrium point of the system is far away from the original equilibrium position. It is found that damage accumulation results in the transition of the equilibrium point.

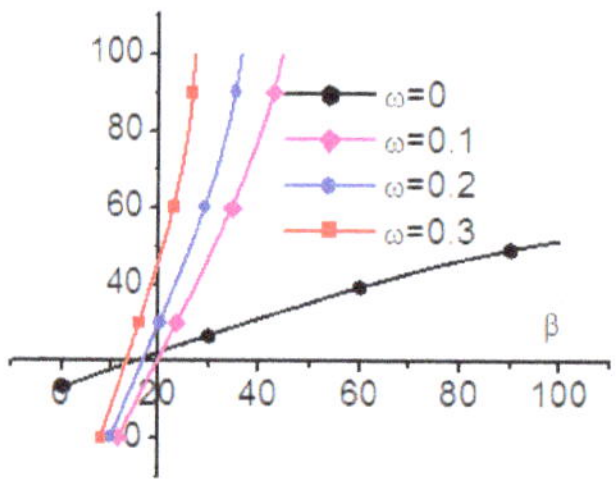

Figure 15. Relative position of equilibrium point $\left\{\frac{\beta'+\sqrt{(\beta')^2-4}}{2},0\right\}$.

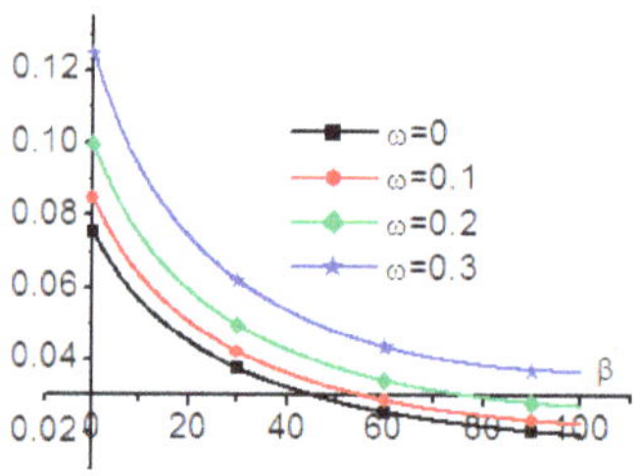

Figure 16. Relative position of equilibrium point $\left\{\frac{\beta-'\sqrt{(\beta')^2-4}}{2},0\right\}$.

7. Conclusions

In this paper, the dynamic stability theory of cylindrical reticulated shells with initial damage is proposed. The damage constitutive relations of the building steels are built. Based on the damage constitutive equation, the nonlinear vibration differential equations of cylindrical reticulated shells with initial damage are established. Then, the nonlinear natural vibration frequency of cylindrical reticulated shells with initial damage is obtained via the energy variation principle. Further, the bifurcation problem of cylindrical reticulated shells with initial damage at the equilibrium point is discussed by using the Flouquet Index, and the stability state of cylindrical reticulated shells with initial damage at the equilibrium point is analyzed numerically in depth.

It is found that the local dynamic stability of the system is determined by its initial condition, geometric parameters, and initial damage via the example of cylindrical reticulated shells. Moreover, the initial damage dominates over other influence factors due to its strong randomness and uncertainty for the same structure. A smaller nonlinear natural vibration frequency Ω^2 of cylindrical reticulated shells is obtained under larger damage accumulation. The damage accumulation results in the transition of its equilibrium point, changing the dynamic stability condition at the equilibrium point. In addition, the nonlinear natural vibration frequency decreases to zero with accumulation of the damage reaching 0.618; the local stability of cylindrical reticulated shells fails and they even lose whole stability. The present study's results provide some theoretical basis for practical engineering.

For the recommendation of future study, the nonlinear dynamic stability of the whole system can be studied systemically via experimental and numerical simulation based on the findings of this paper. The nonlinear vibration differential equations of cylindrical reticulated shells with initial damage can be solved and the dynamic mechanical mechanism of the system should be revealed, as well as the dynamical instability failure assessment of cylindrical reticulated shells with initial damage.

Author Contributions: L.L.: Investigation, Data curation, Software, Writing—original draft. N.U.: Supervision, Writing—review and editing. G.L.: Supervision, Writing—review and editing. L.T.: Supervision, Writing—review and editing. C.B.: Data curation, Writing—review and editing. Z.Z.: Data curation, Writing—review and editing. C.S.: Writing—original draft, Writing—review and editing. All authors have read and agreed to the published version of the manuscript.

Funding: Science Foundation in the Henan province (Grant No: 192102210030, 182102210439 and 232103810099), Key Science Foundation in the Henan province (Grant No: 231111221200) and the National Science Foundation in China (NSFC) (Grant No: U1404524).

Data Availability Statement: Data will be made available on request. The data are not publicly available due to privacy.

Conflicts of Interest: Author Chen Shen was employed by the company China Construction Seventh Engineering Division. The remaining authors declare that the research was conducted in the absence of any commercial or financial relationships that could be construed as a potential conflict of interest.

Nomenclature

R	radius of cylindrical reticulated shells
F	rise of arch of cylindrical reticulated shells
Q	uniformly distributed load
a	chord length of cylindrical reticulated shells
A	chord sectional area of cylindrical reticulated shells
γ	mass unit area of cylindrical reticulated shells
K_x, k_y	the curvature at x, y direction
c	damping coefficient of cylindrical reticulated shells
ω^2	natural vibration frequency of cylindrical reticulated shells
σ_{ij}, ε_{ij}	Cauchy stress, strain tensor
Ψ	Helmholtz free energy
Ω	damage value of materials
Y	strain energy release rate of damage
ψ^*	thermodynamic potential
ψ_0	Helmholtz free energy at original state
C_{ijkl}	elastic tensor of a material
λ, μ	Lamé coefficient, Poisson's ratio
α^n, β^n	dimensionless nonnegative coefficient
$\mu(\omega)$, $\lambda(\omega)$	Lamé coefficient with damage
E	elastic modulus of steel
u, v, w	corresponds to the displacement in the x, y, z direction
ε_d	strain with damage
ε'	strain without damage
Y_0	maximum amplitude of cylindrical reticulated shells
Ω^2	nonlinear natural vibration frequency of cylindrical reticulated shells
λ	characteristic roots
I, l	inertia moment of chord, length of chord
$\bar{e}_x$ $\bar{e}_y$	mean displacement of four edges of cylindrical reticulated shells in x and y direction
p_x, p_y	mean tension of two side edge of cylindrical reticulated shells
α	damping coefficient of dimensionless

References

1. Sharei, E.; Scholzen, A.; Hegger, J.; Chudoba, R. Structural behavior of a lightweight, textile-reinforced concrete barrel vault shell. *Compos. Struct.* **2017**, *171*, 505–514. [CrossRef]
2. Lu, M.; Ye, J. Stability optimization design for single-layer cylindrical domes based on joint well-formedness. *J. Vib. Shock* **2018**, *37*, 6.
3. Zhang, S.; Xu, Z.; Li, X. Static Behavior Analysis of Hinged Six-Bar Tetrahedral Spherical Lattice Shell. *Steel Constr.* **2023**, *38*, 5. [CrossRef]
4. Yu, P.; Yun, W.; Bordas, S.; He, S.; Zhou, Y. Static Stability Analysis of Single-Layer Reticulated Spherical Shell with Kiewitt-Sunflower Type. *Int. J. Steel Struct.* **2021**, *21*, 1859–1877. [CrossRef]
5. Ye, K.; Lu, T.; Yuan, S. A direct method for the computation of critical points in structural geometric nonlinear analysis. *Eng. Mech.* **2010**, *27*, 7.
6. Bulenda, T.; Knippers, J. Stability of grid shells. *Comput. Struct.* **2001**, *79*, 1161–1174. [CrossRef]
7. Sun, C.; Zhou, X.; Zhou, S. Nonlinear responses of suspended cable under phase-differed multiple support excitations. *Nonlinear Dyn.* **2021**, *104*, 1097–1116. [CrossRef]
8. Ye, J.; Lu, M. Design optimization of domes against instability considering joint stiffness. *J. Constr. Steel Res.* **2020**, *169*, 105757. [CrossRef]
9. Amabili, M. Non-linear vibrations of doubly curved shallow shells. *Int. J. Non-Linear Mech.* **2005**, *40*, 683–710. [CrossRef]
10. Chen, L.; Cheng, C.; Zhang, N. Chaotic motion of viscoelastic beams with geometric and physical nonlinearities. *Eng. Mech.* **2001**, *18*, 1–6.
11. Han, Q.; Zhang, N.; Yang, G. Chaotic motion of a nonlinear thermoelastic elliptic plate. *Appl. Math. Mech.* **1999**, *20*, 960–966. [CrossRef]
12. Zhi, X.; Fan, F.; Shen, S. Failure mechanism of latticed shells subjected to harmonic actions. *Earthq. Eng. Eng. Vib.* **2005**, *25*, 76–84.
13. Wang, X.; Liang, C.; Han, M.; Yeh, K.; Wang, G. Nonlinear dynamical behavior of shallow cylindrical reticulated shells. *Appl. Math. Mech.* **2007**, *28*, 151–156. [CrossRef]
14. Wang, X.; Wang, G.; Zhao, Y.; Yeh, K. Nonlinear dynamical stability analysis of the circular three-dimensional frame. *Appl. Math. Mech.* **2004**, *25*, 367–372. [CrossRef]
15. Wang, Y.; Dai, S. Chaotic motion of a heated bimetallic thin circular plate with initial deflection. *J. Tsinghua Univ.* **2004**, *44*, 1134–1137.
16. Li, L.; Hao, J. Nonlinear dynamic characteristic of the shallow cylindrical reticulated shells with initial damage. *Xi'an Jianzhu Keji Daxue Xuebao/J. Xi'an Univ. Archit. Technol.* **2017**, *49*, 76–80. [CrossRef]
17. Huang, Y.; Li, L. Chaos motion of shallow reticulated spherical shells considering initial imperfection. *Spat. Struct.* **2010**, *16*, 13–16.
18. Zeng, Q.; Guo, X.; Yang, X.; Zhu, S.; Li, Z. Constrained Stochastic Imperfection Modal Method for Nonlinear Buckling Analysis of Single-Layer Reticulated Shells. *J. Struct. Eng.* **2023**, *149*, 04022265. [CrossRef]
19. Nie, G.; Li, Z. Nonlinear analysis of imperfect squarely-reticulated shallow spherical shells. *Sci. China Ser. G Phys. Mech. Astron.* **2007**, *50*, 109–117. [CrossRef]
20. He, S.; Jiang, Z.; Cai, J. Investigation on Simulation Methods of Initial Geometric Imperfection Distribution in Elasto-plastic Stability Analysis of Single-layer Reticulated Shells. *KSCE J. Civ. Eng.* **2018**, *22*, 1193–1202. [CrossRef]
21. Wu, Q.; Wang, H.; Qian, H.; Han, K.; Fan, F. Effect of insufficient screwing depth of bolt on mechanical behavior of bolt-ball joint and stability of single-layer reticulated shell. *Eng. Struct.* **2020**, *213*, 110590. [CrossRef]
22. Liu, H.; Li, F.; Yuan, H.; Ai, D.; Xu, C. A Spiral Single-Layer Reticulated Shell Structure: Imperfection and Damage Tolerance Analysis and Stability Capacity Formulation for Conceptual Design. *Buildings* **2021**, *11*, 280. [CrossRef]
23. Liang, J.; Li, Y.; Huang, H.; Fan, F. A new seismic damage assessment method for single-layer spherical reticulated shells based on structural residual displacement under sequential earthquakes. *Structures* **2023**, *54*, 1391–1401. [CrossRef]
24. Sirigiri, V.K.R.; Gudiga, V.Y.; Gattu, U.S.; Suneesh, G.; Buddaraju, K.M. A review on Johnson Cook material model. *Mater. Today Proc.* **2022**, *62*, 3450–3456. [CrossRef]
25. Wierzbicki, T.; Bao, Y.; Lee, Y.-W.; Bai, Y. Calibration and evaluation of seven fracture models. *Int. J. Mech. Sci.* **2005**, *47*, 719–743. [CrossRef]
26. Bai, Y.; Wierzbicki, T. A comparative study of three groups of ductile fracture loci in the 3D space. *Eng. Fract. Mech.* **2015**, *135*, 147–167. [CrossRef]
27. Lemaître, J. *A Course on Damage Mechanics*; Springer Science & Business Media: Berlin/Heidelberg, Germany, 1992.
28. Lemaitre, J. A Continuous Damage Mechanics Model for Ductile Fracture. *J. Eng. Mater. Technol.* **1985**, *107*, 83–89. [CrossRef]
29. Rice, J.R. Inelastic constitutive relations for solids: An internal-variable theory and its application to metal plasticity. *J. Mech. Phys. Solids* **1971**, *19*, 433–455. [CrossRef]
30. Krajcinovic, D. Damage mechanics: Accomplishments, trends and needs. *Int. J. Solids Struct.* **2000**, *37*, 267–277. [CrossRef]
31. Tang, X.; Jiang, C.; Zheng, J. A General Theory for Elastic Damage. *J. Beijing Univ. Aeronaut. Astronaut.* **2001**, *27*, 4.

32. Shilin, D. Continuum Analogy Method for Analyzing Latticed Spherical Shells. *J. Build. Struct.* **1988**, *9*, 1–14. [CrossRef]
33. Tang, X.; Jiang, C.; Zheng, J. General Expressions of Constitutive Equations for Isotropic Elastic Damaged Materials. *Appl. Math. Mech.* **2001**, *22*, 7.

buildings

MDPI

Review

Advanced Composite Materials for Structure Strengthening and Resilience Improvement

Xinghuai Huang [1], Shaoyang Su [2], Zhaodong Xu [2,*], Qisong Miao [3], Wenfeng Li [3] and Lixin Wang [4]

[1] Shenzhen Research Institute, Southeast University, Nanjing 211189, China; huangxh@seu.edu.cn
[2] China-Pakistan Belt and Road Joint Laboratory on Smart Disaster Prevention of Major Infrastructures, Southeast University, Nanjing 211189, China; 220221423@seu.edu.cn
[3] Beijing Institute of Architectural Design, Beijing 100045, China; miaoqisong@biad.com.cn (Q.M.); liwenfeng00@126.com (W.L.)
[4] Shenzhen Academy of Disaster Prevention and Reduction, Shenzhen 518003, China; wlx@szadpr.cn
* Correspondence: xuzhdgyq@seu.edu.cn

Abstract: Advanced composite materials have excellent performance and broad engineering application prospects, and have received widespread attention in recent years. Advanced composite materials can mainly be divided into fiber-reinforced composite materials, laminated composite materials, matrix composite materials, and other composite materials. This article provides a comprehensive overview of the types and characteristics of advanced composite materials, and provides a comprehensive evaluation of the latest research on structural strengthening and resilience improvement in advanced composite materials from the perspectives of new methods, modeling optimization, and practical applications. In the field of fiber-reinforced composite materials, the hybrid technology of carbon fiber and glass fiber can achieve dual advantages in combining the two materials. The maximum increase in mechanical properties of multilayer sandwich RH plate by hybrid technology is 435.4% (tensile strength), 149.2% (flexural strength), and 110.7~114.2% (shear strength), respectively. In the field of laminated composite materials, different mechanical properties of laminated composite materials can be obtained by changing the deposition sequence. In the field of matrix composites, nano copper oxide particles prepared by nanotechnology can increase the hardness and tensile strength of the metal matrix material by 77% and 78%, respectively. In the field of other composite materials, viscoelastic materials and magnetorheological variants have received widespread attention. The development of composite materials benefits from the promotion of new methods and technologies, but there are still problems such as complex preparation, high cost, and unstable performance. Considering the characteristics, application requirements, cost, complexity, and performance of different types of composite materials, further improvements and innovations are needed in modeling and optimization to better meet practical engineering needs, such as the application of advanced composite materials in civil engineering, ships, automobiles, batteries, and other fields.

Keywords: advanced composite materials; structure strengthening; resilience improvement; fiber-reinforced composite materials; laminated composite materials; matrix composite materials

1. Introduction

With the development of science and technology and the diversification of social needs, industrial development has increasingly higher requirements for material properties. Many types of materials have received widespread attention in this field, from commonly used composite materials (such as glass fibers) to advanced composite materials (such as carbon fibers, magnetorheological elastomers, metal substrates, etc.). Advanced composite materials have excellent performance and broad engineering application prospects, and have received widespread attention in recent years. Therefore, increasingly more scholars are beginning to study structural strengthening and resilience improvement in advanced composite materials, which is crucial for improving the safety, reliability, and durability of composite equipment. Advanced composite materials can be mainly divided into four categories: fiber-reinforced composite materials, laminated composite materials, matrix composite materials, and other composite materials. The distinction among these four material categories can be outlined in the following manner: Firstly, fiber-reinforced composite materials are mainly characterized by embedding high-strength fiber materials as reinforcing materials in a continuous matrix. Fibers can be carbon fiber, glass fiber, aramid fiber, etc. Usually, fibers in fiber-reinforced composite materials are distributed in a continuous form in the matrix to provide reinforcement effects. Secondly, from a precise perspective, laminated composite materials are also categorized within the realm of fiber-reinforced composite materials. However, unlike traditional fiber-reinforced composite materials, they are alternately stacked with fiber layers, each with a different fiber direction, and then formed by pressing layer by layer. The arrangement of these laminated fibers in different directions provides excellent anisotropic properties, allowing the material to exhibit excellent mechanical properties in multiple directions. Thirdly, matrix composite materials refer to composite materials formed by the combination of two or more different materials, with at least one material serving as the matrix and the other materials serving as reinforcing or filling materials. The matrix material is usually metal or ceramic, while the reinforcing material can be fiber, particle, or sheet-like material. Lastly, other composite materials do not belong to any of these types, including composite materials with special applications in certain fields such as viscoelastic materials, magnetorheological fluids, cement mortar, etc. The continuous research and improvement in the performance of these four types of advanced composite materials from different research perspectives is still of great significance. This article provides a summary of new methods, modeling and simulation, and new applications for enhancing the structure and resilience of advanced composite materials. The intention is to offer the most recent research advancements in advanced composite materials, potentially benefiting other researchers in acquiring ideas and enhancing their research endeavors.

2. State-of-Art Concerning Fiber-Reinforced Composite Materials in Structural Reinforcement and Resilience Enhancement

2.1. Comparison of Common Fiber-Reinforced Composite Materials and Their Mixing Techniques

One of the biggest problems facing contemporary civil engineering is the strengthening, updating, and transformation of old structures. One of the most promising answers to these needs is the use of fiber-reinforced polymer (FRP). FRP is a composite material composed of synthetic fibers (such as carbon, glass, asbestos, beryllium, molybdenum, and aromatic polyamides) and synthetic polymer matrices (such as epoxy resin, polyester, and vinyl ester) [1]. Glass fiber-reinforced polymer (GFRP) and carbon fiber-reinforced polymer (CFRP) are the most suitable reinforcement materials for various civil engineering applications, as they have the highest temperature resistance, greatest strength, and lowest cost [2]. Bonding carbon fiber-reinforced polymer (CFRP) to the tensile surface of components can significantly improve structural strength. In addition, the type of adhesive material used is important to prevent debonding of the CFRP. Rageh has proposed two unique methods to enhance the adhesion between glass fiber-reinforced polymer (GFRP) and RC concrete surfaces, as shown in Figure 1 [3]. The purpose of Figure 1 is to demonstrate how CFRP is bonded to RC concrete beams through two different bonding materials, EP and GPP. The two adhesive materials are epoxy (EP) and geopolymer (GPP) with different proportions of short glass fiber (SGF). A more conventional method than the two proposed unique methods is to use epoxy resin to bond structural elements to steel plates. Through experiments and ABAQUS simulations, it has been proven that geopolymer slurries with and without short glass fibers in reinforced beams outperform epoxy resins in terms of performance and cost. At the same time, increasing the SGF ratio in GPP has a significant effect on improving the bearing capacity of the beam. GPP as an adhesive material enhances the maximum beam deflection and the beam ductility factor better than EP. At present, CFRP and GFRP composite materials have been applied in fields such as bridge reinforcement and building components due to their excellent performance, as is shown in Table 1. In recent years, the hybrid technology of carbon fiber and glass fiber can achieve the dual advantages of combining the two materials. Xian designed two types of carbon fiber/glass fiber hybrid rods with sandwich structures through pultrusion technology: fiber random hybrid (RH) with multilayer sandwich structures and fiber core–shell hybrid (CH) with single-layer sandwich structures [4]. The experimental results show that the tensile strength, flexural strength, and in-plane shear strength of the multilayer sandwich RH plate are increased by 50%, 30–40%, and 18%, respectively. With the increase in bending load, the flexural strength and in-plane shear strength decreased by 18% and 19%, respectively. By evenly dispersing carbon fibers into glass fibers, the problem of uneven interfacial stress concentration in multilayer sandwich RH panels can be effectively alleviated. The interfacial shear strength is more sensitive to bending load and immersion temperature. The long-term life prediction results show that the maximum increase in mechanical properties of multilayer sandwich RH plates is 435.4% (tensile strength), 149.2% (flexural strength), and 110.7~114.2% (shear strength). The mixing technology of basalt fiber and glass aramid fiber can also improve the mechanical properties of composite materials. Mazur used aramid fiber (AF) and basalt fiber (BF) as reinforcement materials and prepared polylactic acid composite materials through injection molding [5]. The increase in fiber content significantly improved the mechanical properties. AF increased Young's modulus and the tensile strength of the composite material, while BF increased the tensile strength. The trend in thermodynamic testing is similar to that for mechanical testing, where the higher the filler content, the higher the storage modulus. At the same time, adding AF and BF can increase the stiffness of the composite material, with the same effect as adding expensive aramid fibers of the same weight. The addition of fiber did not change the characteristic temperature obtained by differential scanning calorimetry. This work provides the possibility for creating hybrid materials with customizable mechanical and structural properties, and the use of different fibers can adjust the performance of composite materials to make them suitable for specific applications.

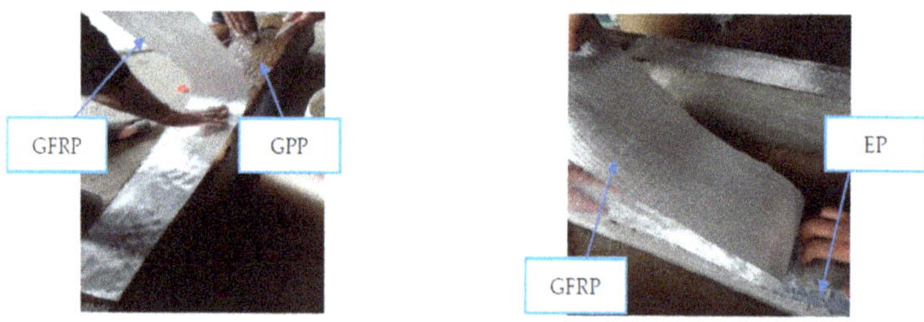

(a) (b)

Figure 1. Strengthening RC beams with (**a**) GFRP and (**b**) EP (adapted from [3]).

Table 1. Performance comparison between CFRP and GFRP.

Performance	CFRP	GFRP
mechanical properties	high strength and stiffness	good flexibility and impact toughness
fatigue performance	higher fatigue strength	poor fatigue performance
creep	good stability at high temperatures	poor stability at high temperatures
durability	good durability	relatively poor CFRP
cost	expensive	less expensive

However, research has shown that there are various types of FRP debonding failures regardless the type of bonding material or bonding method used, including middle crack

debonding, concrete cover separation, and plate end debonding. When FRP degumming fails, the effectiveness of the strengthening mechanism is greatly reduced [6–8]. A solution to this problem may be to combine effective installation procedures with well-defined and controlled prestressing of CFRP materials. An anchor system is one of the commonly used installation procedures and strengthening technologies in construction engineering, which has a wide range of applications in strengthening and reinforcing concrete structures, steel structures, and other components. When FRP debonding fails, this anchoring system may significantly improve the debonding process and interface bonding between FRP and RC concrete surfaces [9,10]. Schmidt studied a new type of prestressed CFRP NSMR reinforcement anchorage system, as shown in Figure 2 [11]. The purpose of Figure 2 is to show the approximate appearance of the anchoring system. The system has undergone five laboratory tests and has been successfully used in pilot projects to reinforce cast-in-place concrete bridges. The study focuses on considering the system's responsiveness to ensure that it can provide consistent load/deformation curves, yield thresholds, and yield mechanisms. All reinforcement systems have undergone in situ anticounterfeiting loading after installation and prestressing procedures have been carried out as required to eliminate any possible failure modes below the anticounterfeiting loading level, as is shown in Figure 3. This study indicates that this new anchoring system has performed well in laboratory testing and pilot projects, and has great potential for application. The fatigue resistance of CFRP and GFRP materials is an important mechanical property. Engineering structures are often affected by cyclic loads, which can lead to material fatigue and adversely affect the service life and damage tolerance of components and connections. Alam detailed the effect of cyclic loading on the fatigue performance of carbon fiber composite materials [12]. This article argues that more targeted research is needed on how the manufacturing of CFRP, inherent material defects, crosslinking density (curing), and material uniformity will affect the dispersion degree of CFRP under cyclic loading. Wu verified their good fatigue resistance by studying the fatigue performance and macroscopic damage in CFRP bending anchor cables [13]. Under stress amplitudes of 500 and 600 MPa, the residual strength retention rates of CFRP rods after two million fatigue cycles are as high as 95.1% and 76.7%. The bending–anchoring CFRP anchor cable can withstand two million fatigue loads with a maximum stress of 0.4 σ Ult, with an amplitude of 500 MPa and without significant fatigue damage. Under more severe fatigue loading conditions, the macroscopic damage of CFRP bars mainly manifests as fiber splitting and compression shear fracture. The additional shear effect is a decisive factor affecting the fatigue resistance of CFRP rods. These research results contribute to optimizing the bending anchorage system, improving its fatigue resistance, and promoting the application and development of CFRP cables and bending anchorage systems in bridge structures.

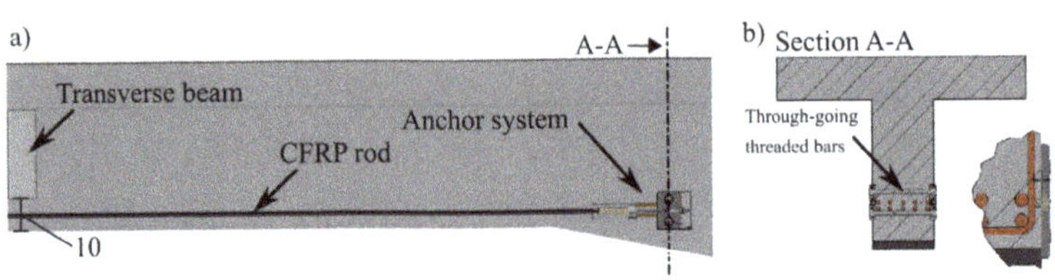

Figure 2. Side view (**a**) and section view (**b**) of the proposed strengthening method (adapted from [11]).

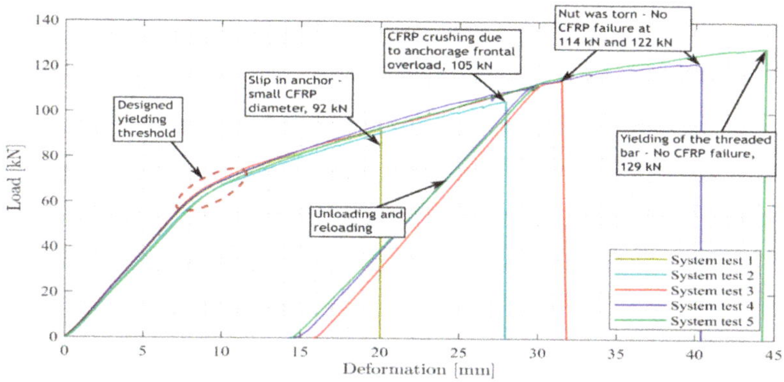

Figure 3. Load deformation test curves (adapted from [11]).

2.2. Methods for Improving the Performance of Fiber-Reinforced Composite Materials

Fiber-reinforced composite materials, as an important structural material, are influenced by various factors in their performance. In order to improve the performance of fiber-reinforced composite materials, many new improvement methods have emerged in recent years, such as surface modification, crosslinking, changing fiber orientation, additives, etc. Yang studied the analysis method of fiber fracture and orientation changes during the manufacturing process of short-fiber-reinforced composite materials [14]. Using a simplified CFD model to study experimental phenomena, the results showed that nozzle diameter has a significant impact on fiber fracture and mechanical properties, with smaller nozzle diameters leading to higher shear rates and more fiber fracture. Fibers of different lengths have different fracture ratios at different printing stages. The direction of the fibers inside the bead is related to the height of the layer. A larger layer height leads to better alignment, but it will reduce mechanical performance. Song adopted a new fiber surface treatment method to improve interface compatibility, and the treated fibers are shown in Figure 4. The use of a low-cost alkyl ketene dimer (AKD) is aimed at replacing relatively expensive silane-based reagents. The effect of fiber surface treatment on the interfacial compatibility and mechanical properties of composite materials was studied [15]. Lu proposed a fiber modification method by forming a thin SiO_2 layer on the surface of carbon fibers and reacting with $Ca(OH)_2$ to form calcium silicate hydrates, resulting in condensation at the fiber matrix interface [16]. The experiment found that the bending strength and tensile strength of carbon fiber-reinforced cement-based composites modified with nano SiO_2 were increased by 24.7% and 25.1%, respectively. At the same time, nanomodification enhances the compactness of the interface transition zone, reduces porosity, and increases the post crack behavior and resilience performance of the material. The research results indicate that this fiber modification method can effectively improve the performance of fiber-reinforced cement-based composites.

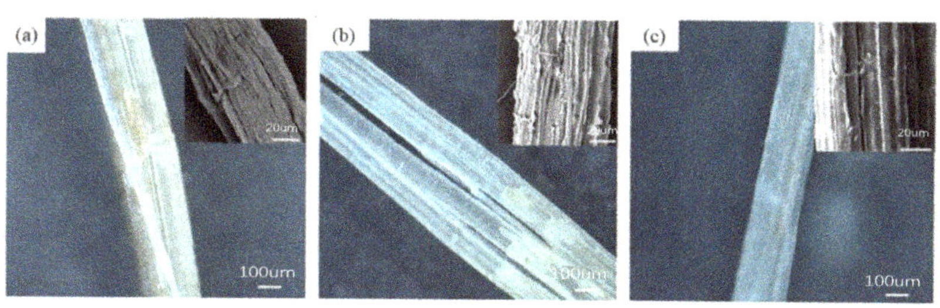

Figure 4. Micromorphology and structure of fibers with different treatments (**a**) H, (**b**) HP, and (**c**) HPM (adapted from [15]).

Lee investigated how fiber orientation and distribution influenced the mechanical behavior of short-fiber-reinforced thermoplastic composites by using synchrotron X-ray imaging combined with in situ tensile testing and digital volume correlation analysis [17]. This research found that fiber orientation is crucial for local compressive strain and anisotropic mechanical behavior, while load-direction strain is greatly affected by local volume fraction. The local volume fraction also determines the stiffness of the subvolumes that can be estimated from the analysis homogenization model. Batu studied the effects of different fiber volume fractions and fiber orientations on the mechanical properties of fake banana/glass fiber-reinforced hybrid composites [18]. The composites were manufactured by hand-lay technology, and the tensile, flexural, and compressive properties were measured by ASTM standard test methods. The effects of volume fraction and fiber orientation on the properties of composite materials were analyzed, and the results showed that both significantly affect the mechanical properties of hybrid composite materials. Zhang prepared five types of paper based composite friction materials with different glass fiber contents through the papermaking process, and studied the effect of glass fiber on material properties [19]. The results indicate that the sample with 10 wt.% glass fiber has the best mechanical properties and wear resistance. With the increase in glass fiber content, the shear strength first increases and then decreases, the compressibility increases, and the recovery rate decreases. Additionally, the friction coefficient first increases and then decreases with the increase in glass fiber content, and the optimal composition effect is for samples with medium glass fiber content. Liu was able to improve the strength of SFRTPCs by adding a small amount of strong materials to the matrix to connect short fibers [20]. In this work, micromechanical analysis was conducted using a finite element model, taking into account the fiber crosslinking effect. The simulation results showed that increasing the mass fraction of the bonding material can improve the mechanical properties of short-fiber-reinforced thermoplastic composites (SFRTPCs); by increasing the strength of the connecting rod material, the tensile strength and resilience of the composite material were especially improved. Shivamurthy studied the possibility of producing jute fiber-reinforced epoxy composites using a mixture of alkaline fibers and cashew shell liquid epoxy resin, and tested the mechanical properties [21]. The results show that the JFRE composite material made from modified matrix and alkaline-treated jute fibers has an average tensile strength of 679 MPa and a bending strength of 88.83 MPa, exhibiting better mechanical properties and can be used for high-strength applications. The emergence of these methods has greatly promoted the development of fiber-reinforced composite materials. However, there are still problems and defects in fiber-reinforced composite materials, such as complex preparation processes, high costs, and poor performance stability, which need to be addressed through technological innovation.

2.3. *Model Algorithm for Optimizing Fiber-Reinforced Composite Materials*

Establishing model optimization algorithms plays an important role in the development of fiber-reinforced composite materials. At present, the commonly used modeling methods include multiscale topology optimization, theoretical model, and so on. Optimization algorithms mainly include genetic algorithms, trial-and-error methods, etc. The application of these models and algorithms can better predict the mechanical properties of composite materials, thereby optimizing the ratio and preparation process of composite materials. Alhaddad uses artificial neural network modeling methods to predict the performance of composite materials, especially the ultimate tensile strength [22]. At the same time, the artificial bee colony algorithm is used to optimize printing and material parameters to achieve maximum tensile strength. The research results indicate that fiber properties, fiber morphology, fiber orientation, and fiber volume fraction have the greatest impact on the tensile strength of composite materials. Duan introduced a novel multiscale and multimaterial anisotropic penalty (MMCAP) model [23]. This model is tailored for exploring the design optimization of composite structures with variable stiffness (VS), reinforced with fibers. The objective is to minimize structural compliance, as demonstrated

in Figure 5. This model utilizes improved solid isotropic material punishment (SIMP) and discrete material optimization (DMO) at the macro- and microscales, respectively, to achieve clear macromultimaterial structure topology and microspecific discrete fiber-laying angle selection. This method can effectively achieve multimaterial and multiscale design optimization of fiber-reinforced composite structures, achieving clear macroscopic topology of multimaterial structures and microscopic fiber-laying angles. This provides a new implementation strategy for lightweight, multimaterial, and multiscale design optimization of composite materials. Obid has proposed a new constitutive model that can describe the tensile and compressive responses of nonlinear anisotropic fiber materials under general stress states [24]. Built upon the Ramberg–Osgood model, this framework is expanded by incorporating triaxial stress invariance as a marker for stress-state classification. This extension aims to account for tension–compression asymmetry within multiaxial stress conditions. Compared to other models, this model provides an elegant formula for describing all phenomena and implements a simple calibration program. Therefore, this method can describe the nonlinear behavior and tension compression asymmetry of fiber-reinforced composite materials, and has better advantages. Liu studied the fracture behavior of fiber-reinforced composite laminates under shear dominant loads through experiments and numerical simulations [25]. The V-notch guide-rail shear testing method was used, and the fracture surface was studied using scanning electron microscopy. At the same time, a high-fidelity, elastic–plastic damage model was developed to predict shear fracture behavior. The experimental results show that the composite sample undergoes a significant nonlinear process and exhibits a series of failure modes. The simulation results exhibit good correlation with the experimental results, demonstrating the predictive capability of the developed model. This study is of great significance for the design of composite material structures. Sakata utilized SEM images of randomly arranged, unidirectional fiber-reinforced composite samples [26]. The apparent strength in each segment was extracted, and the validity of the random field modeling for local strength was examined. Through microstructure simulation, the spatially varying random apparent fracture strength was subsequently calculated. Single scale and multiscale analysis were attempted in this study, and random field and its autocorrelation function were calculated by using the moving window method. The numerical results indicate that a smaller window is preferred for a more accurate evaluation of apparent strength. From the perspective of establishing models and optimizing algorithms, there are still some shortcomings in the development of fiber-reinforced composite materials, such as the lack of effective prediction models and immature optimization algorithms.

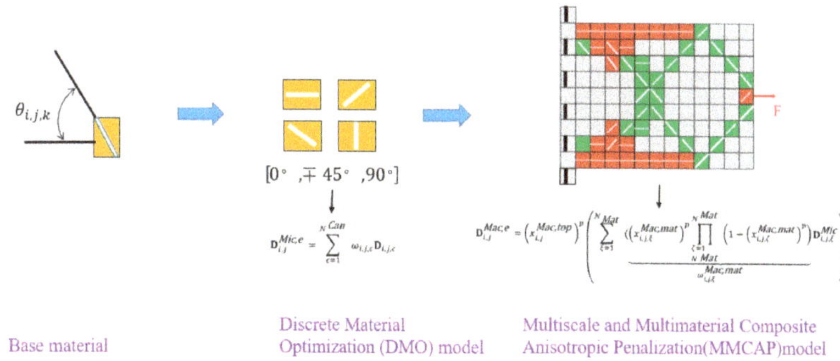

Figure 5. Schematic showing concurrent macroscale and multimaterial topology design optimization for composite laminates with minimum structural compliance.

2.4. Application of Fiber-Reinforced Composite Materials in Practical Projects

However, whether it is new technological methods or innovative modeling and simulation methods, their core goal is to improve practical applications. The mixed fiber technology in technical methods can significantly improve the strength and stiffness of

materials, thereby producing lighter and higher strength materials for practical engineering, such as civil infrastructure reinforcement. The proposal for novel modeling and simulation methods, such as the MMCAP model, provides a new solution for lightweight application and multiscale optimization application of composite materials. Fiber-reinforced composite materials have been widely used in practical engineering and projects, mainly in fields such as aerospace, automotive, shipbuilding, architecture, electronics, etc. Lv used ultrasonic-assisted electroplating technology to evenly coat copper coatings on the surface of carbon fibers to improve their wettability with the aluminum substrate [27]. This research found a linear relationship between the thickness of copper coatings and electroplating time. At the same time, copper coating can promote the infiltration of aluminum melt into carbon fiber bundles, thereby enhancing the mechanical properties of composite materials. The optimal coating thickness is 1.5 microns, and the tensile strength of the composite material is 169 MPa, the elastic modulus is 80 GPa, and the elongation is 4.2%. Therefore, ultrasonic-assisted electroplating technology can improve the performance of carbon fiber-reinforced aluminum matrix composites and enhance their application prospects in industrial and commercial applications. Kumar explored the preparation of carbon fiber powder-reinforced polymer composites and their application in pump impellers [28]. Researchers use specific materials and molds to predict the strength of composite materials through tensile, bending, and impact testing, and use nondestructive testing methods to predict crack formation. The results showed that the strength of composite materials was superior to that of metals, cast iron, and structural steel, and ultimately achieved application in pump impellers. Researchers believe that by changing the thickness and combination of fibers, higher strength and corrosion resistance can still be achieved compared to lightweight materials such as steel and cast iron. In the future, this method can be applied to the field of pump impellers. Jeong investigated the assessment of mechanical performance for cargo sealing systems on liquid natural gas transport vessels [29]. This was accomplished by employing the continuous installation of E-glass fiber-reinforced composite materials with extended structures, as depicted in Figure 6. This research revealed that the material properties associated with this specific length can effectively act as representative material properties for infinite lengths, offering a dependable characterization of the material attributes pertaining to extended structures of infinite nature. By using a Weibull distribution to calculate the failure strength, the study also found that the standard deviation and failure strength of the material increase with the increase in length within a specific length. These findings are of great significance for ensuring the safety of liquid natural gas transport ships.

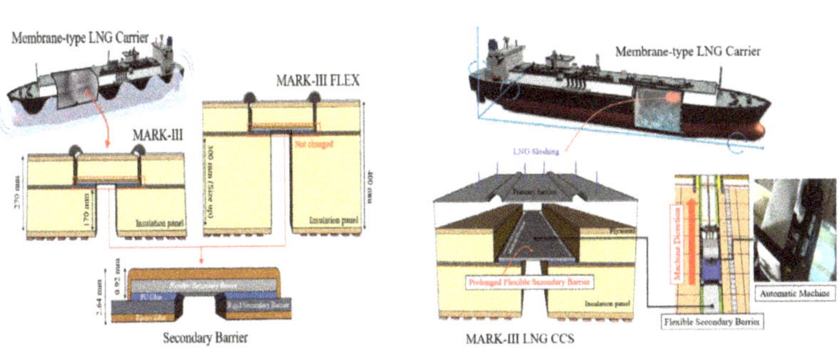

Figure 6. Schematic of the secondary barrier in the MARK-III and FLEX LNG CCS (adapted from [29]).

Saintane prepared two composite samples using carbon-based materials and powder metallurgy methods to manufacture carbon fiber-reinforced polymer composites to improve friction performance [30]. By conducting shore hardness testing and compression testing on the sample, it was proven that Sample 2 had higher hardness and compression strength. The results of this study indicate that the carbon-based polymer friction composite material manufactured has excellent mechanical properties and can be used as a candidate material for commercial brake pads and other application fields. Additionally, to meet the needs of practical engineering and applied projects, it is necessary to further improve specifications and standards, promote and apply mature fiber-reinforced composite material technology, and strengthen research in material testing, performance evaluation, and quality control.

3. State-of-Art Concerning Laminated Composite Materials in Structural Reinforcement and Resilience Enhancement

3.1. Effect of Changing the Deposition Order on the Properties of Laminated Composite Materials

Laminated composite materials are composite materials composed of two or more layers of different materials combined in a specific way, with good mechanical properties and various application fields. In order to further improve the strength and resilience of laminated composite materials, various new methods have been proposed, for example, changing the deposition sequence, adding synthetic fibers, interlayer reinforcement technology, design optimization, explosion-proof welding methods, etc. The deposition sequence and deposition materials are important factors affecting laminated composite materials. In recent years, many researchers have conducted experiments on material manufacturing and deposition sequence. Yang et al. studied 316L stainless steel in laminated composite materials and compared four laser scanning strategies. By using subtraction milling technology, they obtained a smooth surface and analyzed the average density, defects, average hardness, and tensile properties of the material. In addition, due to the clear interface of laminated composite materials, they also studied the possible residual stress, tensile anisotropy surface roughness, and internal defects that may exist during the manufacturing process of laminated composite materials through addition and subtraction hybrid manufacturing (HASM) [31–36]. Bhaburi et al. estimated the surface/interface quality, microstructure, and mechanical properties of hybrid aluminum parts, and improved their microhardness and joint strength through heat treatment [37]. In the field of laminated composite materials, SS316L and Inconel 718 alloy laminated composite materials are manufactured through mixed additives and subtraction to obtain materials with good mechanical and high-temperature properties. In addition, by studying the microstructure near the interface and the tensile properties in the deposition direction, it is possible to further understand the performance and feasibility of laminated composite materials. Li used HASM technology to manufacture laminated composite materials in one step and provides customized performance [38]. The improved additive manufacturing system and auxiliary traction milling can maintain a smooth and clear interface for energy deposition, solving the challenge of connecting different materials. The experimental results indicate that laminated composite materials with different deposition sequences have different microstructures and mechanical properties, among which IN718 laminated composite material deposited on SS316L exhibits the best performance, while new substances are generated at the interface between SS316L and 1040 steel to improve performance, as shown in Figures 7 and 8.

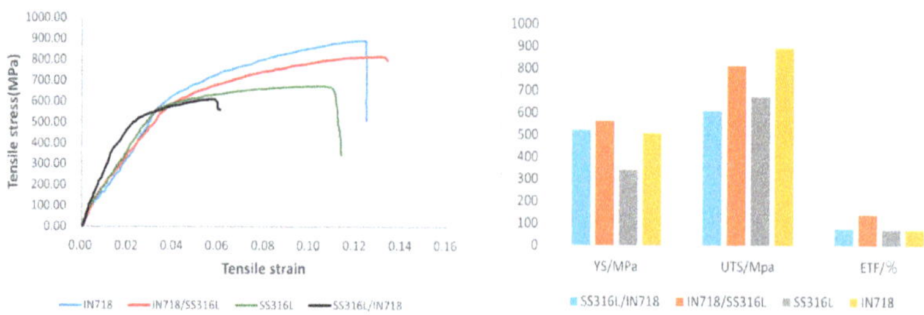

Figure 7. Tensile stress–strain curves for laminated composite specimens with different deposition sequences, manufactured by HASM (the initial calibration length was 20 mm).

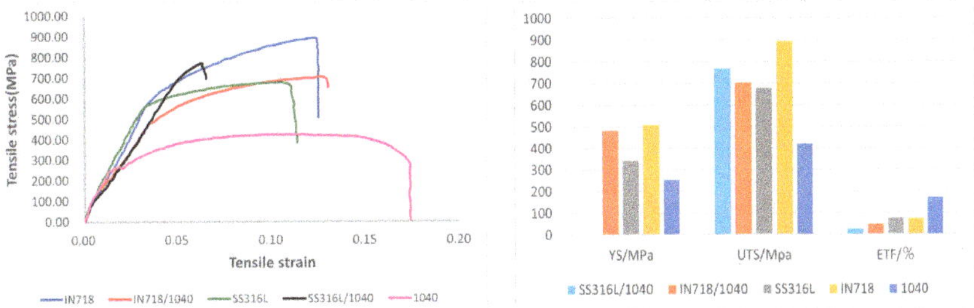

Figure 8. Tensile stress–strain curves for different materials deposited on 40 steel (the initial calibration length was 20 mm).

3.2. Methods for Improving the Performance of Laminated Composite Materials

In addition to the main method of changing the sedimentary sequence, many other new methods have been proposed in recent years. Kumar studied the preparation process for titanium alloy laminates for multilayer composite material input [39]. This study used a single stage process instead of rolling to prepare laminated plates, and investigated the sliding wear behavior of Ti-6Al-4V laminated plates. The research results indicate that composite materials have better wear resistance and lower wear rate than matrix materials, and the hardness value of composite materials is increased by about 25% compared to matrix materials. Biswas used the finite element method to study the effect of fiber orientation on the tensile, bending, and fatigue properties of bidirectional glass and carbon fiber-reinforced composite laminates [40]. The experimental results show that laminates with a laminate angle of (0°/90°) exhibit the maximum tensile strength, bending strength, and fatigue life. The tensile strength, bending strength, and fatigue life decrease with the increase in layer angle, and the layer angle reaches the minimum value (±45°). Therefore, the fiber orientation of composite materials can affect the quality of laminated materials. Liu introduced a test method for characterizing continuous fiber-reinforced composite materials, the deflection test, and its application in comparing the shear deformation of cross-unidirectional prepregs and woven prepregs [41]. This research has shown that material composition and different process parameters play a crucial role in the shear properties of composite materials, while the metal layer support can improve the shear deformation capability of fibers. This experimental method helps to better understand the process design in manufacturing hybrid metal composite laminates. Slattery emphasized the importance of robust residual strength testing methods for evaluating the effectiveness of composite laminate repair techniques [42]. The study used four different mechanical tests to study the strength of original, damaged, and repaired solid laminated composite materials, including post impact compression (CAI), tensile, and four-point bending tests. This research found that the different testing methods can affect the strength recovery capability of repair methods, and the four-point bending test is considered the best preliminary testing method for screening repair techniques. However, when selecting testing methods to evaluate the effectiveness of repairs, design standards should always be considered. Maithil introduced a methodology aimed at enhancing the mechanical properties of naturally reinforced fiber–polymer composites [43]. This approach primarily involves the synthesis of composite materials by combining reinforced fibers with polymer matrices. The purpose of the research was to improve the tensile strength of composites, and to achieve this goal by blending synthetic fibers into natural fiber. Using the ABAQUS software package for numerical construction and tensile simulation, the tensile behavior of composite material samples was evaluated, and the surface treatment was noted to improve the adhesion between fibers and matrix. The results show that the tensile strength of jute fiber is lower than that of hybrid fiber-reinforced composite materials, and the addition of carbon fiber can improve the strength of the laminate. Mortell described two highly coupled damage mechanisms in laminated composite materials—transverse microcrack growth and delamination [44]. This article investigates the effect of changes in transverse crack density on delamination growth. The results indicate that increasing the number of transverse layers can significantly improve the load-bearing capacity of the laminate. In addition, the article also discusses the use of new in situ scanning electron microscopy micromechanical testing and acoustic damage detection techniques to determine the initial load, density, and local differential layer length of transverse cracks, together with the performance and damage behavior of laminates under different loads. Zuo explored how to optimize the effective stiffness design for laminated composite materials [45]. By solving a constrained optimization problem, the laminated composites are designed by finding the best material properties of the constituent phases and their volume fractions. The findings of the research suggest that by identifying the ideal volume fraction for the constituent phases, it is possible to attain significant longitudinal and transverse stiffness. Additionally, enhancing Young's modulus in a specific stage leads to the attainment of the highest effective Young's modulus. These results indicate that the design and manufacturing of laminated composite materials have special effective stiffness, which can be easily achieved through the proposed methods and modern manufacturing technologies. Liu introduced an innovative technique that effectively examines interface strength within layered composite materials through the utilization of viscous elements [46]. This method also offers improved precision in forecasting the extent of the fracture process zone, particularly when employing coarse grids. This research has shown that the constant cohesion law is more suitable for simulation layering of coarse grids than the bilinear cohesion law. This method is applicable to different structural thicknesses and material properties. Li described that TC4/TA1/Ti6321 titanium alloy laminated composites were successfully fabricated by explosive welding, and the interfacial properties and dynamic mechanical properties were studied, as shown in Figure 9 [47]. The results show that there are high-density crystal defects at the interface, and the composite plate shows a hard/soft/hard hardness distribution, and the microhardness near it increases. Owing to the presence of interfaces, dynamic mechanical properties exhibit different results, and the failure mode is adiabatic shear failure mode. However, there are still some shortcomings in the new method for laminating composite materials. For example, although new technologies can improve the performance and reduce costs of composite materials, their actual preparation process may be limited by factors such as material cost, processing difficulty, and stability.

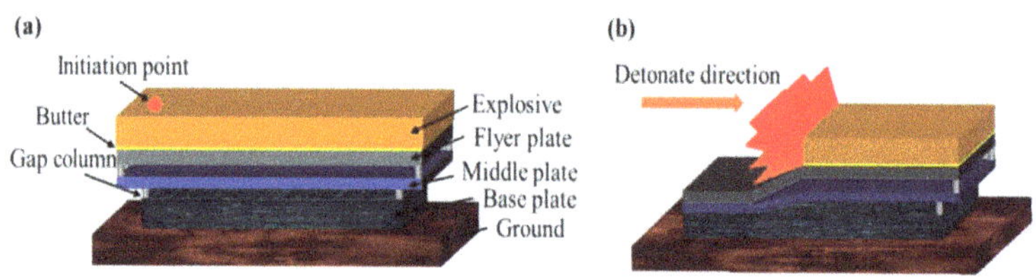

Figure 9. Schematic diagram showing the EXW experiment: (**a**) welding process design; (**b**) welding process at some point (adapted from [47]).

3.3. *Modeling and Simulation for Optimization of Laminated Composite Materials*

At present, models and software simulations have been widely established and applied toward laminated composite materials. Among them, finite element analysis and the genetic algorithm are the most commonly used methods, which can simulate the mechanical behavior of composite materials. In addition, multiscale simulation methods based on microstructure are gradually receiving attention, which can more realistically reflect the microstructure and mechanical properties of composite materials. Coelho described a computational model for designing bimaterial composite laminates using a multiscale topology optimization model to optimize the design of structures and materials [48]. The model is based on a hierarchical optimization strategy, taking into account the manufacturing process and basic characteristics, and using a mixed set of micro- and macrodesign variables to characterize the distribution of the two materials. This method has yielded better design results, which helps to provide a deeper understanding of the effectiveness of the microstructure characteristics of composite laminates. Shi employed the LS-DYNA software to simulate the dynamic and progressive failure process in laminated composite materials [49]. This simulation utilized the MAT_162 material models, encompassing factors such as strain rate, damage evolution, and anisotropy effects. To determine the model parameters, traditional standard and nonstandard testing methods were used, including double shear and Brazilian testing. Through experiments and numerical simulation, the fidelity of the developed material model parameters was verified, including the modulus and strength in plane and full thickness directions, fiber matrix splitting and shear fracture, and other failure modes. Nastos introduced a strength prediction method for composite laminates based on a combination of numerical and nondestructive testing, aiming to predict strength through nondestructive testing, as shown in Figure 10 [50]. In order to manage the difference between the maximum load of 20% and 100% in the sample, a high-fidelity physics-based numerical model, deep learning technology, and noncatastrophic experiments were used. This application is trained using convolutional neural networks and Monte Carlo dropout techniques, and validated using probabilistic failure analysis datasets generated by stochastic finite element methods.

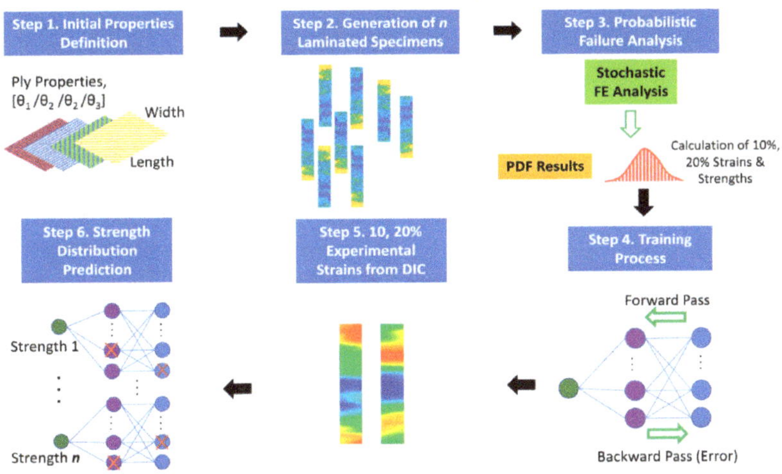

Figure 10. The concept of the nondestructive methodology for predicting strength in composite laminates (adapted from [50]).

Yashiro conducted a mode II interlaminar fracture resilience test (DENT) and examined it through elastic–plastic finite element analysis [51]. The aim was to explore the effect of material nonlinearity on the assessment of interlaminar fracture resilience within composite materials. The study revealed that the plastic zone size in the DENT test remains unaffected by the crack length. Furthermore, the disparity between the estimated energy release rate and the integration within the plastic zone size is relatively minor when compared to other quasi-static mode II interlaminar fracture resilience tests. The CC method is an effective way to evaluate the resilience of adhesives, and even significant plastic deformation does not affect the test results. These findings can be useful in evaluating the fracture resilience of laminated composite materials. Torabi described conducting experimental and theoretical crack investigations on laminated glass/epoxy composites to evaluate their load-bearing capacity [52]. By using a new concept called virtual isotropic material (VIMC) and combining predictions from two brittle fracture models, namely maximum tangential stress (MTS) and average stress (MS), the fracture behavior of laminated materials can be predicted. This prediction is necessary because laminated materials have complex structures and predicting their load-bearing capacity is difficult. The research results indicate that using VIMC and LENFM filters to predict the load-bearing capacity of laminated materials can be directly, quickly, and conveniently predicted without the need for complex analysis. Lu proposed a three-dimensional separable cohesive element (SCE) that can simulate the interaction between matrix cracking and interface delamination in laminated fiber-reinforced composites [53]. The results indicate that SCE can effectively simulate stress concentration caused by matrix cracks and load transfer from solid components with cracks to interface viscous components. The application of SCE to the progressive failure model of composite laminates is consistent with experimental results. Lin proposed a framework for fast computation of low-speed impact analysis and compressive composite material impact response, using intelligent grid mode, effective modeling strategy, and the damage transfer algorithm [54]. This framework can significantly reduce computational time while maintaining accuracy, helping to expand design space and generate large databases. Compared with traditional methods, using this framework can reduce computational time by 67% and obtain prediction results within 6% error compared to experimental values. This framework can be used to optimize the impact resistance of laminated composite materials. However, there are also some limitations in establishing models and software simulations. First, the establishment of the model requires a large amount of experimental data for verification, which may be a significant challenge for composite materials. Second, existing models and software simulation methods often only consider the influence of a single factor and cannot fully consider the composite effects of multiple factors. Therefore, in the future, it is necessary to further develop and improve models and software simulation methods to better meet the needs in practical engineering.

3.4. *Application of Laminated Composite Materials in Practical Engineering Projects*

Changing the deposition sequence in technical methods can produce laminates with different resilience and tensile strength, and laminates manufactured using nanotechnology can be applied to very precise electrical components. The combination of new numerical values and nondestructive testing for strength prediction methods is of great significance for testing the performance of laminates, which is of great significance for practical applications. In recent years, there have been many new technological breakthroughs and practical applications in laminated composite materials. Among them, nanotechnology has been widely developed. One of the most effective uses of nanofibers or nanoparticles is as a staggered reinforcement for composite laminates, which can be embedded between two layers of laminates to study their effect on the overall performance of composite laminates. Bodduru studied the performance changes when E-glass fiber and sisal fiber were mixed together to manufacture laminates reinforced with epoxy polymer composites [55]. Researchers treated sisal fibers with alkali (NaOH) and made laminated samples using the manual lamination method. Additionally, they explored the effect of introducing carbon nanotubes (CNTs) and MXene ($Ti_3C_2T_x$) nanoparticles on hybrid composite materials. Nanofillers are mixed with pure epoxy resin and dispersed using ultrasound. The multiscale structure of MXene-dispersed hybrid plates was examined using scanning electron microscopy. The results indicate that by enhancing with 42% MXene, the tensile strength of sisal/glass fiber-laminated materials is enhanced. In addition, when MXene nanoparticles were used to enhance the matrix, the bending strength and modulus of sisal/glass fiber-laminated materials increased by 34% and 30%, respectively. This research found that the incorporation of nanofillers significantly improves the load transfer efficiency between the fiber material and the epoxy matrix, resulting in better elastic properties. Abot proposed the advantages and problems of fiber-reinforced polymer composites, namely the possibility of interlayer stress leading to premature failure [56]. To solve this problem, a composite material can be established between layers of laminated composite materials to maximize shear stress transfer. The dense arrangement of multiwalled carbon nanotube arrays is considered a nanoparticle-reinforced material that can improve the interlayer shear modulus of layered composite materials. The preliminary results indicate that this method is feasible. Ipackchi uses PVB electrospun nanomesa to toughen phenolic resin/glass fiber composite materials [57]. The experiment found that the thickness of the interlayer is the most effective factor in improving the resilience of type I and type II. Additionally, the decrease in nanofiber diameter and the increase in impact strength also contribute to the improvement in fracture resilience. The arrangement of nanofibers has little effect on the fracture of composite materials. High-tensile, ultrathin electrospun polyamide thermoplastic nanofibers were created, which were then incorporated into layers of glass fiber-reinforced woven phenolic preimpregnated composites. The experimental results indicate that an appropriate density of electrospun nanofibers can significantly improve the structural integrity of composite laminates without causing weight gain. There is a threshold for the density of nanofibers; exceeding this threshold will reduce the bonding performance between glass fibers and phenolic matrix [58]. Ji introduced a method of preparing nonwoven cellulose acetate nanofiber felt using electrospinning and adding it to the core layer of high-pressure laminates [59]. When the CA concentration is 16 wt.%, the morphology of the prepared CA nanofiber felt is the best. After adding CA nanofiber felt, the mechanical properties of the polymer composite material were significantly improved, including increased tensile strength and elongation at break. Kubin prepared low-density polyvinylidene fluoride (PVDF) nanocomposites containing cubic symmetric nano-$BaTiO_3$ using the electrospinning method, and the electron microscope image is shown in Figure 11 [60]. Brugo integrates nanopiezoelectric sensors into composite laminates to achieve self-sensing monitoring [61]. The composite laminate used is a composite material called glass-laminated aluminum-reinforced epoxy resin (GLARE). This nanostructured hybrid laminate is itself a piezoelectric sensor capable of detecting the effects on its entire surface. Nondestructive impact testing and research and optimization of circuit electrical parameters are used to evaluate the performance of self-sensing laminates, including linearity and spatial uniformity.

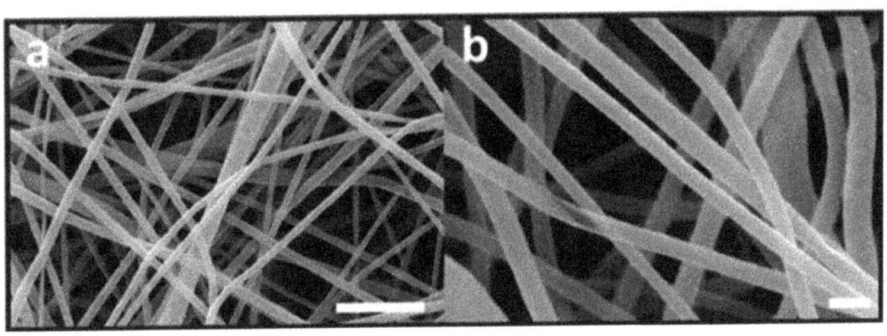

Figure 11. SEM images of electrospun PVDF/$BaTiO_3$ nanocomposite fibers: ES-PVDF (**a**,**b**). (adapted from [60]).

Palazzetti et al. reviewed the latest composite laminate technology, which interweaves electrospun nanofibers to enhance the material. The research in this literature adopts mechanical methods and focuses on the main load types. Researchers conducted in-depth research on the role and working mechanism of nanofibers and summarized the main results. The research results indicate that the nanofiber interlayer between the layers of the laminate can bring significant benefits from a structural and load-bearing perspective, while its impact on the weight and size of the laminate is minimal and even negligible. Compared with basic materials, the interlaced laminated sheets of nanofibers can significantly improve the mechanical properties of the material [62,63]. Xu introduced a multiunit-structured battery composite laminate composed of three advanced structural battery composite units in series [64] and studied its electrochemical and mechanical properties. The experimental results indicate that the capacity of the multiunit-structure battery composite laminate is slightly affected by tensile load. Additionally, after calculation and estimation, the mechanical properties of the multicore-structure battery laminated plate are equivalent to a traditional glass fiber composite multiaxis-laminated plate. Yoo examined the fundamental attributes and magnetoelectric potentials of a piezoelectric composite material consisting of 15 modules [65]. Additionally, Yoo crafted a structure with a magnetostrictive piezoelectric laminated plate to induce shear stress, employing the aforementioned material. The investigation also delved into the material's magnetoelectric properties. The research results indicate that this piezoelectric composite material has great potential for application in high-performance magnetic and electrical components. Although there are many new technologies and potential applications for laminated composite materials, there are still some limitations in practical applications. Compared to traditional materials, the damage and wear of laminated composite materials may lead to more complex repairs and higher maintenance costs. Therefore, it is necessary to comprehensively consider these issues when promoting and applying laminated composite materials.

4. State-of-Art Concerning Matrix Composite Materials in Structural Reinforcement and Resilience Enhancement

Matrix composite materials are composite materials composed of two or more different materials. One type of material is a matrix, and the other is a reinforcing material. This usually provides better performance and characteristics than a single material. Matrix composite materials are mainly divided into metal matrix composite materials and ceramic matrix composite materials.

4.1. Methods for Improving the Properties of Metal Matrix Composite

The strength and resilience improvement in metal matrix composite is one of the current research hotspots. Its new methods mainly focus on the stirring–casting method, material microstructure design, processing technology optimization, surface modification, and addition of appropriate second equalities. Rajaram studied the effect of particle weight fraction on the mechanical properties of metal matrix composite (MMC) by stirring–casting [66]. Metal matrix was prepared using aluminum (Al7075) and quarry rock dust powder, with 0%, 5%, and 7.5% rock powder and Al7075 added. Hardness, bending, and wear tests were conducted, and machining was carried out according to ASTM standards. This research found that the quarry stone powder of Al7075 has a higher hardness and lower wear rate, and the metal matrix containing 7.5 wt.% has the highest bending degree. Therefore, sample 2 has higher hardness and wear resistance. Rajaram studied the preparation of AA6063/ZrO_2 metal matrix composite by stirring–casting technology, and studied the mechanical properties of the composite through hardness, UTS, pin-on-disk wear, and other testing methods [67]. This research found that in the AA6063 sample of 4 wt.% ZrO_2 composite material, the maximum hardness is 71.0 HV and the maximum ultimate strength is 168.384 MPa, which is about 12% higher than the unreinforced AA6063 composite material. The results of the pin-plate wear test indicate that the lower wear rate identified in the AA6063 test sample with 4 wt.% ZrO_2 composite material is 2.5×10^{-6} mg/m. SEM microscopic testing confirmed that ZrO_2 particles were uniformly distributed in the test samples of 2 wt.% and 4 wt.% AA6063 composite materials. The research results indicate that as the weight percentage of ZrO_2 increases, the hardness and UTS of AA6063/ZrO_2 composite material significantly increase, while the wear rate significantly decreases. Chinababu successfully prepared two kinds of metal matrix composite, which are composed of different proportions of TiB_2 and CoCrFeMnNi high-entropy alloys [68]. The microstructure and morphology of the composites were studied by TEM and SEM. Compared with traditional ceramic particle composite materials, composite materials with added reinforcing phases have better mechanical properties, with LM25-2 wt.% TiB_2-3 wt.% HEA composite materials exhibiting the highest tensile strength. It was observed that with the increase in HEA content, the cleavage planes on the fracture surface changed. Xu successfully prepared carbon nano-onion (CNO)-reinforced AZ31B magnesium alloy composite MMC through friction stir treatment (FSP), and studied their microstructure and mechanical properties [69]. The experimental results indicate that FSP CNO/AZ31B MMC achieved a uniform fine-grained structure in the stirring zone, and the grain refinement mechanism is attributed to the continuous and discontinuous dynamic recrystallization, together with the key role played by CNOs in activating twinning behavior. The MMC exhibits enhanced yield strength and improved strain hardening ability, owing to a comprehensive effect of grain boundary strengthening, dislocation strengthening, Orowan strengthening, and the load-bearing mechanism of the reinforced material. Compared with the matrix material, the yield strength and elongation at break increased by 14% and 11.8%, respectively, indicating that both strength and ductility were improved due to the addition of CNO-reinforcement material. Zhai has developed a new method to manufacture metal matrix composite, that is, to use a three-dimensional metal glass lattice to strengthen metal [70]. This processing route overcomes the limitation in using only discontinued crystalline ceramic particles as reinforcing materials in existing methods. In the experiment, stainless steel matrix composites reinforced with 30% by volume, three-dimensional metallic glass in the form of a body-centered cubic lattice exhibited excellent mechanical properties. The enhanced mechanical properties mainly come from well-designed structures, good bonding between hard three-dimensional metallic glass and soft stainless steel, and their mutual reinforcement mechanisms. This study provides a novel design concept and manufacturing method that can be extended to other alloy systems to achieve improved mechanical properties. Kumar prepared Si_3N_4-reinforced Al7075 metal matrix composite (AMMC) by stirring–casting technology [71]. The results show that AMMC reinforced with Si_3N_4 has better hardness, tensile strength, compressive strength, and impact strength than the base alloy. The Si_3N_4 reinforced sample with a 9% weight percentage showed the best performance, and the predicted results of the regression equation were in good agreement with the experimental results. Therefore, stirring–casting is an effective method for preparing AMMC. Patil introduced a new microfilling material in metal matrix composite, namely fly ash [72]. This research has shown that adding different contents of fly ash can significantly improve the mechanical properties and wear resistance of aluminum alloy composite materials. SEM analysis also demonstrated the important role of the dispersion characteristics of fly ash in improving the performance of composite materials. These results show that it is feasible to use fly ash as a reinforcing material in metal matrix composite, and it has application prospects. Kumar has prepared powder reinforced metal matrix composite using traditional powder metallurgy methods, in which Al powder and ceramic particle reinforcement are an attractive combination [73]. The composite material prepared was subjected to hardness, compression, and tensile tests, and the seventh batch of samples was identified to have the highest strength. The aluminum grade, silicon carbide powder size, and weight fraction have an impact on the effective manufacturing of aluminum-based powder samples. Al SiC Gr is considered as a special composite combination with high strength to weight ratio and is an important metal matrix composite. Thirupathi compared the physical and mechanical properties under two different sintering cycles, and studied the density and microhardness [74]. By mixing 1% NbC powder with pure aluminum and compacting through conventional sintering and the Chu two-step sintering process (TSS-C), four samples were obtained, two of which were pure aluminum and two were Al-1% NbC. The results show that the mechanical properties of the TSS-C sample are better than the other samples, with higher density and microhardness, which is of great significance for the application of metal matrix composite in the automotive and aircraft industries. Sachin [75] studied the mechanical properties and specific wear rate of aluminum alloy Al6061 reinforced with different weight percentages of silicon carbide (SiC). Metal matrix composite were prepared by stirring–casting, and two SiC particles with different weight percentages were added into Al6061. The tests include tensile, hardness, and dry sliding wear, among which the C2 composite material shows higher wear resistance than the C1 composite material. Scanning electron microscopy was used to observe the microstructure of the composite material and determine the distribution of the reinforcement. The results indicate that SiC particles can adapt to specific wear applications, and the reinforcement has a strong impact on the matrix material. Meher prepared a TiB_2-reinforced magnesium RZ5 alloy matrix–in situ metal matrix composite by self-propagating high-temperature synthesis [76]. TiB_2 particles were nearly evenly distributed in the magnesium RZ5 alloy matrix. The solid solution heat treatment improved the mechanical properties of the material, with a maximum tensile strength of 178.7 MPa. Adding 8 wt.% TiB_2 can reduce the wear of the material. The microstructure of the composite material was observed through optical microscopy and field-emission scanning electron microscopy, and the wear surface of the composite material was analyzed. Gayathri introduced a new preparation method for metal matrix composite, which uses waste catalyst and copper oxide nanoparticles to strengthen the pure aluminum matrix [77]. The composite material was prepared by the double stirring–casting method, and the influence of different nano copper oxide contents on the mechanical properties of the composite material was investigated. The optimal composition was ultimately determined to be Al10% SAC1% nano copper oxide, which increased the hardness and tensile strength by 77% and 78%, respectively. Rahul studied the microstructure, mechanical properties, and texture evolution of AA6061-T6 metal matrix composites reinforced with silicon carbide (SiC) and zinc (Zn) particles during friction stir welding (FSW). The fracture diagrams of two particle-reinforced specimens are shown in Figure 12 [78]. Kumar studied the effects in the use of exfoliated graphite nanosheets and thermally sealed nanofillers on the mechanical properties of copper-based composites [79]. This research has shown that adding nanofillers can improve relative density, hardness, and wear resistance, while reducing porosity. The hardness and tensile strength of Cu-2 wt.% xGnP composite materials are higher than those of pure copper and Cu-MWCNT composite materials. In addition, the uniform distribution of xGnP and good adhesion to the copper matrix contribute to improving the mechanical properties and wear behavior of the composite material. Adding nanofillers can also change the crystal structure of composite materials. Although these new methods have improved the performance of metal matrix composite to a certain extent, there are still some problems and challenges, such as the loss of resilience while improving the strength, and the increase in material cost and manufacturing complexity due to alloying.

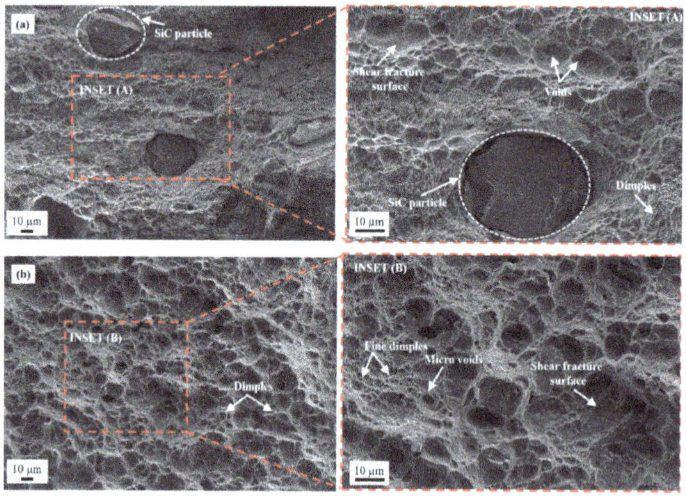

Figure 12. Fractography images of (**a**) SiC-particle reinforced and (**b**) Zn-particle reinforced tensile failure specimens (adapted from [78]).

4.2. Methods for Improving the Properties of Ceramic Matrix Composite

The strength and resilience improvement in ceramic matrix composite is one of the research focuses in the field of composites. Among them, using new methods to enhance strength and resilience is one of the main directions of current research, and common methods include nanoparticle reinforcement, adding new elements, microstructure regulation, and powder metallurgy, among others. Li found that the Nb content and sintering temperature significantly affect the phase formation and microstructure of the sample, while the width of the Ti_2AlC layered structure increases with the increase in sintering temperature [80]. The optimal compressive strength of Ti_2AlC MAX phase ceramic matrix composites was obtained under optimized Nb content (1 wt.%) and sintering temperature (1300 °C). This work demonstrates that LPBF and sintering methods can be used to design composite materials, and explores the properties of Ti_2AlC MAX phase ceramics from the perspective of microalloying. Xi investigated the effect of mixed ceramic reinforcement content on the mechanical properties of aluminum matrix composites [81]. They used laser powder bed fusion to prepare composites reinforced with different contents of (ZrC TiC) ceramic fractions. The results showed that with the increase in mixed ceramic content, the laser absorption behavior of the composite material is enhanced, and the fraction of nanoparticles increases. The composites containing 20 wt.% ceramics exhibited high micro- and nanohardness, and the elastic modulus and tensile strength of 15 wt.% (ZrC + TiC)/Al composite material were significantly higher than the unreinforced Al matrix, due to the formation of nanoprecipitates and coherent binding at the reinforcement/matrix interface. Yao predicted the effective elastic properties of conductivity of 3D printing materials and the mechanical properties of ceramic matrix composite using micromechanical models and resistance network models [82]. The study considers the hierarchical structure of two scales, and uses an appropriate model to homogenize each scale, which can be applied to alumina ceramic matrix composite containing multiwall carbon nanotubes. Through parameterization research, it was found that the electromechanical properties of composite materials are sensitive to geometric features related to machining. This study provides a method for predicting effective electromechanical performance, which can evaluate the electromechanical performance of composite material parts. Galizia compared the microstructure and mechanical properties of different ZrB_2-based composites prepared by slurry infiltration sintering (SIS), polymer infiltration pyrolysis (PIP), and radio-frequency chemical-vapor infiltration (RF-CVI) processes, and evaluated the structural properties of these composites [83]. High temperature tensile tests were conducted on an ultrahigh temperature hot carrier trap, and the results showed that different porosity and fiber properties have varying degrees of influence on the mechanical properties of the composite material. All ZrB_2-based CMCs exhibit excellent structural performance at high temperatures. Additionally, the role of high-level residual thermal stress was also discussed. Wang introduced a new method to predict the fracture resilience of granular ceramic matrix composite at high temperatures [84]. This method uses the stress intensity factor method and high-temperature fracture strength theory, taking into account the effects of internal and surface defects, laminated structures, and thermal residual stresses inside particles and laminated materials, without the need for traditional destructive experiments. The predicted results are in good agreement with the measured data, therefore providing an effective method to obtain the high-temperature fracture resilience and fracture strength for multilayer and particle composites. Wang prepared three types of ceramic particle-reinforced iron matrix composites using the powder metallurgy method, namely alumina/Fe45, ZTA (zirconia-toughened alumina ceramic particles)/Fe45, and zirconia/Fe45 [85]. This paper studied their mechanical properties, impact abrasive wear resistance, and impact abrasive wear mechanism, and found that ZTA/Fe45 composite materials have better wear resistance. In the phase transition analysis of composite materials and ceramic particles, the t-ZrO_2 to m-ZrO_2 phase transition increases the resilience of ZTA and improves the wear resistance of ZTA/Fe45 composite materials. Xiang used the HQEM method to predict the working pressure of unidirectional fiber-reinforced ceramic matrix composite, and determined the CMC interface properties damaged by micromachining by analyzing sensitivity and optimal parameters [86]. The results indicate that compared to the classical shear model, the HQEM method has higher accuracy in verifying pressure distribution. Additionally, the study analyzed micromechanical behavior through three typical cases and provided insights into the microstructure behavior related to composite material failure characteristics and global failure mechanisms. This study provides a promising method for effectively analyzing the fracture of CMC. Yu studied transverse tensile mechanical experiments using ceramic matrix microcomposite materials (CMMC) [87]. Digital optoelectronic technology was used in the experiment to analyze the evolution process of the strain field and study the uniformity changes in the transverse tensile strain field. There is a strong correlation between the uniformity of the strain field and the degree of damage, which can sensitively reflect the damage evolution process of materials. The relationship between local damage and overall mechanical response, along with the localization and nonuniformity characteristics of damage evolution, were discussed. Additionally, the influence of microstructure on the transverse tensile modulus of CMMC was analyzed, and a prediction model for transverse tensile modulus of CMMC considering microstructure was established. Chen prepared Ti_3SiC_2/SiC ceramic matrix composite through molten salt synthesis and subsequent polymer infiltration and pyrolysis techniques, as shown in Figure 13 [88]. Compared with PDC SiC, this composite material has higher mechanical properties, electromagnetic interference shielding effect, and thermal conductivity, due to its preformed Ti_3SiC_2 connecting to the network skeleton. The flexural strength and fracture resilience of the composites have good properties.

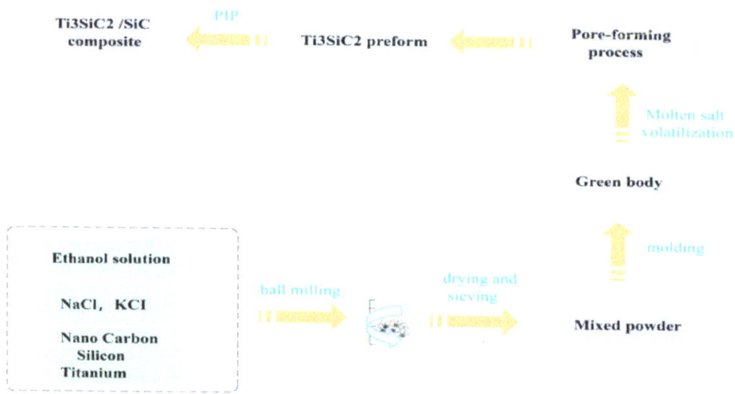

Figure 13. The preparation process diagram for Ti_3SiC_2/SiC composites.

Li successfully synthesized ceramic-reinforced 316L/IN718 matrix composites in situ through directed energy deposition technology [89]. The study compared the microstructure, microhardness, tensile properties, and wear resistance of metal matrix composite materials with unreinforced samples. It was found that when 10% Ti was added, the microhardness increased to 629.6 HV, the ultimate tensile strength increased to 717.9 MPa, and the friction coefficient decreased to 0.688. This study provides a practical and economical method for preparing 316L/IN718-Ti composite materials as potential high-strength and wear-resistant materials. Farvizi studied the use of YSZ particles with the coefficient of thermal expansion close to NiTi alloy as reinforcement materials to improve the wear resistance of NiTi alloy [90]. By using YSZ instead of monoclinic zirconia as a reinforcing agent, lower mismatch stress can be generated. Therefore, there is still a high proportion of austenite phase in the matrix, which has strain recovery ability and is beneficial for the application of tribology. Therefore, NiTi YSZ samples with a higher proportion of austenite and better mechanical properties have superior wear resistance. The wear mechanism is mainly dominated by abrasive, delamination, and adhesion mechanisms. The research results indicate that ceramic-based materials can provide excellent performance to meet various application needs. In general, there are still some defects in the strength and resilience improvement in ceramic matrix composite, such as high processing difficulty and high cost.

4.3. Application of Matrix Composite Materials in Practical Engineering Projects

Matrix composite materials have a wide range of applications due to their special properties, and have been applied in recent years in biomedical, biodegradable, brake pads, radiation shielding, and other related fields. Manso studied how to improve the wear resistance of metal bone implants by adding hard reinforcements [91]. Studying the use of nontoxic and nonallergenic β, two different in situ composite materials were prepared by adding NbC powder to Ti-Nb alloy as the matrix material. This strategy can achieve the synthesis of reinforcing phases during the manufacturing process and achieve strong interface bonding through high chemical compatibility. The results showed that the friction and corrosion properties of both composite materials were improved, and the wear amount was less than 50%. These results indicate that the matrix composite material is a potential bone implant material that can improve its wear resistance and degradation process. Kabir's study introduced a zinc-based composite material for biodegradable implant materials, which added graphene nanosheets to improve its mechanical properties and corrosion behavior [92]. The research results show that the composite material has high mechanical properties and good biocompatibility, making it a potential biodegradable implant material. In the past few years, carbon ceramic brakes have been favored for their light weight, high-temperature resistance, and long lifespan. Bianchi's research analyzed the environmental impact of carbon ceramic brakes using a lifecycle assessment method, and the specimen implementation program is shown in Figure 14. The results show that although cast iron brakes are more sustainable in the production stage, carbon ceramic brakes prepared from recycled prepreg waste have lower environmental impacts in long-term use [93]. Bhakuni described the development of granular ceramic matrix composite (PCMC) for automobile brake pads prepared by two powder metallurgy methods [94]. The first type shows good density, thermal stability, and low noise, while the second type shows excellent properties such as high scratch hardness. This study indicates that these two PCMC samples outperform traditional commercial brake pads in terms of performance, demonstrating potential effective performance in practical use. Suna introduced a preparation method for low-density polyethylene (LDPE)-based composites using boron carbide, dysprosium tetraborate, dysprosium oxide borate, and boron carbide dysprosium compounds as reinforcing materials [95]. The chemistry, mechanics, morphology, thermal properties, and neutron radiation shielding performance of composite materials were studied in this work. The research results indicate that composite materials not only have effective neutron shielding performance, but also have performance comparable to pure boron carbide LDPE-based composite materials. This composite material has potential for safe applications that require protection from neutrons and gamma rays.

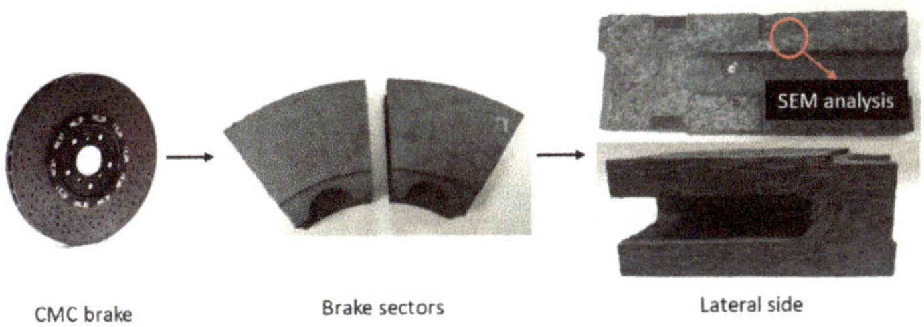

Figure 14. Procedure for specimen realization for SEM analysis (adapted from [93]).

5. State-of-Art Concerning Other Composite Materials in Structural Reinforcement and Resilience Enhancement

In addition to the advanced composite materials mentioned above, there are also many other advanced composite materials, such as viscoelastic materials, magnetorheological fluids, and concrete mortar. These advanced composite materials have excellent applications in certain single fields. Viscoelastomers and magnetorheological variants can be used to make dampers for vibration reduction control of structures, and improved concrete mortar can be used for reinforcement and resilience improvement in building structures.

5.1. Theoretical Research and Practical Application of Viscoelastic Materials

Viscoelastic materials have characteristics such as elastic behavior, noninstantaneous deformation, and viscosity, making them the main materials for manufacturing viscoelastic dampers. Xu developed viscoelastic materials based on different matrix rubbers and conducted experiments [96]. The results indicate that the viscoelastic damper based on a nitrile rubber matrix has high energy dissipation capacity, while the viscoelastic damper based on a silicone rubber matrix has stable performance under different working conditions. To elucidate its mechanical properties, an equivalent high-order fractional derivative model considering temperature and frequency effects was proposed, and numerical results consistent with experimental results were obtained. Xu introduced the application of viscoelastic dampers in seismic reduction [97]. Firstly, the mathematical model of VE dampers and the dynamic analysis of structures with VE dampers were introduced. Then, an equivalent standard solid model is used to describe the effect of temperature on the energy absorption characteristics of VE dampers. Finally, the response of a three-story reinforced-concrete frame structure with and without VE dampers was studied through elastic–plastic time field analysis, frequency field analysis, and shaking table tests. The results showed that VE dampers can be modeled using an equivalent standard solid model, and are effective in reducing structural seismic response. Xu introduced the preparation and performance testing of acrylic viscoelastic dampers [98]. On the basis of acrylic rubber molecules, the optimal formulation of viscoelastic dampers was prepared and mechanical properties were tested. The performance of the damper under different temperatures, excitation frequencies, and displacement amplitudes was studied, and it was found that it has good damping performance and is influenced by environmental temperature and excitation frequency, while the energy dissipation capacity of a single circuit is significantly affected by displacement amplitudes. In order to accurately represent these effects, a modified fractional derivative equivalent model was introduced and compared with experimental data to verify the correctness of the mathematical model. Xu proposed a high order fractional derivative model and temperature frequency equivalence principle to characterize the effects of frequency and temperature, and introduced internal variable theory to consider the effect of internal/microscale structure evolution on capture displacement [99]. The mechanical properties of a plate shear-type viscoelastic damper under sinusoidal displacement excitation at room temperature have been tested, as shown in Figure 15, and it has been proved that the high-order fractional derivative model modified using the internal variable theory and the temperature frequency (0.1–25 Hz) equivalence principle (ITHF) is sufficiently accurate in describing the dynamic behavior of viscoelastic dampers with different frequencies and displacement amplitudes.

$$G_1 = k_1 \left\{ \begin{array}{l} \mu_1 + \eta_1 (\alpha_T \omega)^\gamma \cos(\gamma\pi/2) \\ + \frac{\mu_2 c_2 (\alpha_T \omega)^\beta \cos(\beta\pi/2) + \mu_2 (\alpha_T \omega)^{\alpha+\beta} \cos[(\alpha-\beta)\pi/2]}{c_2^2 + 2c_2 (\alpha_T \omega)^\alpha \cos(\alpha\pi/2) + (\alpha_T \omega)^{2\alpha}} \end{array} \right\} \frac{1}{1+(\varepsilon/\varepsilon_c)^{2m}+b_1}$$

$$G_2 = k_2 \left\{ \begin{array}{l} \eta_1 (\alpha_T \omega)^\gamma \sin(\gamma\pi/2) \\ + \frac{\mu_2 c_2 (\alpha_T \omega)^\beta \sin(\beta\pi/2) + \mu_2 (\alpha_T \omega)^{\alpha+\beta} \sin[(\beta-\alpha)\pi/2]}{c_2^2 + 2c_2 (\alpha_T \omega)^\alpha \cos(\alpha\pi/2) + (\alpha_T \omega)^{2\alpha}} \end{array} \right\} \frac{2(\varepsilon/\varepsilon_c)^m}{1+(\varepsilon/\varepsilon_c)^{2m}+b_2}$$

The higher-order fractional derivative model modified after the Kraus model and temperature–frequency equivalence principle (KTHF) is shown above.

Xu introduced the common applications of viscoelastic dampers as passive control devices for structural vibration reduction, and mentioned that their energy dissipation characteristics are influenced by temperature and frequency [100]. To describe the complex characteristics of this damper, a new model, the equivalent standard solid model, was proposed and applied to the comparison of experimental and numerical data. At the same time, the author applied 52 viscoelastic dampers to the seismic reinforcement design of Xi'an Petroleum Hotel, and conducted finite element analysis on the structures with and without dampers. The results indicate that viscoelastic dampers can effectively reduce seismic response and are a high-performance energy dissipation device, as shown in Figure 16.

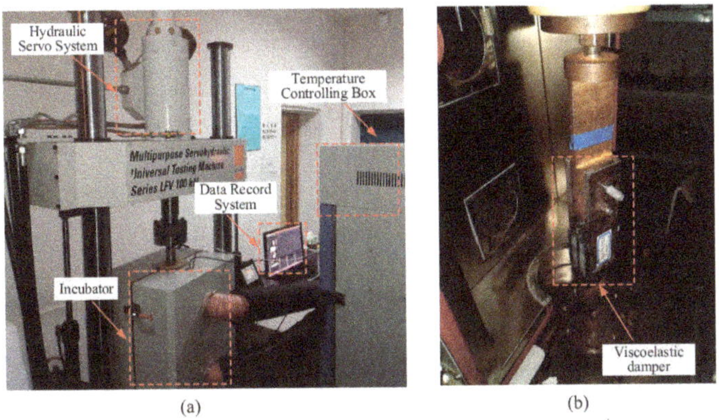

Figure 15. Equipment for testing the dynamic properties of a viscoelastic damper: (**a**) schematic image; (**b**) viscoelastic damper inside the incubator (adapted from [95]).

Figure 16. Viscoelastic vibration reduction device for the Xi'an Fujin JiaYuan building apartment.

5.2. Theoretical Research and Practical Application of Magnetorheological Materials

Magnetorheological fluid has the characteristics of adjustability, fast response speed, small size, and light weight, making it the main material for manufacturing magnetorheological dampers. Xu conducted research on magnetorheological fluids and their composite materials [101]. Magnetorheological fluid is an intelligent material widely used for structural vibration reduction, consisting of iron particles, carrier fluid, and additives. One of the research hotspots is the expectation that ferromagnetic particles have low density and high magnetic properties. This article uses ultrasonic and mechanical stirring–grafting techniques to coat multiwalled carbon nanotubes (MWNT) onto carbonyl iron (CI) particles;

an electron microscope image is shown in Figure 17. In addition, shear yield stress tests were conducted on self-made composite materials, and the test results were compared with the theoretical values of the single- and double-chain micromechanical model to verify the effectiveness and accuracy of the proposed model. Xu discussed how to control the vibration response in aerospace or precision instrument platforms in the frequency range of 0 to 500 Hz [102]. A new scheme using vibration isolation and damping devices, magnetorheological dampers, and viscoelastic dampers has been proposed. The dynamic response of these dampers under different excitations was calculated by coupling the motion equations of the vibration model, and the multistate control concept was adopted to control the input current of the magnetorheological damper. The experimental results show that the scheme is effective and suitable for vibration control of the platform [103]. Xu designed and manufactured a new type of high-damping vibration isolation and damping device, aimed at reducing the dynamic response of the platform over a wide frequency range. After testing, the effect of excitation frequency and amplitude on the properties of the device was considered. Additionally, a platform dynamics model for vibration isolation and damping devices was established, and the dynamic response of the damping platform under different excitations was calculated. By comparing the time-domain and frequency-domain responses of damped and undamped structures, it was concluded that these devices can significantly reduce the dynamic response of the platform.

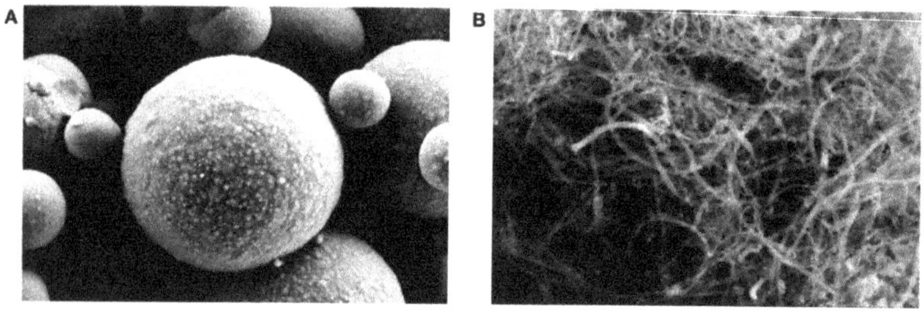

Figure 17. The electron micrographs of (**A**) original CI particles and (**B**) MWNTs (adapted from [101]).

Xu introduced a new hybrid testing system based on Matlab OpenSees combination programming, which is used for hybrid testing of seismic response of structures [104]. The feasibility and accuracy of the hybrid experimental system have been demonstrated by verifying the performance of the testing equipment and Matlab OpenSees system, and testing the mechanical properties of viscoelastic dampers under different conditions. In addition, through mixed experiments on three-dimensional viscoelastic damping structures, it was verified that viscoelastic dampers have good damping effects on structures under seismic excitation. Compared with previous hybrid testing systems, this system has a wider range of applications and can achieve more accurate seismic response testing. Xu uses magnetorheological dampers to carry out semiactive control of buildings and structures to reduce earthquake damage [105]. The article proposes a Bingham model for magnetorheological dampers and proposes a relationship between the yield shear stress and control current of magnetorheological dampers that matches the experimental data. Then, an online real-time control method for semiactive control of MR damper structures was introduced, which can solve the problem of structural response distortion. Finally, an example was used to verify the effectiveness of online real-time control. Xu introduced the characteristics of magnetorheological dampers as a semiactive control device, particularly highlighting the application of magnetorheological elastomeric composites, as shown in Figure 18 [106]. The magnetic saturation problem of magnetorheological dampers is an important research direction. This article describes experimental and simulation research on shear valve-type magnetorheological dampers, proposes a mathematical model of magnetic saturation, and compares it with other models. The magnetic-saturation mathematical model can describe the effects of input current, displacement amplitude, and excitation frequency on the property and magnetic saturation performance of magnetorheological dampers, providing important references for the development and application of magnetorheological dampers.

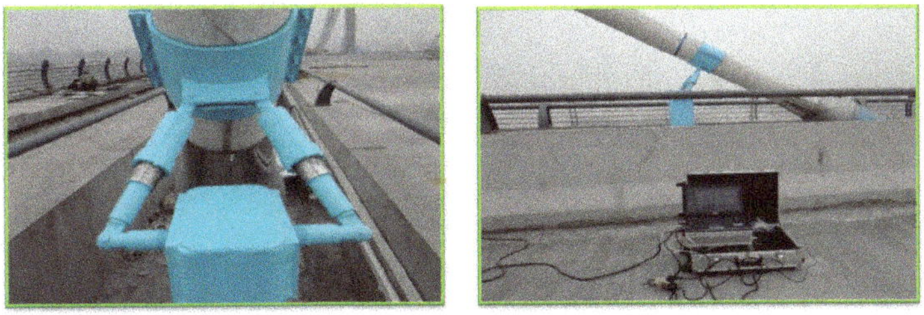

Figure 18. MR dampers for Hanzhong City West Second Ring Bridge (images by Zhao-Dong Xu).

However, there are still some shortcomings in the current research. First, the preparation process and technology of composite materials need to be further optimized and improved to increase their performance. Second, the viscoelastic and magnetorheological properties of composite materials have a significant impact on their preparation and application, and require in-depth research.

5.3. *Performance Improvement and Practical Application of Concrete Mortar Composite Materials*

Concrete and mortar composite materials are widely used in the construction industry, and their strength and resilience are key factors affecting their performance. In recent years, researchers have proposed various new methods to improve the strength and resilience of concrete and mortar composite materials, for example, the addition of nanomaterials, the use of additives, and microstructure regulation. Malakopoulos and Salifoglou studied the properties of mortars composed of Portland limestone cement, calcium carbonate, butyl stearate, and oleic acid [107]. The results indicate that compared to ordinary Portland limestone cement mortar, adding calcium carbonate, butyl stearate, and oleic acid to the mortar mixture can improve durability. The use of alkaline activators to chemically polymerize aluminosilicate materials for the production of alkaline-activated materials can effectively reduce greenhouse gas emissions from ordinary Portland cement (OPC) (approximately 73%). In order to solve the treatment problem concerning silicon manganese slag, Luo et al. used silicon manganese slag in alkali-activated materials to replace the original slag composition; an electron microscope image is shown in Figure 19 [108]. Through research, it has been proven that the flowability, setting time, compressive strength, flexural strength, microstructure, and freeze–thaw resistance of alkali-activated silicon manganese slag concrete (AASSC) have been improved under different ratios, steel-fiber volume fraction, and alkali-activation modulus (MS). Cheng studied the effects of various toughening materials on the resilience of concrete and mortar [109]. A variety of toughening materials, including PVA fiber, rubber particles, and polymer latex powder, were prepared using specific specimens, and their bending and compression properties were tested. The results indicate that mixing multiple toughening materials can greatly improve the bending resilience for mortar and concrete compared to adding only PVA fibers. From the scanning electron microscope images, it can be seen that the incorporation of mixed toughening materials can promote the formation of network structures and improve crack resistance. Ohemeng used waste concrete powder and fly ash as cement substitutes to produce mortar, exploring their combined effects, environmental impact, and cost analysis [110]. The study was conducted in two stages. The first stage replaced cement with 0%, 15%, 40%, 60%, and 100% waste concrete powder, while the second stage added 0% to 25% fly ash at a ratio of 40% waste concrete powder and 60% cement. The experimental results indicate that the performance of the produced mortar improved and met the strength requirements for masonry engineering. Additionally, the use of fly ash can reduce costs and electrical resistivity, and has good economic and environmental benefits.

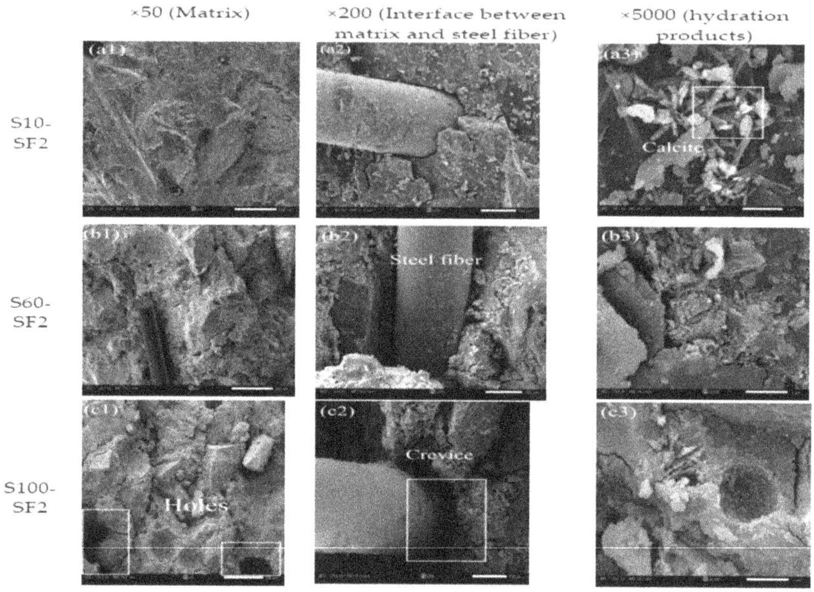

Figure 19. SEM images of AASSC with different substitution ratios of silicomanganese slag: (**a1–a3**) S10-SF2 (**b1–b3**) S60-SF2 (**c1–c3**) S100-SF2 (adapted from [108]).

Dobiszewska studied the effect of concrete production on the environment and used waste byproducts as substitutes for clinker [111]. Rock dust is considered a potential alternative material for the production of cement composite materials. However, there are conflicting findings in the literature regarding the effect of partial substitution of cement on the physical and mechanical properties and durability of cement composite materials. The effect of using rock dust instead of cement is mainly related to the filler effect. The higher the amount of replacement cement, the lower the mechanical properties and durability of cement composite materials. However, replacing cement with up to 10–15% rock powder will not affect the performance of cement composite materials. Sara uses construction and demolition (C&D) waste instead of natural sand, and uses ground and granulated blast furnace slag (GGBFS) instead of cement to manufacture self-compacting mortar [112]. The research results indicate that using recycled concrete sand (RCS) instead of natural sand can produce self-compacting mortar, with an optimal yield of 50%. Replacing cement with GGBFS can offset the negative effects of mortar porosity and capillary water absorption, and improve long-term compressive strength. By adding hydrated lime to mortar, Zhang studied the effect of hydrated lime on the mechanical properties of ultrahigh-performance concrete mortar and analyzed its impact mechanism through microstructure testing [113]. The results show that under the conditions of hot-water curing and dry-air heating, different proportions of hydrated lime can improve the mechanical properties of mortar, because the generated (C-(A)-S-H) gel is transformed into crystalline gabbro and xonotlite, and the microstructure is refined. Additionally, when the volcanic ash reaction and cement hydration reach an optimal balance, the mortar containing higher crystalline phases has a denser microstructure and better mechanical properties. Pan uses an encapsulated silane coupling agent (SCA) to modify the surface of waste polyvinyl chloride (PVC) particles to improve the performance of concrete and mortar [114]. The results showed that after SCA modification, the compressive strength and flexural strength of concrete and mortar were both improved. Scanning electron microscopy observation showed that SCA improved the bonding between PVC particles and cementitious materials, thereby improving the mechanical properties of the mortar. Liu recycled the ground concrete waste powder to prepare a green binder, and prepared a highly sustainable metakaolin-based polymer mortar [115]. Adding concrete waste powder into metakaolin can improve the dry shrinkage and compressive strength, but also affect the microstructure and the amount of the cement-based composite geopolymer slurry product. By optimizing the type and amount of concrete waste powder, the sustainable metakaolin-based polymer mortar with high strength and poor conveying performance can be prepared. Although these methods have achieved certain research results, there are still some shortcomings in practical applications, such as the high expense of the added materials, complex production processes, and the ability to only improve a single performance.

6. Conclusions

Advanced composite materials, such as fiber-reinforced composites, laminated composites, matrix composites, and other advanced variants, possess distinct advantages in enhancing structural reinforcement and improving resilience. The main representatives of fiber-reinforced composite materials, CFRP and GFRP, are characterized by their light weight and high strength, and are widely used in civil engineering, aerospace, and automotive industries. Laminated composite materials can exhibit excellent crack resistance and resilience by changing the deposition sequence and optimization modeling, and are widely used in fields such as electrical components and piezoelectric sensors. The addition of nanoparticles or fiber-reinforced materials to the matrix enhances the performance of the matrix itself. It has good structural reinforcement effect, while also providing a certain degree of resilience and durability, and is widely used in fields such as construction and shipbuilding. Other advanced composite materials, such as viscoelastic materials and magnetorheological fluids, are very good energy dissipation materials with good resilience, and are widely used in the production of dampers and applied in other engineering fields. They are of great significance for seismic reduction and reinforcement of buildings.

(1) For fiber-reinforced composite materials, current research mainly focuses on developing new methods, establishing models, optimizing algorithms, and creating new applications. The new methods mainly focus on controlling and enhancing fiber orientation, modifying surfaces, crosslinking, using additives, and applying mixed fibers. The models and algorithms focus on multiscale topology optimization, genetic algorithms, prediction models, etc. New applications are concentrated in fields such as aerospace, construction, electronics, shipbuilding, etc., such as bonding with concrete beams as reinforcement materials, pump impellers, natural gas transport ships, brake pads, etc.

(2) For laminated composite materials, the current research mainly focuses on developing new methods, establishing models, and optimizing algorithms. The new methods mainly focus on changing the deposition sequence, using finite element comparative analysis, applying interlayer reinforcement technology, and, especially, adding nanofibers, which has become a new research hotspot. The establishment of models and optimization algorithms mainly focus on finite element analysis, multiscale simulation, genetic algorithms, etc.

(3) For matrix composite materials, matrix materials can be mainly divided into metal and ceramic matrix composites. At present, research on new methods mainly focuses on stirring–casting, secondary addition of new elements, microstructure regulation and optimization, and nanoparticle reinforcement. Additionally, quantitative analysis and performance optimization of matrix composite materials can also be achieved through methods such as establishing models and optimization algorithms.

(4) For other composites, these mainly include materials such as magnetorheological fluids, viscoelastic materials, concrete mortar, etc. The main application of magnetorheological and viscoelastic materials is to produce dampers for vibration reduction and seismic resistance in buildings and various precision instruments. The strength and durability of concrete mortar has been greatly improved by including additives

and toughening materials, which are of great significance for their application in construction engineering.

The development of advanced composite materials in structural reinforcement and resilience improvement has made much progress, but there are still some developmental deficiencies, for example, their high cost, immature processes, and practical application in extreme environments. Further research and development are still needed in the future.

Author Contributions: Conceptualization, X.H., Q.M. and L.W.; methodology, X.H., Z.X., Q.M., W.L. and L.W.; software, S.S.; validation, Q.M. and L.W.; formal analysis, S.S.; data curation, S.S.; writing—original draft preparation, S.S.; writing—review and editing, Z.X. and W.L.; supervision, X.H.; project administration, W.L.; funding acquisition, Z.X. All authors have read and agreed to the published version of the manuscript.

Funding: This research was funded by the Shenzhen Sustainable Development Science and Technology Special Project (KCXFZ20211020165543004), National Natural Science Foundation of China (52178463, 52130807).

Data Availability Statement: The data presented in this study are available in the respective references.

Conflicts of Interest: The authors declare no conflict of interest.

References

1. Codispoti, R.; Oliveira, D.V.; Olivito, R.S.; Lourenço, P.B.; Fangueiro, R. Mechanical performance of natural fiber-reinforced composites for the strengthening of masonry. *Compos. Part B* **2015**, *77*, 74–83. [CrossRef]
2. Aravind, N.; Samanta, A.K.; Thanikal, J.V.; Roy, D.K.S. An experimental study on the effectiveness of externally bonded corrugated GFRP laminates for flexural cracks of RC beams. *Constr. Build. Mater.* **2017**, *136*, 348–360. [CrossRef]
3. Rageh, B.O.; El-Mandouh, M.A.; Elmasry, A.H.; Attia, M.M. Flexural Behavior of RC Beams Strengthened with GFRP Laminate and Retrofitting with Novelty of Adhesive Material. *Buildings* **2022**, *12*, 1444. [CrossRef]
4. Xian, G.; Rui, G.; Li, C.; Hong, B. Mechanical properties of carbon/glass fiber reinforced polymer plates with sandwich structure exposed to freezing-thawing environment: Effects of water immersion, bending loading and fiber hybrid mode. *Mech. Adv. Mater. Struct.* **2023**, *30*, 814–834. [CrossRef]
5. Mazur, K.; Siwy, Z.S.; Adamczyk, A.; Kuciel, S. Synergistic effect of aramid and basalt fibers on mechanical, thermal and dynamic properties of polylactide hybrid composites. *Ind. Crops Prod.* **2023**, *198*, 116630. [CrossRef]
6. Radfar, S.; Foret, G.; Saeedi, N.; Sab, K. Simulation of concrete cover separation failure in FRP plated RC beams. *Constr. Build. Mater.* **2012**, *37*, 791–800. [CrossRef]
7. De Maio, U.; Fabbrocino, F.; Greco, F.; Leonetti, L.; Lonetti, P. A study of concrete cover separation failure in FRP-plated RC beams via an inter-element fracture approach. *Compos. Struct.* **2019**, *212*, 625–636. [CrossRef]
8. Fu, B.; Teng, J.G.; Chen, J.F.; Chen, G.M.; Guo, Y.C. Concrete Cover Separation in FRP-Plated RC Beams: Mitigation Using FRP U-Jackets. *J. Compos. Constr.* **2017**, *21*, 04016077. [CrossRef]
9. Yazdani, N.; Aljaafreh, T.; Beneberu, E. Concrete beam flexural strengthening with anchored pre-saturated CFRP laminates. *Compos. Struct.* **2020**, *235*, 111733. [CrossRef]
10. Grelle, S.V.; Sneed, L.H. Review of Anchorage Systems for Externally Bonded FRP Laminates. *Int. J. Concr. Struct. Mater.* **2013**, *7*, 17–33. [CrossRef]
11. Schmidt, J.W.; Sørensen, J.D.; Christensen, C.O. In Situ Concrete Bridge Strengthening Using Ductile Activated NSMR CFRP System. *Buildings* **2022**, *12*, 2244. [CrossRef]
12. Alam, P.; Mamalis, D.; Robert, C.; Floreani, C.; Brádaigh, C. The fatigue of carbon fibre reinforced plastics—A review. *Compos. Part B* **2019**, *166*, 555–579. [CrossRef]
13. Wu, J.; Zhu, Y.; Li, C. Experimental Investigation of Fatigue Capacity of Bending-Anchored CFRP Cables. *Polymers* **2023**, *15*, 2483. [CrossRef] [PubMed]
14. Yang, Z.; Yang, Z.; Chen, H.; Yan, W. 3D printing of short fiber reinforced composites via material extrusion: Fiber breakage. *Addit. Manuf.* **2022**, *58*, 103067. [CrossRef]
15. Song, H.; Liu, T.; Gauvin, F.; Brouwers, H.J.H. Improving the interface compatibility and mechanical performances of the cementitious composites by low-cost alkyl ketene dimer modified fibers. *Constr. Build. Mater.* **2023**, *132186*, 395. [CrossRef]
16. Lu, M.; Xiao, H.; Liu, M.; Feng, J. Carbon fiber surface nano-modification and enhanced mechanical properties of fiber reinforced cementitious composites. *Constr. Build. Mater.* **2023**, *370*, 130701. [CrossRef]
17. Lee, S.; Hong, C.; Ji, W. In situ micromechanical analysis of discontinuous fiber-reinforced composite material based on DVC strain and fiber orientation fields. *Compos. Part B* **2022**, *247*, 110361. [CrossRef]
18. Batu, T.; Lemu, H.G. Investigation of mechanical properties of false banana/glass fiber reinforced hybrid composite materials. *Results Mater.* **2020**, *6*, 100131. [CrossRef]

19. Zhang, X.; Li, K.Z.; Li, H.J.; Fu, Y.W.; Fei, J. Tribological and mechanical properties of glass fiber reinforced paper-based composite friction material. *Tribol. Int.* **2014**, *69*, 156–167. [CrossRef]
20. Liu, P.; Zhong, Y.; Pei, Q.-X.; Sorkin, V.; Zhang, Y.-W. Simultaneously enhancing the strength and toughness of short fiber reinforced thermoplastic composites by fiber cross-linking. *Compos. Sci. Technol.* **2022**, *217*, 109076. [CrossRef]
21. Shivamurthy, B.; Naik, N.; Thimappa, B.; Bhat, R. Mechanical property evaluation of alkali-treated jute fiber reinforced bio-epoxy composite materials. *Mater. Today Proc.* **2020**, *28*, 2127–2130. [CrossRef]
22. Wa, A.; Envelope, M.; Yh, B.; Kyma, A. Optimizing the material and printing parameters of the additively manufactured fiber-reinforced polymer composites using an artificial neural network model and artificial bee colony algorithm. *Structures* **2022**, *46*, 1781–1795.
23. Duan, Z.; Liu, Y.; Fan, J.; Long, K.; Xu, B.; Zhu, J.; Yan, J. Concurrent multi-material and multi-scale design optimization of fiber-reinforced composite material and structures for minimum structural compliance. *Compos. Struct.* **2023**, *311*, 116796. [CrossRef]
24. Obid, Š.; Halilovič, M.; Urevc, J.; Starman, B. Non-linear elastic tension-compression asymmetric anisotropic model for fibre-reinforced composite materials. *Int. J. Eng. Sci.* **2023**, *185*, 103829. [CrossRef]
25. Liu, H.; Zhou, J.; Kong, X.; Li, S. Fracture behaviour of fibre-reinforced composite materials subjected to shear loading: An experimental and numerical study. *Int. J. Lightweight Mater. Manuf.* **2023**, *6*, 108–116. [CrossRef]
26. Sakata, S.; Stefanou, G.; Araki, S. Random field modeling of local strength of a randomly arranged unidirectional fiber-reinforced composite material under transverse tensile loading. *Mater. Today Commun.* **2022**, *33*, 104783. [CrossRef]
27. Lv, Z.Z.; Wang, J.; Guo, Y.C.; Dong, S.Q.; Sha, J.J.; Cheng, X.P. Effect of Cu coating thickness on carbon fiber surface on microstructure and mechanical properties of carbon fiber reinforced aluminum matrix composites. *Mater. Today Commun.* **2023**, *34*, 105424. [CrossRef]
28. Kumar, T.V.; Prasad VM, M.; Santhosh, D.; Prasanth, C.; Ranjith, K. Evaluation of mechanical and interfacial properties of carbon fiber reinforced polymer (CFRP) composite materials. *Mater. Today Proc.* **2020**, *21*, 477–482. [CrossRef]
29. Jeong, Y.-J.; Kim, H.-T.; Kim, J.-D.; Kim, J.-H.; Kim, S.-K.; Lee, J.-M. Evaluation of mechanical properties of glass fiber-reinforced composites depending on length and structural anisotropy. *Results Eng.* **2023**, *17*, 101000. [CrossRef]
30. Saindane, U.V.; Soni, S.; Menghani, J.V. Studies on mechanical properties of brake friction materials derived from carbon fibers reinforced polymer composite. *Mater. Today Proc.* **2021**, *47*, 5760–5765. [CrossRef]
31. Maji, A.; Mahato, P.K. Development and applications of shear deformation theories for laminated composite plates: An overview. *J. Thermoplast. Compos. Mater.* **2022**, *35*, 2576–2619. [CrossRef]
32. Li, P.; Gong, Y.; Xu, Y.; Qi, Y.; Sun, Y.; Zhang, H. Inconel-steel functionally bimetal materials by hybrid directed energy deposition and thermal milling: Microstructure and mechanical properties. *Arch. Civ. Mech. Eng.* **2019**, *19*, 820–831. [CrossRef]
33. Yang, Y.; Gong, Y.; Qu, S.; Xie, H.; Cai, M.; Xu, Y. Densification, mechanical behaviors, and machining characteristics of 316L stainless steel in hybrid additive/subtractive manufacturing. *Int. J. Adv. Manuf. Technol.* **2020**, *107*, 177–189. [CrossRef]
34. Zhang, S.; Gong, M.; Zeng, X.; Gao, M. Residual stress and tensile anisotropy of hybrid wire arc additive-milling subtractive manufacturing. *J. Mater. Process. Technol.* **2021**, *293*, 117077. [CrossRef]
35. Guo, W.; Wu, C.; Ding, Z.; Zhou, Q. Prediction of surface roughness based on a hybrid feature selection method and long short-term memory network in grinding. *Int. J. Adv. Manuf. Technol.* **2021**, *112*, 2853–2871. [CrossRef]
36. Ye, Z.P.; Zhang, Z.J.; Jin, X.; Xiao, M.Z.; Su, J.Z. Study of hybrid additive manufacturing based on pulse laser wire depositing and milling. *Int. J. Adv. Manuf. Technol.* **2017**, *88*, 2237–2248. [CrossRef]
37. Bhaduri, D.; Penchev, P.; Essa, K.; Dimov, S.; Carter, L.N.; Pruncu, C.I.; Pullini, D. Evaluation of surface/interface quality, microstructure and mechanical properties of hybrid additive-subtractive aluminium parts. *CIRP Ann.* **2019**, *68*, 237–240. [CrossRef]
38. Li, P.; Zhou, J.; Li, L.; Gong, Y.; Lu, J.; Meng, X.; Zhang, T. Influence of depositing sequence and materials on interfacial characteristics and mechanical properties of laminated composites. *Mater. Sci. Eng. A* **2021**, *827*, 142092. [CrossRef]
39. Kumar, P.; Joshi, R.S.; Singla, R.K. Investigation of tribological and metallurgical effects on Ti–6Al–4V laminates made by LSEM for Ti/GFRP stacked composite. *Wear* **2023**, *524–525*, 204860. [CrossRef]
40. Biswas, R.; Sharma, N.; Singh, K.K. Numerical analysis of mechanical and fatigue behaviour of glass and carbon fibre reinforced polymer composite. *Mater. Today Proc.* **2023**, *3*, 479. [CrossRef]
41. Liu, S.; Sinke, J.; Dransfeld, C. A modified bias-extension test method for the characterisation of intra-ply shear deformability of hybrid metal-composite laminates. *Compos. Struct.* **2023**, *314*, 116964. [CrossRef]
42. Slattery, P.G.; McCarthy, C.T.; O'Higgins, R.M. Assessment of residual strength of repaired solid laminate composite materials through mechanical testing. *Compos. Struct.* **2016**, *147*, 122–130. [CrossRef]
43. Maithil, P.; Gupta, P.; Chandravanshi, M.L. Study of mechanical properties of the natural-synthetic fiber reinforced polymer matrix composite. *Mater. Today Proc.* **2023**, *1*, 245. [CrossRef]
44. Mortell, D.J.; Tanner, D.A.; McCarthy, C.T. In-situ SEM study of transverse cracking and delamination in laminated composite materials. *Compos. Sci. Technol.* **2014**, *105*, 118–126. [CrossRef]
45. Zuo, Z.H.; Xie, Y.M. Maximizing the effective stiffness of laminate composite materials. *Comput. Mater. Sci.* **2014**, *83*, 57–63. [CrossRef]

46. Liu, W.; Chen, P. A novel method for the determination of the interface strength with coarse meshes for laminated composite materials. *Eng. Fract. Mech.* **2021**, *242*, 107469. [CrossRef]
47. Li, Y.; Wang, L.; Zhu, L.; Li, Y.; Yan, Z.; Song, Y.; Cheng, X. Investigation of interfacial structure and dynamic mechanical behavior of titanium alloy laminated composites. *J. Mater. Res. Technol.* **2022**, *21*, 5111–5120. [CrossRef]
48. Coelho, P.G.; Guedes, J.M.; Rodrigues, H.C. Multiscale topology optimization of bi-material laminated composite structures. *Comp. Struct.* **2015**, *132*, 495–505. [CrossRef]
49. Shi, C.; Guo, B.; Sarkaya, M.; Elik, M.; Güden, M. Determination of the material model and damage parameters of a carbon fiber reinforced laminated epoxy composite for high strain rate planar compression. *Int. J. Impact Eng.* **2021**, *149*, 103771. [CrossRef]
50. Nastos, C.; Komninos, P.; Zarouchas, D. Non-destructive strength prediction of composite laminates utilizing deep learning and the stochastic finite element methods. *Compos. Struct.* **2023**, *311*, 116815. [CrossRef]
51. Yashiro, S.; Yamasaki, T.; Nagai, H.; Yoshimura, A. Effect of material nonlinearity on the Resilience evaluation in quasi-static mode ii interlaminar fracture Resilience tests of composite laminates. *Eng. Fract. Mech.* **2021**, *253*, 107879. [CrossRef]
52. Torabi, A.R.; Pirhadi, E. Notch failure in laminated composites under opening mode: The Virtual Isotropic Material Concept. *Compos. Part B* **2019**, *172*, 61–75. [CrossRef]
53. Lu, X.; Chen, B.Y.; Tan, V.; Tay, T.E. A separable cohesive element for modelling coupled failure in laminated composite materials. *Compos. Part A Appl. Sci. Manuf.* **2018**, *107*, 387–398. [CrossRef]
54. Lin, S.; Waas, A.M. Accelerating computational analyses of low velocity impact and compression after impact of laminated composite materials. *Compos. Struct.* **2020**, *260*, 113456. [CrossRef]
55. Bodduru, K.; Singh, L.K.; Amrita, M.; Anbumani, P.; Singh, A.P. Influence of MXene and carbon nanotube on the mechanical properties of sisal fibre/glass fibre reinforced epoxy composite laminates. *Mater. Today Proc.* **2023**, *02*, 446. [CrossRef]
56. Abot, J.L.; Song, Y.; Schulz, M.J.; Shanov, V.N. Novel carbon nanotube array-reinforced laminated composite materials with higher interlaminar elastic properties. *Compos. Sci. Technol.* **2008**, *68*, 2755–2760. [CrossRef]
57. Ipackchi, M. Improvement of interlaminar fracture toughness of phenolic laminates interleaved with electrospun polyvinyl butyral nanofibers. *Theor. Appl. Fract. Mech.* **2020**, *105*, 102406. [CrossRef]
58. An, D.; Lotfian, S.; Mesbah, D.; Ayre, D.; Yoosefinejad, A.; Thakur, V.K.; Nezhad, H.Y. Ultra-thin electrospun nanofibers for development of damage-tolerant composite laminates. *Mater. Today Chem.* **2019**, *14*, 100202. [CrossRef]
59. Ji, Y.; Xia, Q.; Cui, J.; Zhu, M.; Han, S. High pressure laminates reinforced with electrospun cellulose acetate nanofibers. *Carbohydr. Polym.* **2021**, *254*, 117461. [CrossRef]
60. Kubin, M.; Makreski, P.; Zanoni, M.; Gasperini, L.; Selleri, G.; Fabiani, D.; Gualandi, C.; Bužarovska, A. Effects of nano-sized $BaTiO_3$ on microstructural, thermal, mechanical and piezoelectric behavior of electrospun PVDF/$BaTiO_3$ nanocomposite mats. *Polym. Test.* **2023**, *126*, 108158. [CrossRef]
61. Brugo, T.M.; Maccaferri, E.; Cocchi, D.; Mazzocchetti, L.; Zucchelli, A. Self-sensing hybrid composite laminate by piezoelectric nanofibers interleaving. *Compos. Part B Eng.* **2021**, *212*, 108673. [CrossRef]
62. Zucchelli, A.; Focarete, M.; Gualandi, C.; Ramakrishna, S. Electrospun nanofibers for enhancing structural performance of composite materials. *Polym. Adv. Technol.* **2011**, *22*, 339–349. [CrossRef]
63. Palazzetti, R.; Zucchelli, A. Electrospun nanofibers as reinforcement for composite laminates materials—A review. *Compos. Struct.* **2017**, *182*, 711–727. [CrossRef]
64. Xu, J.; Geng, Z.; Johansen, M.; Carlstedt, D.; Duan, S.; Thiringer, T.; Liu, F.; Asp, L.E. A multicell structural battery composite laminate. *Ecomat* **2022**, *4*, e12180. [CrossRef]
65. Yoo, K.H. 15-mode piezoelectric composite and its application in a magnetoelectric laminate structure. *J. Alloys Compd.* **2018**, *767*, 61–67. [CrossRef]
66. Rajaram, S.; Kanimurugan, K.; Kumaresan, N.; Mohemed Naveed, S.; Praveena, P. Effect of rock powder on the mechanical properties of AL7075 metal matrix composites. *Mater. Today Proc.* **2023**, *74*, 85–90. [CrossRef]
67. Rajaram, S.; Karthikeyan, P.; Surya, T.J.; Kaniskar, K.; Pazhanivel, V. Experimental investigation of mechanical properties of AA6063/ZrO_2 metal matrix composites. *Mater. Today Proc.* **2023**, *74*, 105–109. [CrossRef]
68. Chinababu, M.; Rao, E.B.; Sivaprasad, K. Fabrication, microstructure, and mechanical properties of Al-based metal matrix-TiB2-HEA hybrid composite. *J. Alloys Compd.* **2023**, *947*, 169700. [CrossRef]
69. Xu, N.; Ren, Z.; Qiu, Z.; Lu, Z.; Wang, D.; Shen, J.; Song, Q.; Zhao, J.; Bao, Y. Microstructure and mechanical properties of carbon nano-onions reinforced Mg-based metal matrix composite prepared by friction-stir processing. *Mater. Today Commun.* **2023**, *35*, 105552. [CrossRef]
70. Zhai, J.T.; Gao, W.J.; Dong, H.K.; Hu, Y.C.; Zhang, T.; Zhu, X.G.; Zhang, W.W.; Yang, C.; Liu, L.H. Novel metal matrix composites reinforced with Zr-based metallic glass lattices. *Appl. Mater. Today* **2022**, *29*, 101649. [CrossRef]
71. Kumar, R.A.; Prasad, B.I.; Bibin, C.; Thomas, I.J.; Soundararajan, G.; Darshan, P.R.; Arunkumar, S. Physical and mechanical characterization of Al7075/Si_3N_4 metal matrix composite prepared by stir casting technique. *Mater. Today Proc.* **2022**, *11*, 096.
72. Patil, A.; Banapurmath, N.R.; Hunashyal, A.M.; Hallad, S. Enhancement of mechanical properties by the reinforcement of fly ash in aluminium metal matrix composites. *Mater. Today Proc.* **2020**, *4*, 487. [CrossRef]
73. Mvpka, B.; Xxs, B. Dual matrix and reinforcement particle size (sps and dps) composites: Influence on mechanical behavior of particulate aluminum-sic-gr metal matrix composites. *Mater. Today Proc.* **2020**, *24*, 1644–1653.

74. Thiruppathi, K.; Mohan, R.R.; Kumar, S.Y.; Raghuraman, S.; Balaji, K. Experimental study on the effect of sintering cycle and reinforcement of 1% niobium carbide (nbc) in aluminium metal matrix composite (al-mmc). *Mater. Today Proc.* **2021**, *37*, 75–79. [CrossRef]
75. Sachin, D.; Uday, K.N.; Rajamurugan, G.; Krishnasamy, P. Effect of sic reinforcement on the mechanical and tribological behaviour of al6061 metal matrix composites. *Mater. Today Proc.* **2021**, *46*, 7082–7089. [CrossRef]
76. Meher, A.; Mahapatra, M.M.; Samal, P.; Vundavilli, P.R. Study on effect of tib2 reinforcement on the microstructural and mechanical properties of magnesium rz5 alloy based metal matrix composites. *J. Magnes. Alloys* **2020**, *8*, 780–792. [CrossRef]
77. Gayathri, J.; Elansezhian, R. Influence of dual reinforcement (nano cuo + reused spent alumina catalyst) on microstructure and mechanical properties of aluminium metal matrix composite. *J. Alloys Compd.* **2020**, *829*, 154538. [CrossRef]
78. Kesharwani, R.; Jha, K.K.; Imam, M.; Sarkar, C.; Barsoum, I. Comparison of microstructural, texture, and mechanical properties of SiC and Zn particle reinforced FSW 6061-T6 aluminum alloy. *J. Mater. Res. Technol.* **2023**, *26*, 3301–3321. [CrossRef]
79. Lka, B.; Sks, B.; Sna, B. Effect of xgnp/mwcnt reinforcement on mechanical, wear behavior and crystallographic texture of copper-based metal matrix composite. *Mater. Sci. Eng. B* **2019**, *263*, 114718.
80. Li, S.; Zhang, K.; Yan, Z.; Liu, T.; Zhu, Z.; Xiong, Z.; Zou, Z.; Li, J.; Liao, W. Effect of Nb addition on the microstructure and mechanical properties of additively manufactured Ti_2AlC MAX-phase ceramic matrix composites. *Addit. Manuf.* **2023**, *67*, 103488. [CrossRef]
81. Xi, L.; Feng, L.; Gu, D.; Prashanth, K.G.; Kaban, I.; Wang, R.; Xiong, K.; Sarac, B.; Eckert, J. Microstructure formation and mechanical performance of micronanoscale ceramic reinforced aluminum matrix composites manufactured by laser powder bed fusion. *J. Alloys Compd.* **2023**, *939*, 168803. [CrossRef]
82. Koutsawa, Y.; Rauchs, G.; Fiorelli, D.; Makradi, A.; Belouettar, S. A multi-scale model for the effective electro-mechanical properties of short fiber reinforced additively manufactured ceramic matrix composites containing carbon nanotubes. *Compos. Part C* **2022**, *7*, 100336. [CrossRef]
83. Galizia, P.; Contardi, M.; Benedetti, E.; Gasparotto, G.; Montecchi, M.; Zanotto, F.; Iucolano, F.; Gisario, A. Elevated temperature tensile and bending strength of ultra-high temperature ceramic matrix composites obtained by different processes. *J. Eur. Ceram. Soc.* **2023**, *43*, 2533–2542. [CrossRef]
84. Wang, R.; Zhou, S.; Li, W. Modeling effects of temperature, laminated structure, flaws and residual stress on fracture resilience of multilayered and particulate ceramic matrix composites. *Ceram. Int.* **2023**, *49*, 18412–18420. [CrossRef]
85. Wang, Y.; Fan, L.; Chen, H. Effect of strength and resilience of ceramic particles on impact abrasive wear resistance of iron matrix composites. *Mater. Today Commun.* **2023**, *35*, 105689. [CrossRef]
86. Xiang, W.; Li, X.; Ni, H.; Liu, B. Micromechanical analysis of fiber-reinforced ceramic matrix composites by a hierarchical quadrature element method. *Compos. Struct.* **2022**, *300*, 116143. [CrossRef]
87. Yu, G.; Jia, Y.; Xie, C.; Du, J.; Gao, X.; Song, Y.; Wang, F. Transverse tensile mechanical experimental method and behavior of ceramic matrix mini-composites. *Compos. Struct.* **2022**, *297*, 115923. [CrossRef]
88. Chen, J.; Ye, F.; Cheng, L.; Yang, J.; Chen, X. Preparation and properties of Ti_3SiC_2 preform reinforced SiC ceramic matrix composites. *J. Eur. Ceram. Soc.* **2023**, *43*, 3146–3157. [CrossRef]
89. Li, P.; Yang, Q.; Li, L.; Gong, Y.; Zhou, J.; Lu, J. Microstructure evolution and mechanical properties of in situ synthesized ceramic reinforced 316L/IN718 matrix composites. *J. Manuf. Process.* **2023**, *93*, 214–224. [CrossRef]
90. Farvizi, M.; Bahamirian, M.; Faraji, A.; Kim, H.S. Influence of ceramic reinforcement characteristics on the microstructure and wear behavior of NiTi matrix composites. *Mater. Today Proc.* **2023**, *28*, 101726. [CrossRef]
91. Gonçalves, V.R.M.; Corrêa, D.R.N.; Grandini, C.R.; Pintão, C.A.F.; Afonso, C.R.M.; Filho, P.N.L. Assessment of improved tribocorrosion in novel in-situ Ti and β Ti–40Nb alloy matrix composites produced with NbC addition during arc-melting for biomedical applications. *Mater. Chem. Phys.* **2023**, *301*, 127597. [CrossRef]
92. Kabir, H.; Munir, K.; Wen, C.; Li, Y. Microstructures, mechanical and corrosion properties of graphene nanoplatelet-reinforced zinc matrix composites for implant applications. *Acta Biomater.* **2023**, *157*, 701–719. [CrossRef] [PubMed]
93. Bianchi, I.; Forcellese, A.; Simoncini, M.; Vita, A.; Delledonne, L.; Castorani, V. Life cycle assessment of carbon ceramic matrix composite brake discs containing reclaimed prepreg scraps. *J. Clean. Prod.* **2023**, *413*, 137537. [CrossRef]
94. Bhakuni, H.; Muley, A.V.; Ruchika. Fabrication, testing & analysis of particulate ceramic matrix composite for automotive brake pad application. *Mater. Today Proc.* **2023**, *1*, 413.
95. Avcioğlu, S. LDPE matrix composites reinforced with dysprosium-boron containing compounds for radiation shielding applications. *J. Alloys Compd.* **2022**, *927*, 166900. [CrossRef]
96. Xu, Z.-D.; Liao, Y.-X.; Ge, T.; Xu, C. Experimental and theoretical study on viscoelastic dampers with different matrix rubbers. *J. Eng. Mech. (ASCE)* **2016**, *142*, 04016051. [CrossRef]
97. Xu, Z.-D. Earthquake mitigation study on viscoelastic dampers for reinforced concrete structures. *J. Vib. Control.* **2007**, *13*, 29–45. [CrossRef]
98. Xu, Z.-D.; Ge, T.; Liu, J. Experimental and theoretical study of high energy dissipation viscoelastic dampers based on acrylate rubber matrix. *J. Eng. Mech. ASCE* **2020**, *146*, 04020057. [CrossRef]
99. Xu, Y.; He, Q.; Guo, Y.-Q.; Huang, X.-H.; Dong, Y.-R.; Hu, Z.-W.; Kim, J. Experimental and Theoretical Investigation of Vis-coelastic Damper by Applying Fractional Derivative Method and Internal Variable Theory. *Buildings.* **2023**, *13*, 239. [CrossRef]

100. Xu, Z.-D.; Wang, D.-X.; Shi, C.-F. Model, tests and application design for viscoelastic dampers. *J. Vib. Control.* **2011**, *17*, 1359–1370. [CrossRef]
101. Guo, Y.-Q.; Sun, C.-L.; Xu, Z.-D.; Jing, X. Preparation and Tests of MR Fluids with CI Particles Coated with MWNTs. *Front. Mater.* **2018**, *5*, 50. [CrossRef]
102. Xu, Z.-D.; Xu, F.-H.; Chen, X. Intelligent vibration isolation and mitigation of a platform by using MR and VE devices. *J. Aerosp. Eng. ASCE* **2016**, *29*, 04016010. [CrossRef]
103. Xu, Z.-D.; Xu, F.-H.; Chen, X. Vibration suppression on a platform by using vibration isolation and mitigation devices. *Nonlinear Dyn.* **2016**, *83*, 1341–1353. [CrossRef]
104. Xu, Z.-D.; Dong, Y.-R.; Chen, S.; Guo, Y.-Q.; Li, Q.-Q.; Xu, Y.-S. Development of hybrid test system for three-dimensional viscoelastic damping frame structures based on Matlab-OpenSees combined programming. *Soil Dyn. Earthq. Eng.* **2021**, *144*, 106681. [CrossRef]
105. Xu, Z.-D.; Shen, Y.-P.; Guo, Y.-Q. Semi-active control of structures incorporated with magnetorheological dampers using neural networks. *Smart Mater. Struct.* **2003**, *12*, 80–87. [CrossRef]
106. Xu, Z.-D.; Jia, D.-H.; Zhang, X.-C. Performance tests and mathematical model considering magnetic saturation for magnetorheological damper. *J. Intell. Mater. Syst. Struct.* **2012**, *23*, 1331–1349. [CrossRef]
107. Malakopoulos, A.; Salifoglou, A. Assessment of Durability Indicators for Service Life Prediction of Portland Limestone Cementitious Systems Produced with Permeability-Reducing Admixtures. *Buildings* **2022**, *12*, 1712. [CrossRef]
108. Luo, B.; Wang, D.; Mohamed, E. Study on Mechanical Properties and Durability of Alkali-Activated Silicomanganese Slag Concrete (AASSC). *Buildings* **2022**, *12*, 1621. [CrossRef]
109. Cheng, Z.; Yang, K.; Tang, Z.; Ge, F.; Zhou, X.; Zeng, X.; Ma, K.; Long, G. Experimental investigation on flexural and compressive Resilience of mortar and concrete with hybrid toughening materials. *Structures* **2022**, *43*, 1592–1599. [CrossRef]
110. Ohemeng, E.A.; Naghizadeh, A. Alternative cleaner production of masonry mortar from fly ash and waste concrete powder. *Constr. Build. Mater.* **2023**, *374*, 130859. [CrossRef]
111. Dobiszewska, M.; Bagcal, O.; Beycioğlu, A.; Goulias, D.; Köksal, F.; Płomiński, B.; Ürünveren, H. Utilization of rock dust as cement replacement in cement composites: An alternative approach to sustainable mortar and concrete productions. *J. Build. Eng.* **2023**, *69*, 106180. [CrossRef]
112. Sara, B.; Mhamed, A.; Otmane, B.; Karim, E. Elaboration of a Self-Compacting mortar based on concrete demolition waste incorporating blast furnace slag. *Constr. Build. Mater.* **2023**, *366*, 130165. [CrossRef]
113. Zhang, G.; Peng, G.-F.; Zuo, X.-Y.; Niu, X.-J.; Ding, H. Adding hydrated lime for improving microstructure and mechanical properties of mortar for ultra-high performance concrete. *Cem. Concr. Res.* **2023**, *167*, 107130. [CrossRef]
114. Pan, Z.; Chen, J.; Zhan, Q.; Wang, S.; Jin, R.; Shamass, R.; Rossi, F. Mechanical properties of PVC concrete and mortar modified with silane coupling agents. *Constr. Build. Mater.* **2022**, *348*, 128574. [CrossRef]
115. Liu, M.; Hu, R.; Zhang, Y.; Wang, C.; Ma, Z. Effect of ground concrete waste as green binder on the micro-macro properties of eco-friendly metakaolin-based geopolymer mortar. *J. Build. Eng.* **2023**, *68*, 106191. [CrossRef]

buildings

MDPI

Article

Mechanical Property Research for CSIP Thin-Wall Box-Beams

Lei Li [1,*], Nasim Uddin [2], Xianxian Zhao [1] and Limin Tian [3]

1 School of Civil Engineering, Zhengzhou University of Aeronautics, Zhengzhou 450015, China
2 Department of Civil Construction and Environmental Engineering, University of Alabama at Birmingham, Birmingham, AL 35294, USA
3 Shaanxi Key Lab of Structure and Earthquake Resistance, School of Civil Engineering, Xi'an University of Architecture and Technology, Xi'an 710055, China
* Correspondence: lilei@zua.edu.cn

Abstract: Composite structural insulated panels (CSIPs) are eco-friendly, high-performance materials, which not only good have mechanical properties, but also good waterproof, moisture-proof, fire-proof, and anti-corrosion characteristics, so they have been used to build envelope structures in recent years. However, how to improve stiffness of CSIPs remains unsolved. The poor stiffness is one of the biggest obstacles for the application of CSIPs in the load-bearing members of civil engineering. In this study, the layout of glass–polypropylene (PP) laminate layers is designed to enhance its stiffness, and this study applies CSIPs as load-bearing members of civil engineering for the first time. Thus, the bend model of CSIP thin-wall box-beams under uniform loading is built, based on Timochenko's theory. The deflection curve equation is presented, considering shearing deformation. The expressions for the bending of normal strain flanges of the beam and the equation considering principal shearing strain at the beam's web are obtained, respectively. Finally, mechanical properties of the thin-wall box-beam under uniformly distributed loads were performed by FE. FE results are entirely consistent with the theoretical results, thereby making the theoretical method applicable for the design of thin-wall box-beams, which are made of composite materials. Different from other beams, the shearing deformation is a critical factor that influences the deformation of thin-walled box-beams.

Keywords: thin-wall box-beam; CSIPs; mechanical property; shearing deformation

1. Introduction

Conventional building materials encounter challenges, such as high self-weight, poor corrosion, poor fireproofing, poor waterproofing, poor moisture-proofing, and high maintenance costs. Nowadays, more and more researchers are committed to investigating innovative new material substitutes. Thus, extensive research on CSIPs has been conducted at the University of Alabama at Birmingham. Professor Uddin's research team presented composite structural insulated panels (CSIPs). The designs for the section of the CSIPs were proposed by Vaidva et al. [1]. (Figure 1), such as the thickness of the glass–polypropylene (PP) face sheets (3.04 mm) and EPS core (140 mm). The face sheets of CSIPs exhibit high impact-resistance and strength, good waterproofing, moisture-proofing, fire-proofing, anti-corrosion, and eco-friendly characteristics, whereas the EPS middle sheets bear the shearing stress, and two outer PP faces resist buckling and improve its stiffness. Thus, as the structural layout of CSIPs, the use of glass–polypropylene (PP) sheets is appropriate because it provides a high ratio of strength to weight, fine impact resistance, and good durability. Meanwhile, EPS foam is so light weight that it is used for the CSIP core. Thus, it has an excellent performance in things such as heat insulation, excellent impact-resistance, anti-bending, fire-proofing, and keeping warm [1]. Therefore, it is important that CSIPs' mechanical properties are studied.

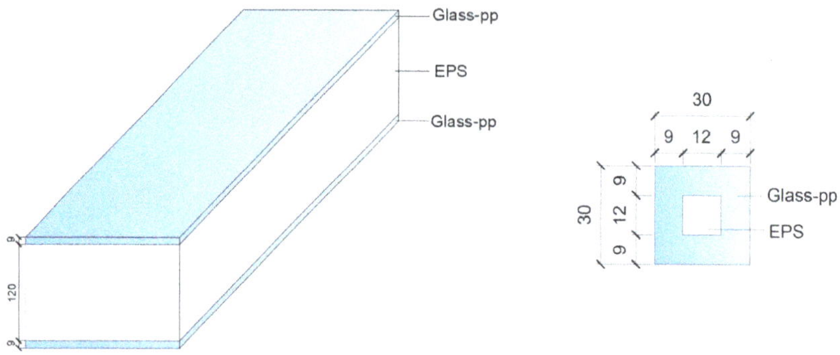

Figure 1. CSIP Sketch (unit: mm).

In recent years, considerable research has been devoted to developing sandwich composite materials. Wang et al. discussed the shearing of beams and panels with higher order hypotheses [2]. Liu et al. proposed layer-wise shear deformation theory of composite laminated plates by DQFEM [3]. Pawlus solved the stability of asymmetrical three-layered annular plates under loads in a plane [4]. The influence of a thin middle sheet of bending on the sandwich beams was presented the by Iaccarino et al. [5]. Grygorowicz and Lewiński presented a three-point bending of an I-beam considering the transverse shear effect [6]. Under the higher-order zigzag principle, static FE analyses of sandwich beams in both outer sheets and soft cores were presented by Chakrabarti et al. [7]. Wittenbeck et al. studied the three-point bending of sandwich beams with corrugated cores [8]. Roque et al. presented the bending of simply supported laminated composite beams subjected to transverse loads by a modified couple stress theory and a meshless method [9]. Batra and Xiao gave finite transient deformations of a curved laminated beam composed of a St. Venant–Kirchhoff material [10]. De Santis et al. obtained the behavior effects of composite glulam beams with circular holes [11]. Rocha et al. presented the structural concept of post-tensioned Fe-SMAs glass systems [12]. Bui et al. investigated transient responses and natural frequencies of sandwich beams with inhomogeneous functionally graded (FG) cores [13]. Marczak and Jędrysiak built a model of sandwich structures with inert cores [14,15]. Matuszewska and Strek studied a vibration analysis of a beam by an auxetic cross-section [16]. Grygorowicz and Magucka-Blandzi gave static and dynamic stabilities of a simply supported three-layered beam with a metal foam core [17]. Jopek and Strek presented the torsion of a two-phased composite bar with helical distribution of constituents by the FEM method [18]. Teter showed that simply supported columns are subjected to in-plane pulse loading at the loaded end by the analytical–numerical method (ANM), using Koiter's perturbations method [19]. The study of the failure mechanisms of sandwich beams with horizontal loads was performed by Steeves et al. and Qin et al. [20,21]. Tian et al. proposed the instability of structural members under axial compression causes [22] The bending problems of sandwich beams made of a metal foam core were presented by Magnucka-Blandzi et al. [23,24]. The connections between foam sheets was gained considering cracks in the connections of the fibers and foam sheets by Zenkert. [25]. The effect of the fiber transverse dimension of the binding layer under the deformation of beam bending was presented by Magnucki et al. [26]. Mori et al. gained insight into the mechanisms of deformation and the mechanics of sandwich structures subjected to underwater shock loading [27]. A formal engineering method of the mechanics of thin-walled laminated beams was obtained under kinematic assumptions by Barbero et al.; it has a good agreement with Timoshenko's beam theory [28]. Yuan et al. proposed the bending influence of shear deflection on laminated sandwich box cantilever beams [29]. Pawar demonstrated that the thin-walled composite box beams by 3D Finite Element Modeling were a feasible solution [30].

Uddin et al. investigated the behavior of the whole bending of CSIP wall panels (CSIPs) under concentric and eccentric load in a plane [31,32]. Du and Uddin presented concepts of the new CSIP thin-walled laminated shell, which it is different from the traditional concrete thin shell [33]. Due to the advantages of CSIP plates and shells, such as a high strength–weight ratio, excellent impact resistance, and high durableness, they are used as enclosure members in civil engineering. However, previous research did not focus on the stress, strain, and deflection in each laminated layer of the sandwich plates and beams.

With an excellent strength–weight ratio and good anti-impact properties, civil engineering will apply CSIPs to load-bearing structures. The stiffness of CSIPs is so small that the load-bearing members made of CSIP composite materials have a great challenge, which is the large deflection. Consequently, focusing research on enhancing CSIPs' stiffness is essential. In the present study, the new innovative CSIPs improved the stiffness of the materials, which are made of E-glass (PP) layers as the surface, and a polystyrene (EPS) foam core. Innovative CSIPs are derived from sandwich panels, which have a foam core with a thickness of 120 mm and a surface with thickness of 9 mm. On the other hand, the box-beams are used because of their excellent mechanical properties—a larger rigidity and strong torsion resistance. These innovations can be developed to enhance efficiency through special designs. In this study, stress, strain, and deflection of each of the face sheets of the CSIP thin-wall box-beam will be carried out by theoretical analysis. The FE model of the CSIP thin-wall box-beam was constructed using the ANSYS Workbench. Finally, by comparing the theoretical results with the FE results, the feasibility of the theoretical model is evaluated.

2. Design of the CSIP Thin-Wall Box-Beam

2.1. Cross Sectional Shapes of CSIP Box-Beams

The section design of the girder is important in building design. The section format of girders is mainly things such as box-beams, T-beams, and I-beams. In this study, the box-beam is formed (Figure 2), because it has a much greater torsional stiffness and strength than an open section. The closed cell is made of EPS foam, its web and flange are made of 12 mm and 9 mm glass PP, respectively.

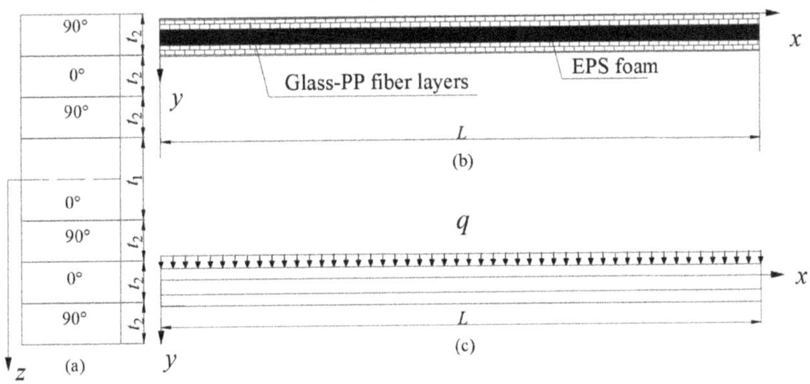

Figure 2. CSIP thin-wall box-beam. (**a**) Cross-section of the laminated composites. (**b**) The configuration vertical section. (**c**) A simply supported beam with a uniform load diagram.

2.2. Design of CSIPs

A cheap orthotropic thermoplastic (TP) glass PP surface and EPS foam core sheets are the CSIP composite materials. Bending stresses are borne by face sheets. The foam core has an important function, the shear load is resisted by the foam core. On the other hand, it improves bending performance of the CSIP girder [34]. A CSIP was developed, which was made of low-cost TP glass PP laminate as a face sheet and EPS as a core (Figure 1).

The PP resin impregnated E-glass fibers, then the glass PP surface was made of 70%, bidirectionally. The TP polymers have some advantages, such as high-temperature, short-time processing, good durability, being inexpensive, good toughness, excellent impact ability, good plasticity, and high recyclability and reusability for the thermosetting wastes. A direct reinforcement fabrication technology (DRIFT) process is used to turn out glass PP surfaces [35]. The manufacturer (Crane Composites, Inc., Channahon, IL, USA) can produce glass PP laminated layers (3.04 mm thickness). In this study, the glass PP mechanical performance was as described in Table 1. At the same time, due to the low price and low weight of EPS foam, a structure made of CSIPs has low weight, high thermal resistivity, and good impact ability, and is very good thermal insulation material. Due to its low cost, the core sheet comprised 140 mm EPS foam (Table 1). The manufacturer (Universal Packaging, Inc., Lutherville, MD, USA) supplied the EPS foam materials, its performance is as described in Table 2. Above all, CSIPs are high performance materials [36]. In this research, the new design method for CSIPs has been presented to improve its stiffness.

Table 1. Properties of the glass PP face sheets [36].

Items	Value
Nominal thickness (t)	3.04 mm
Density (ρ_f)	980 kg/m^3
Weight percentage of glass fiber	70%
Longitudinal modulus (Ex)	15,169 MPa
Transverse modulus (Ey)	15,169 MPa
Thickness direction modulus (Ez)	1050 MPa
Poisson's ratio (ν_{xy}, ν_{yz}, ν_{xz})	0.11, 0.22, 0.22
Shear modulus (G_{xy}, G_{yz}, G_{xz})	1800 MPa, 1800 MPa, 750 MPa
Tensile strength	690 MPa
Compression strength	317 MPa

Table 2. Properties of EPS foam core [36].

Items	Value
Nominal thickness (t)	140 mm
Density (ρ_f)	16 kg/m^3
Elastic modulus (Ec)	1.2–1.5 MPa
Poisson's ratio (ν)	0.25
Shear modulus (G)	1.9–2.2 MPa
Tensile strength	0.11–0.14 MPa
Compression strength	0.07–0.1 MPa

The CSIPs are made of laminate surfaces where glass fiber is 70% of the weight, and the EPS foam core the remainder. The new innovative composite structure for insulated panels (Figure 1), which is derived from the sandwich composite structure of a lightweight thick core (120 mm) and two stronger and thinner surfaces (9 mm), improves the stiffness of the materials. The box-beams are used because of their excellent mechanical properties, such as large rigidity and strong torsion resistance.

2.3. CSIP Thin-Wall Box-Beam Geometry

The mechanical behavior of a beam depends on its span–height ratio. Generally, the span–height ratio of a simply supported beam is from 10 to 20. The a span –height ratio is 20 in this study. The span of the box-beam is 600 mm, the cross section is a box-beam that is a square of about 30 mm on a side (Figure 2). To improve the stiffness of the CSIP thin-wall box-beams, the thickness of the flange is the same as in a beam web, 9 mm. The E-glass/PP layup orientation of the webs is consistent with those of the flanges.

3. Static Analysis of a CSIP Thin-Wall Box-Beams

3.1. Strain Stress of a CSIP Thin-Wall Box-Beams

Based on the hypothesis of the plain section, CSIP thin-wall box-beams with simple support will be studied under uniform loads. Regardless of the shear strain, the strain of a CSIP thin-wall box-beam can be calculated using Equation (1):

$$\varepsilon_k = \frac{M(x)}{D_{11}} y_k \tag{1}$$

where $M(x)$ is the bending moment of the cross-section at x point, D_{11} is the bending stiffness of the cross-section of the beam, and y_k is the distance from the neutral layer to the kth layer of box-beam.

The bending stress at any point of its cross-section is presented using Equation (2):

$$\sigma_x^k = \frac{M(x)}{D_{11}} \overline{Q}_{11}^k y_k \tag{2}$$

where σ_x^k is the bending stress at the kth glass–PP layers of the cross section, $\overline{Q}_{11}^k$ is the stiffness coefficient at the kth glass-PP layers of the cross-section, the flexural rigidity of the cross-section is $D_{12} = \frac{E_c b_c t_c^3}{12(1-\mu_c^2)}$, where E_c is the elasticity modulus of the EPS foam, and μ_c is the Poisson's ratio of EPS foam.

The flexural rigidity of the web (D_{13}) is given as $D_{13} = 2\sum_{i=1}^{12} \overline{Q}_{11}^i \frac{t_{fi} b_{fi}^3}{12}$, where t_{fi} is web thickness at i^{th} layer, and b_{fi}^3 is the web height at the ith layer. The flexural rigidity of the upper or lower flanges of the box-beam cross-section is obtained by $D_{14} = 2\sum_{i=1}^{12} \overline{Q}_{11}^i t_{yi} b_{yi} z_{yi}^2$, where b_{yi} is the flange width at the ith layer, t_{yi} is flange thickness at i^{th} layer, and z_{yi} is distance from the middle surface of the ith layer flange to the neutral axis.

The plane stress constitutive equation for orthotropic material is given as:

$$\begin{pmatrix} \varepsilon_1 \\ \varepsilon_2 \\ \varepsilon_3 \end{pmatrix} = \begin{bmatrix} s_{11} & s_{12} & 0 \\ s_{21} & s_{22} & 0 \\ 0 & 0 & s_{GG} \end{bmatrix} \begin{Bmatrix} \sigma_1 \\ \sigma_2 \\ \sigma_3 \end{Bmatrix} = [s]\{\sigma\} \tag{3}$$

where S_{ij} is the flexibility coefficient, and readily calculated by

$$s_{11} = \frac{1}{E} \tag{4}$$

$$s_{12} = -\frac{\mu}{E} \tag{5}$$

$$s_{GG} = \frac{1}{G} = \frac{2(1-\mu)}{E} \tag{6}$$

The other constitutive equation is obtained from Equation (3)

$$\begin{pmatrix} \sigma_1 \\ \sigma_2 \\ \sigma_3 \end{pmatrix} = \begin{bmatrix} \overline{Q}_{11} & \overline{Q}_{12} & \overline{Q}_{1G} \\ \overline{Q}_{21} & \overline{Q}_{22} & \overline{Q}_{2G} \\ \overline{Q}_{1G} & \overline{Q}_{2G} & Q_{GG} \end{bmatrix} \begin{Bmatrix} \varepsilon_x \\ \varepsilon_y \\ \gamma_{xy} \end{Bmatrix} = [\overline{Q}]\{\varepsilon\} \tag{7}$$

where $\left[\overline{Q}_{ij}\right]$ is the stiffness matrix, given as:

$$\overline{Q}_{11} = Q_{11}\cos^4\theta + 2(Q_{12} + 2Q_{GG})\sin^2\theta\cos^2\theta + Q_{22}\sin^4\theta \tag{8}$$

$$\overline{Q}_{12} = (Q_{11} + Q_{22} - 4Q_{GG})\sin^2\theta\cos^2\theta + Q_{12}(\sin^4\theta + \cos^4\theta) \tag{9}$$

$$\overline{Q}_{22} = Q_{11}\sin^4\theta + 2(Q_{12} + 2Q_{GG})\sin^2\theta\cos^2\theta + Q_{22}\cos^4\theta \tag{10}$$

$$\overline{Q}_{1G} = (Q_{11} - Q_{12} - 2Q_{GG})\sin\theta\cos^3\theta + (Q_{12} - Q_{22} + 2Q_{GG})\sin^3\theta\cos\theta \tag{11}$$

$$\overline{Q}_{2G} = (Q_{21} - Q_{12} - 2Q_{1G})\sin^3\theta\cos\theta + (Q_{12} - Q_{22} + 2Q_{GG})\sin\theta\cos^3\theta \tag{12}$$

$$\overline{Q}_{GG} = (Q_{11} + Q_{22} - 2Q_{12} - 2Q_{GG})\sin^2\theta\cos^2\theta + Q_{GG}(\sin^3\theta + \cos^4\theta) \tag{13}$$

where "θ" is ply orientation angle, and Q_{ij} is the stiffness coefficient, which is calculated using the following formulae:

$$Q_{11} = \frac{E_1}{1 - \mu_{12}\mu_{21}} \tag{14}$$

$$Q_{22} = \frac{E_1}{1 - \mu_{12}\mu_{22}} \tag{15}$$

$$Q_{12} = \frac{E_1\mu_{21}}{1 - \mu_{12}\mu_{21}} = \frac{E_1\mu_{12}}{1 - \mu_{12}\mu_{21}} \tag{16}$$

$$Q_{GG} = G_{12} \tag{17}$$

3.2. *Deflection Curve of CSIP Thin-Wall Box-Beams*

On the basis of the plane-section hypothesis of the girder theory, the bending problem is studied in a simply supported CSIP thin-wall box-beam under a uniformly distributed of load magnitude q. When the system is in equilibrium, the flexural rigidity of the CSIP thin-wall box-beam is calculated using Equation (18):

$$EI_y = \int E_x Z^2 dA \tag{18}$$

The CSIP thin-wall box-girder is made of an EPS foam core and glass–PP fiber sheets. Because the ply orientation angle and thickness of the glass–PP fiber layers are the same above and below the EPS foam, the laminate is called the symmetric laminate (Figure 2). And the flexural rigidity of the cross-section of the beam is obtained by

$$\overline{EI_y} = (E_x I_y)_1 + (E_x I_y)_2 \tag{19}$$

where $(E_x I_y)_1$ is the flexural rigidity of the girder flange, which is obtained by $(E_x I_y)_1 = \sum_{i=1}^{n} E_x^i I_x^i$ and $(E_x I_y)_2$ is the flexural rigidity of the beam web that is given as $(E_x I_y)_2 = \sum_{j=1}^{n} E_x^j I_x^j$.

For the CSIP thin-wall box-girder, its shearing modulus is very small; that is, from 1.9 Mpa to 2.2 Mpa. Therefore, effect of the shearing deformation should be considered in the thin-wall box beam design. The shearing modulus of the girder is obtained by

$$\overline{GA} = (GA)_1 + (GA)_2 \tag{20}$$

When a Timoshenko beam is subject to bending moment, owing to shear stress, the additional deflections y_1 and y_2 occur to the beam. Thus, the deformation curvature is given by

$$\frac{d^2y}{dx^2} = -\frac{1}{EI}(M + \frac{KEI}{AG})w \quad (21)$$

where K is the shear correction factor, A is the cross-section area, and G is the shear modulus.

In this study, the simply supported CSIP box-beam carried the uniformly distributed of load magnitude q (Figure 2), where L is the box-girder span. Thus, the shearing and moment of the cross-section of the beam are given as

$$w(x) = \frac{q}{2}(1 - 2x) \quad (22)$$

$$M(x) = \frac{qx}{2}(1 - x) \quad (23)$$

Substituting Equations (22) and (23) into Equation (21), the deflection curvature of the thin-wall box-beam is obtained by

$$W = \frac{q}{EI}(-\frac{Lx^3}{12} + \frac{x^4}{24} + \frac{L^3x}{24}) + \frac{kq}{2AG}(Lx - x^2) \quad (24)$$

where the mid-span deflection is maximum deflection ($W_{\max}$), which is obtained by

$$W_{\max} = \frac{5qL^4}{384EI} + \frac{3kqL^2}{8AG} \quad (25)$$

Because the shear rigidity of the EPS foam is very small, the effect of the EPS foam on shear resistance ability is not considered.

4. Comparison between the Result of the Theoretical and FE Analyses

4.1. FE Model

In this study, the FE model of the simply supported CSIP thin-wall box-beam is based on theory-deduced calculations. Thus, the FE analysis is studied in the CSIP thin-wall box-beam subjected to uniform load. The FE analysis was performed with the FE commercial package ANSYS 18.0. Table 3 describes the element types for the research.

Table 3. Element Type Model.

Material Type Element	ANSYS
Glass–PP fiber sheet faces, EPS foam core	Solide186
Contact	Conta174

In this study, a CSIP thin-wall beam is a laminated sandwich thin shell structure, so element type selects the shell181, which it behaves as a four-node element, one node has six degrees of freedom, and is applied to laminated modeling structures. On the other hand, the CSIP thin-wall beam is made of the glass–PP surfaces and the EPS foam cores, where the cores use the solider45 element as its element type, which behaves as an element with an 8 nodes, a node has three degrees of freedom. It is applied to plastic deformation and cracking. At the same time, the adhesive between glass–PP laminated layers and EPS foam core is glued together for the CSIP box-beam models. In this study, Tables 1 and 2 show separately the performance of the CSIP materials.

The CSIP thin-wall box girder was built by FE (Figure 3). The span of the girder is 600 mm, and its cross-section is a square of about 30 mm on a side. When the aspect ratio of the rectangular unit is close to 1, it is high-precision. In this study, the rectangular unit is taken as finite element model mesh. The FE beam model is describe in Figure 3.

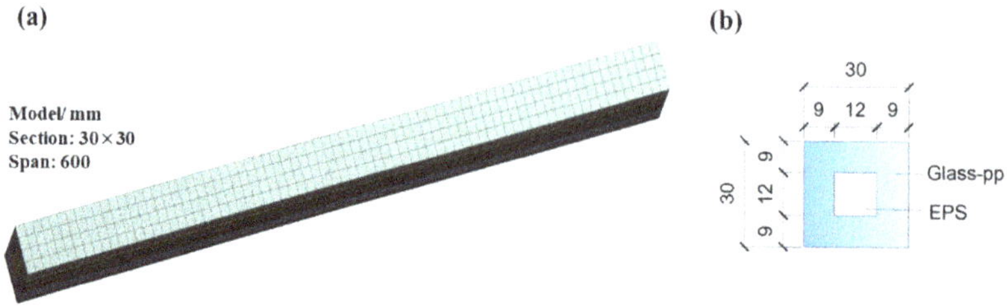

Figure 3. Diagram of the CSIP thin-wall box-beam (mm). (**a**) FE model. (**b**) Section of the beam.

4.2. FE Analysis

The CSIP thin-wall box-beam was subjected to the dead weight that is the uniform load, with a load intensity of 2.581 kN/m^2. In this study, the CSIP thin-wall box-beam is a simple supported beam. The boundary conditions of nodal points are as follows: where X = 0, all its deflections of nodal points are zero; where X = L, its X and Y deflections of nodal points are 0; and Z deflections of nodal points are free.

The stress distribution of each layer of the cross-section of CSIP thin-wall box-girder is shown in Figure 3. The maximum bending stress 6.76 MPa occurs in the outermost layer in a section of the middle of the span of the CSIP thin-wall box-girder. Further, the minimum bending stress exists at the neutral axis of ship section of the girder (Figure 4 and Table 4).

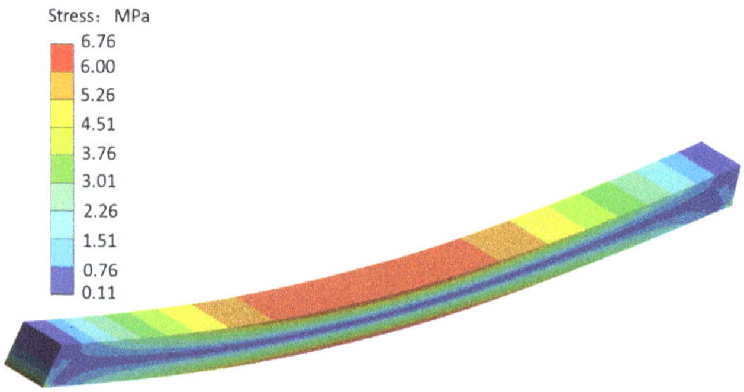

Figure 4. Stress of the CSIP thin-walled box-beam.

Figure 5 shows the total deformation of the CSIP thin-wall box-beam. The maximum deformation occurs at the mid-span of the CSIP thin-wall box-beam section, and its value is 0.83 mm. The FE results meet the requirements of the steel structure design code.

Figure 5. Total deformation of the CSIP thin-wall box-girder.

Table 4. Comparison of theoretical and FE results for the CSIP thin-wall box-beam.

E-Glass Sheets Stress of Mid-Span of Beam (Mpa)				
Section	**Fibrous Layer**	**Ansys**	**Theory**	**Error (%)**
$Flange_{top}$	first	−6.38	−5.93	7.50
	second	−5.53	−4.81	7.20
	third	−4.59	−4.29	7.00
	fourth	−4.14	−3.91	5.88
Web_{left}	first	3.59	−3.45	4.06
	second	−2.21	−2.35	5.60
	third	−1.59	−1.65	4.75
	fourth	−1.34	−0.91	3.64
	fifth	−0.20	−0.19	5.26
	sixth	1.34	0.91	3.64
	seventh	1.59	1.65	4.75
	eighth	2.21	2.35	5.60
	ninth	3.59	3.45	4.06
$Flange_{bottom}$	first	4.14	3.91	5.88
	second	4.59	4.29	7.00
	third	5.53	4.81	7.20
	fourth	6.38	5.93	7.50

A layer strain of the CSIP thin-wall box-girder is described in Figure 6. The maximum bending strain of 0.33 millimetres occurs the outermost layer in a section of the middle of span of the CSIP thin-wall box-girder. Meanwhile, the minimum bending strain exists at the neutral axis of that section of the girder. The strain values of each layer on the CSIP thin-wall box-beam are obtained (Figure 6 and Table 5).

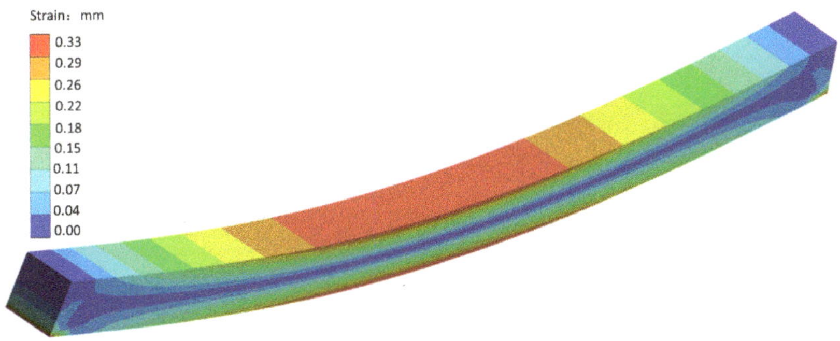

Figure 6. Strain of the CSIP thin-wall box-beam.

Table 5. Comparison of theoretical and FE solution for the CSIP thin-wall box-beam.

	Deflection Mid-Span of Beam		
	Ansys	**Theory**	**Error (%)**
Deflection (mm)	0.83	0.836	0.7

4.3. Comparison of Theoretical and FE Results for the CSIP Thin-Wall Box-Beam

4.3.1. Comparison of Stress Theoretical and FE Results for the CSIP Beam

Table 4 shows the FE and theoretical stress values in the mid-span section of a simply supported CSIP thin-wall box-beam under uniform load. Table 4 shows that the FE results are generally consistent with the theoretical values. The error between the FE and the theoretical results is within 10%, where the maximum error is 7.5%. Thus, the theory results are verified in terms of feasibility and reliability. The theory results meet the requirements of practical engineering.

4.3.2. Comparison of Theoretical and FE Results for the CSIP Beam

Based on Timoshenko's beam theory, the CSIP thin-wall box-beam is concerned with the bending problem. However, the influence on the shear deformation is ignored in the bending problem of the CSIP thin-wall box-beam. Its theoretical deformation model was presented on the basis of shearing deformation [Equation (24)]. The related theoretical results are compared with the FE solution, the maximum deflection is at the middle of the CSIP thin-wall box-gird span. Therefore, the mid-span deflection is the object of research. Table 5 shows the theoretical and FE mid-span deformation results.

The error of the total deformation between theoretical analysis and FE results is 0.7% (Table 5), thus the theory results of the total deformation are verified in terms of feasibility and reliability.

Meanwhile, the study results show that shearing deformation greatly affects the total deformation of the CSIP thin-walled box-beam. Therefore, shearing deformation should be considered in the engineering applications of the CSIP thin-walled box-beam. In addition, different strains of each E-glass layer are shown in Table 6, which contains the FE strain results and theory strain results.

Table 6. Comparison of theoretical and FE results for CSIP thin-wall box-beam.

E-Glass Sheets Strain of Mid-Span of Beam (mm)				
Section	**Fibrous Layer**	**Ansys**	**Theory**	**Error (%)**
$Flange_{top}$	first	−0.31	−0.322	3.72
	second	−0.28	−0.293	4.44
	third	−0.25	−0.263	4.94
	fourth	−0.20	−0.212	3.73
Web_{left}	first	−0.17	−0.181	5.56
	second	−0.13	−0.135	3.70
	third	−0.09	−0.089	1.12
	fourth	−0.06	−0.065	7.69
	fifth	−0.02	−0.021	4.76
	sixth	0.06	0.065	7.69
	seventh	0.09	0.089	1.12
	eighth	0.13	0.135	3.70
	ninth	0.17	0.181	5.56
$Flange_{bottom}$	first	0.20	0.212	3.73
	second	0.25	0.263	4.94
	third	0.28	0.293	4.44
	fourth	0.31	0.322	3.72

The errors between the theoretical and FE results are within 7.69%. The theory results meet requirements of the finite-element analysis values. Thus, the theory results for a CSIP thin-wall box-beam are feasible.

5. Conclusions

This study presents the theory model of a CSIP thin-wall box-beam, and the theory results are verified by the finite element method. The research findings are as follows:

1. The shear deformation is not ignorable in the CSIP thin-wall box-beam on bend problem. In this study, the deformation equations of the CSIP thin-wall box-beam considering the effects of the shear deformation are developed, and this model is verified by the numerical analysis. In addition, the effectiveness of the theorical method is certified.

2. The strain and stress expressions of the CSIP thin-wall box-beam for each E-glass layer are proposed based on Timoshenko's beam theory, and numerical simulations confirm the validity and correctness of the presented theory.
3. The bend theory of the CSIP thin-wall box-beam is presented for the first time. This theoretical model is built to improve the mechanical properties of the CSIP thin-wall box-beam by changing the value of the web width–thickness ratio.

In summary, the results of this study are very useful for the static and dynamic study of the CSIP thin-wall box-beam, and the theoretical deformation model of the CSIP thin-wall box-beam presented in this study may push the applications of CSIP as the main bearing members of civil engineering. Future research is required into nonlinear dynamic characteristics of the CSIP thin-wall box-beam.

Author Contributions: L.L.: Conceptualization, Supervision, Investigation, Formal analysis, Writing—original draft. N.U.: Conceptualization, Supervision, Writing—Review and Editing. X.Z.: FE analysis, Data curation. L.T.: Supervision, Writing—Review and Editing. All authors have read and agreed to the published version of the manuscript.

Funding: Science Foundation in the Henan province, China (Grant No. 192102210030, Grant No. 212102310957, and 182102210439) and NSFC (Grant No. U1404524) and Tackle-Key-Program of S&T Committee of Henan Province China (Grant No. 202102210209) and Key-Research-Projects of Higher Education Institutions of Henan Province, China (Grant No. 23A590002) and Key R&D Special Projects of Henan Province, China (Grant No. 221111230700).

Data Availability Statement: The data will be available as requested.

Conflicts of Interest: The authors declare no conflict of interest.

References

1. Vaidya, A.; Uddin, N.; Vaidya, U. Structural Characterization of Composite Structural Insulated Panels for Exterior Wall Applications. *J. Compos. Constr.* **2010**, *14*, 464–469. [CrossRef]
2. Wang, C.M.; Reddy, J.N.; Lee, K.H. *Shear Deformable Beams and Plates*; Elsevier: Amsterdam, The Netherlands; Laussane, Switzerland; New York, NY, USA; Oxford, UK; Shannon, Ireland; Singapore; Tokyo, Japan, 2000.
3. Liu, B.; Ferreira, A.J.M.; Xing, Y.F.; Neves, A.M.A. Analysis of composite plates using a layerwise theory and a differential quadrature finite element method. *Compos. Struct.* **2016**, *156*, 393–398. [CrossRef]
4. Pawlus, D. Stability of Three-Layered Annular Plate with Composite Facings. *Appl. Compos. Mater.* **2016**, *24*, 141–158. [CrossRef]
5. Iaccarino, P.; Leone, C.; Durante, M.; Caprino, G.; Lamboglia, A. Effect of a thin soft core on the bending behavior of a sandwich with thick CFRP facings. *J. Sandw. Struct. Mater.* **2011**, *13*, 159–175. [CrossRef]
6. Grygorowicz, M.; Lewiński, J. Bending of I-beam with the transvers shear effect included—FEM calculated. In Proceedings of the International Conference of Numerical Analysis and Applied Mathematics 2015 (Icnaam 2015), Rhodes, Greece, 22–28 September 2015.
7. Chakrabarti, A.; Chalak, H.D.; Ashraf Iqbal, M.; Sheikh, A.H. A new FE model based on higher order zigzag theory for the analysis of laminated sandwich beam with soft core. *Compos. Struct.* **2011**, *93*, 271–279. [CrossRef]
8. Wittenbeck, L.; Grygorowicz, M.; Paczos, P. Numerical analysis of sandwich beam with corrugated core under three-point bending. In Proceedings of the International Conference on Numerical Analysis and Applied Mathematics 2014 (Icnaam-2014), Rhodes, Greece, 22–28 September 2014.
9. Roque, C.M.C.; Fidalgo, D.S.; Ferreira, A.J.M.; Reddy, J.N. A study of a microstructure-dependent composite laminated Timoshenko beam using a modified couple stress theory and a meshless method. *Compos. Struct.* **2013**, *96*, 532–537. [CrossRef]
10. Batra, R.C.; Xiao, J. Finite deformations of curved laminated St. Venant–Kirchhoff beam using layer-wise third order shear and normal deformable beam theory (TSNDT). *Compos. Struct.* **2013**, *97*, 147–164. [CrossRef]
11. De Santis, Y.; Pasca, D.P.; Aloisio, A.; Stenstad, A.; Mahnert, K.-C. Experimental, analytical and numerical investigation on the capacity of composite glulam beams with holes. *Eng. Struct.* **2023**, *285*, 115995. [CrossRef]
12. Rocha, J.; Pereira, E.; Sena-Cruz, J. Feasibility of mechanical post-tensioning of annealed glass beams by activating externally bonded Fe-SMA reinforcement. *Constr. Build. Mater.* **2023**, *365*, 129953. [CrossRef]
13. Bui, T.Q.; Khosravifard, A.; Zhang, C.; Hematiyan, M.R.; Golub, M.V. Dynamic analysis of sandwich beams with functionally graded core using a truly meshfree radial point interpolation method. *Eng. Struct.* **2013**, *47*, 90–104. [CrossRef]
14. Marczak, J.; Jędrysiak, J. Some remarks on modelling of vibrations of periodic sandwich structures with inert core. *Compos. Struct.* **2018**, *202*, 752–758. [CrossRef]
15. Marczak, J.; Jędrysiak, J. Tolerance modelling of vibrations of periodic three-layered plates with inert core. *Compos. Struct.* **2015**, *134*, 854–861. [CrossRef]

16. Matuszewska, A.; Stręk, T. Vibration Properties of Auxetic Beam. *Vib. Phys. Syst.* **2018**, *29*, 2018031.
17. Grygorowicz, M.; Magnucka-Blandzi, E. Mathematical modeling for dynamic stability of sandwich beam with variable mechanical properties of core. *Appl. Math. Mech.* **2016**, *37*, 1361–1374. [CrossRef]
18. Jopek, H.; Strek, T. Torsion of a Two-Phased Composite Bar with Helical Distribution of Constituents. *Phys. Status Solidi (B)* **2017**, *254*, 1700050. [CrossRef]
19. Teter, A. Dynamic critical load based on different stability criteria for coupled buckling of columns with stiffened open cross-sections. *Thin-Walled Struct.* **2011**, *49*, 589–595. [CrossRef]
20. Qin, Q.H.; Wang, T.J. An analytical solution for the large deflections of a slender sandwich beam with a metallic foam core under transverse loading by a flat punch. *Compos. Struct.* **2009**, *88*, 509–518. [CrossRef]
21. Steeves, C.A.; Fleck, N.A. Collapse mechanisms of sandwich beams with composite faces and a foam core, loaded in three-point bending. Part II: Experimental investigation and numerical modelling. *Int. J. Mech. Sci.* **2004**, *46*, 585–608. [CrossRef]
22. Tian, L.-M.; Jin, B.-B.; Li, L. Axial Compressive Mechanical Behaviors of a Double-Layer Member. *J. Struct. Eng.* **2023**, *149*, 04023110. [CrossRef]
23. Magnucka-Blandzi, E.; Magnucki, K. Effective design of a sandwich beam with a metal foam core. *Thin-Walled Struct.* **2007**, *45*, 432–438. [CrossRef]
24. Magnucki, K.; Jasion, P.; Szyc, W.; Smyczynski, M.J. Strength and buckling of a sandwich beam with thin binding layers between faces and a metal foam core. *Steel Compos. Struct.* **2014**, *16*, 325–337. [CrossRef]
25. Zenkert, D. Strength of sandwich beams with interface debondings. *Compos. Struct.* **1991**, *17*, 331–350. [CrossRef]
26. Magnucki, K.; Smyczyński, M.J.; Jasion, P. Deflection and strength of a sandwich beam with thin binding layers between faces and a core. *Arch. Mech.* **2013**, *65*, 301–311.
27. Mori, L.; Lee, S.; Xue, Z.; Vaziri, A.; Queheillalt, D.; Dharmasena, K.; Wadley, H.; Hutchinson, J.; Espinosa, H. Deformation and fracture modes of sandwich structures subjected to underwater impulsive loads. *J. Mech. Mater. Struct.* **2007**, *2*, 1981–2006. [CrossRef]
28. Barbero, E.J.; Lopez-Anido, R.; Davalos, J.F. On the Mechanics of Thin-Walled Laminated Composite Beams. *J. Compos. Mater.* **2016**, *27*, 806–829. [CrossRef]
29. Yuan, D.C.; Xu, W.M.; Jiang, Z.G.; Meng, S.Y. Flexibility effect of shearing deformation on cantilever box girder of sandwich composite materials. *Guofang Keji Daxue Xuebao/J. Natl. Univ. Def. Technol.* **2009**, *31*, 33–36.
30. Pawar, D.P.M. Validation of the thin-walled composite box beams using FEM. *IOSR J. Mech. Civ. Eng.* **2012**, *1*, 45–49. [CrossRef]
31. Uddin, N.; Du, W. New thin shells made of composite structural insulated panels. *J. Reinf. Plast. Compos.* **2014**, *33*, 1954–1965. [CrossRef]
32. Mousa, M.A.; Uddin, N. Structural behavior and modeling of full-scale composite structural insulated wall panels. *Eng. Struct.* **2012**, *41*, 320–334. [CrossRef]
33. Du, W.; Uddin, N. Innovative composite structural insulated panels (CSIPs) folded shell structures for large-span roofs. *Mater. Struct.* **2016**, *50*, 51. [CrossRef]
34. Zenkert, D. *An Introduction to Sandwich Structures*; Engineering Materials Advisory Service, Ltd.: Worcester, UK, 1995.
35. Hartness, T.; Husman, G.; Koenig, J.; Dyksterhouse, J. The characterization of low cost fiber reinforced thermoplastic composites produced by the DRIFT™ process. *Compos. Part A Appl. Sci. Manuf.* **2001**, *32*, 1155–1160. [CrossRef]
36. Mousa, M.A.; Uddin, N. Performance of Composite Structural Insulated Panels after Exposure to Floodwater. *J. Perform. Constr. Facil.* **2013**, *27*, 424–436. [CrossRef]

buildings

MDPI

Article

Research on Fatigue Performance of Shape-Memory Alloy Bars under Low Cyclic Loading

Lei Li [1,*], Xianxian Zhao [2] and Junwei Cheng [2]

[1] School of Civil Engineering Architecture, Zhengzhou University of Aeronautics, Zhengzhou 450046, China
[2] School of Materials, Zhengzhou University of Aeronautics, Zhengzhou 450046, China
* Correspondence: lileixiaozhou@163.com

Abstract: In recent years, shape memory alloys (SMAs) have been applied in the vibration control of engineering structures due to their special properties such as super elasticity and high damping, and the study of the performance of SMA wires has been relatively comprehensive, while research on the fatigue performance of SMA bars via cyclic tensile tests has been pretty rare, and low-cycle fatigue test has not been reported. However, the damage to building structures caused by earthquakes is of high-strain, low-cycle fatigue; therefore, in order for SMA bars to be used in seismic design, low-cycle fatigue tests were conducted on SMA bars with a diameter of 14 mm in this paper. Firstly, specimens were heat treated at a constant temperature of 350 °C for 30 min; other specimens were heat treated at a constant temperature of 400 °C for 15 min, while the rests were heat treated under a constant temperature of 400 °C for 30 min. Secondly, the energy dissipation capacity and residual strain of the SMA bar specimens were determined using the low-cycle fatigue test, in which the strain amplitudes were 2.5%, 3.5% and 3.75%. Additionally, the stress–strain relationship for SMA bars under cyclic loading was given. Finally, low-cycle fatigue properties of SMA bars were numerically simulated in the comparison analysis with the experimental results to verify their feasibility. Thus, it is proved that SMA bars can be recommend for seismic design building structures.

Keywords: shape-memory alloy; low-cycle fatigue; superelastic effect; re-centering effect

1. Introduction

In recent years, earthquakes have occurred frequently, such as the Wenchuan earthquake in 2008, the Great Intelligence Earthquake in 2010, the Honshu Earthquake in 2011, and the Great Turkey Earthquake in 2023; they caused significant casualties and economic losses, and tens of thousands of buildings were damaged and collapsed. In the Turkey earthquake, buildings with strict seismic designs survived two consecutive 7.8 magnitude earthquakes, thus proving that the damage to buildings can be prevented and reduced. Under the action of high-strain and low-circumference earthquakes, fatigue damage occurs in a structure, and an earthquake causes high residual deformation or direct damage to traditional ductile members, making the structure difficult to recover after the earthquake, and even causing collapse damage to occur. Therefore, it is imperative to explore ductile and high-strength building materials and design functional members with recoverable performances. The early research on SMAs mainly focused on small-diameter wire: Graesser [1] studied the hysteresis model and cyclic tests of shape-memory alloys and modified the hysteresis model of SMAs. Piedboeuf [2] proposed a third harmonic Fourier sinusoidal series model to simulate the dynamic behavior of shape-memory alloys. Dolce [3] conducted cyclic tensile tests on SMA filaments to observe the changes in properties such as cut-line stiffness, energy loss, equivalent damping and residual strain during the tests. Gong [4,5] conducted an experimental study on the superelastic behavior of SMA wires under cyclic loading conditions with different temperatures and strain rates and suggested that pre-training can make the superelastic properties of SMA wires become stable. Sadiq [6] investigated the effect of heat treatment parameters on the mechanical behavior of NiTi shape-memory alloy wires. Tamai [7] proposed a shape-memory alloy wire-based seismic member and studied the restoring force characteristics of the SMA wire. Zuo [8] investigated the effects of ambient temperature, strain amplitude, number of cycles and loading rate on the mechanical properties of SMA wire in a superelastic state. Qian [9] studied the mechanical properties of SMA wires, showing that the damping energy dissipation characteristics and self-resetting characteristics of SMA wires with diameters of 0.5 mm are better than SMA wires with diameters of 1.2 mm and 2.0 mm. Li [10] discussed the effects of strain amplitude and number of cyclic loadings on the maximum phase change strain of SMA wire. Yan [11] investigated the superelastic properties of SMA wires via cyclic tensile tests and compared the experimental results with Grasser's theoretical model.

However, due to the restoring force provided by SMA wire being limited and the application of bar-shaped members such as steel bars being more extensive in engineering practice, scholars at home and abroad have investigated the effects of different heat treatment conditions, strain amplitudes and loading rates on the mechanical properties of large-diameter SMA bars through torsional and cyclic tensile tests in recent years: Chen [12] studied the stress–strain behavior of shape-memory alloy bars at cyclic loading rates of 0.001–750 s^{-1} and investigated the effect of loading rates on the mechanical properties of shape-memory alloys. Dolce [13] performed torsion tests on SMA bars. Desroches [14] conducted tests on SMA bars under quasi-static and dynamic loading, and the effect of bar size and loading history on its strength and equivalent viscous damping was investigated. Wang [15] discussed issues related to the heat treatment, mechanical property evaluation and joint design of SMA bars with a diameter of 14 mm. Zheng [16] obtained the relationship between the hyperelastic properties of SMA bars and the loading rate via cyclic tensile tests. Ren [17] found that the superelasticity of SMA bars decreases in cyclic loading and unloading, but the output force of 14 mm diameter bars is large. Dong [18] conducted experimental research on the heat treatment and material properties of SMA bars with a diameter of 14 mm and used it to develop a 25 mm diameter SMA self-resetting device. Kang [19] investigated the mechanical properties of shape-memory alloy bars with a diameter of 14 mm for different heat treatment processes. Wang [20] conducted superelasticity tests on SMA bars of different diameters with different loading regimes and proposed that the shape-memory alloy material can be made fully superelastic through "exercise". Zhuang [21] conducted mechanical tests on SMA bars of two chemical compositions and also simulated the mechanical properties with finite element software.

At present, most scholars have studied the fatigue properties of SMA via cyclic tensile tests, but the low-cycle fatigue properties of Ni-Ti shape-memory alloys are rarely studied. Therefore, in this paper, low-cycle fatigue tests are conducted on large-diameter SMA bars to study the effects of different heat treatment conditions and strain amplitudes on the energy dissipation capacity and recovery performance of SMA bars, which lay the foundation for the establishment of the fatigue damage model of shape-memory alloys.

2. Materials and Methods

2.1. Specimen Design and Experimental Materials

The test was conducted using SMA bars with a diameter of 20 mm provided by Baoji Hai peng Metal Materials Co. The component content of the material is shown in Table 1. The critical phase transition temperature of the specimen was Af = 10 °C ± 5 °C, and the experimental temperature was 20 °C ± 3 °C, so the specimen was in an austenite state.

Table 1. Component content of SMA materials.

Ti	Ni	Co	Cu	Cr	Fe	Nb	C	H	O	N
Remaining	55.90	0.005	0.005	0.005	0.012	0.005	0.032	0.001	0.039	0.001

The SMA bars were designed and processed into the specimen shown in Figure 1 in accordance with GBT3075-2008 "Metallic Materials-Fatigue testing-Axial force-controlled method", and the specimens were heat treated at 350 °C (30 min), 400 °C (15 min) and 400 °C (30 min).

(a) (b)

Figure 1. Specimen for low circumference fatigue test. (**a**) Specimen size. (**b**) Specimen.

2.2. Experimental Equipment

The test equipment was a model 370.25 MTS Landmark electro-hydraulic servo mechanics test system with a maximum rated load of 250 KN. The equipment is shown in Figure 2.

Figure 2. Test equipment.

2.3. Low-Cycle Fatigue Test Scheme

The low-cycle fatigue test was carried out in accordance with GB/T 15248-2008 "The test method for axial loading constant-amplitude low-cycle fatigue of metallic materials". In order to study the effect of different strain amplitudes and heat treatment conditions on the fatigue performance of shape-memory alloys, the experimental parameters are listed in Table 2, and the test scheme was formulated as follows:

Table 2. Low-period fatigue experimental parameters.

Specimen Number	Diameter (mm)	Heat Treatment Temperature (°C)	Heat Treatment Time (min)	Strain Amplitude (%)	Load Rate (Hz)
1	14	350	30	±2.5	0.01
2	14	350	30	±3.5	0.01
3	14	350	30	±3.75	0.01
4	14	400	15	±2.5	0.01
5	14	400	30	±2.5	0.01

I: Three shape-memory alloy specimens (specimen 1–3) were taken and cyclically loaded at strain amplitudes of ±2.5%, ±3.5% and ±3.75% 30 times after being held at 350 °C for 30 min to analyze the low-cycle fatigue performance of the shape-memory alloy at different strain amplitudes.

II: The bars (specimen 4–5) held at 400 °C for 15 min and 400 °C for 30 min were cyclically loaded for 30 cycles with the strain amplitude of ±2.5%, and the test results were compared with the test on bars held at 350 °C for 30 min to analyze the effect of different heat treatment conditions on the fatigue performance of the shape-memory alloy.

To reduce the measurement error caused by the loading rate, the loading frequency of the test was 0.01 Hz, and the waveform was a sine wave (as shown in Figure 3). According to the comparison of the specimen before and after loading (as shown in Figure 4), there was no obvious buckling of the specimen.

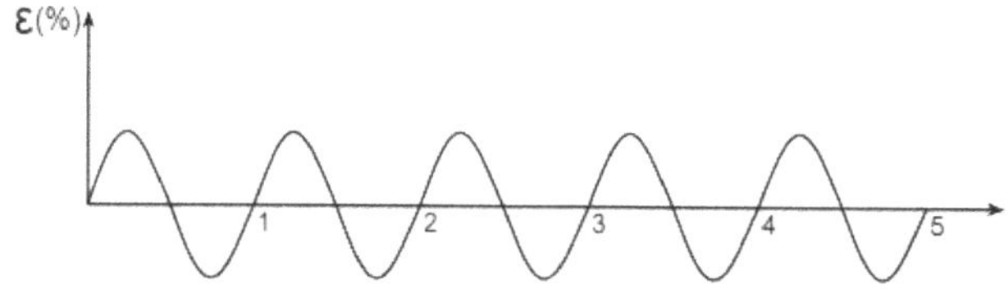

Figure 3. Loading waveform.

Figure 4. Specimen for low-circumference fatigue test. (**a**) Before loop loading. (**b**) After 30 cycles.

3. Static Tensile Test

In this study, the peak tensile stress and the stress–strain relationship of the specimens were required for the low-circumference fatigue test study of the SMA bars. Therefore, a static tensile test was required. The test equipment was a MTS Landmark® electro-hydraulic servo test system, and the test was conducted in accordance with the requirements of GB/T228.1—2010 "Metallic materials-Tensile testing Part 1: Method of test at room temperature".

The static tensile test results of the SMA bar are shown in Table 3, and the stress–strain curves are shown in Figure 3. From Table 3 and Figure 5, it can be seen that the peak tensile stress of the SMA bar is 682.68 MPa and the elongation is 14.52%, which is within the 12% error with the quality inspection report data provided by the manufacturer. It can be seen that the SMA bar has stable performance, so the test data have good reliability and can be used as the basis for determining the bar's fatigue test parameters.

Table 3. Monotonic tensile test data of SMA bar.

Items	Value
Yield strength (f_y)	432.48 MPa
Yield strain (ε_y)	3.75%
Elastic modulus (E_s)	29,454 MPa
Strain of the intensive start phase (ε_{sh})	4.24%
Elastic modulus of the intensive start phase (E_{sh})	3292 MPa
Peak stress (f_u)	682.68 MPa
Strain corresponding to the peak stress (ε_u)	14.52%

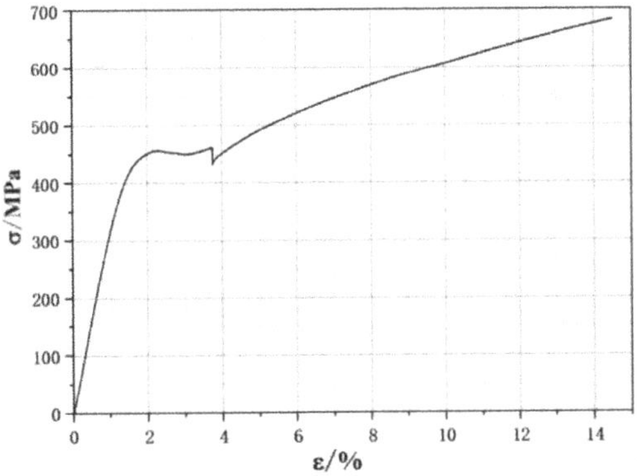

Figure 5. Monotonic tensile stress–strain curve of SMA bar.

4. Analysis of Low-Cycle Fatigue Test Results

4.1. Influence of Strain Amplitude on Superelasticity and Energy Consumption of SMA Bar

Three shape-memory alloy bars (specimens 1–3) with the diameter of 14 mm and heat treatment parameters of being held at 350 °C for 30 min were selected for a low-cycle fatigue test with strain amplitudes of 2.5%, 3.5% and 3.75%, The stress–strain curves of the three specimens after 10 cycles at different strain amplitudes are shown in Figure 6. From Figure 6a–c, it can be seen that with the increase in strain amplitude, the area of hysteresis curve becomes larger, and the shape becomes full, which means that the energy dissipation capacity of the specimen is enhanced.

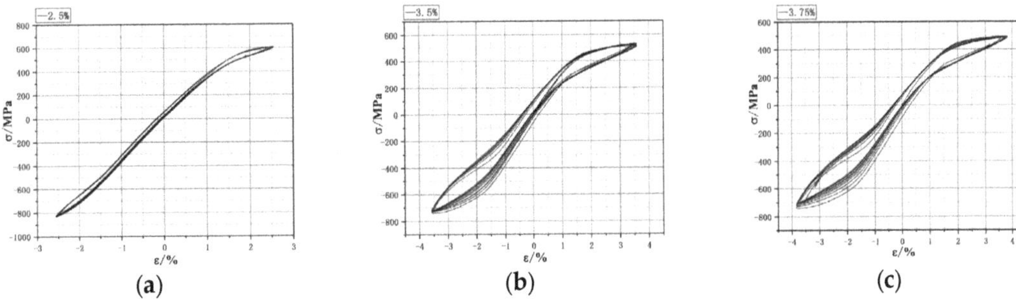

Figure 6. Effect of strain amplitude on stress–strain relationship: (**a**) Strain amplitude of ±2.5%; (**b**) strain amplitude of ±3.5%; (**c**) strain amplitude of ±3.75%.

To investigate the changes in the residual strain and energy dissipation level of SMA bars with the increase in cycle times at different strain amplitudes, the residual strain and energy dissipation capacity of specimens under different strain amplitudes were analyzed by taking cycle times as abscissa (as shown in Figure 7). Figure 7 shows that when the strain amplitude is 2.5%, the energy dissipation and residual strain are very small and almost keeps constant with the increase in cycle times. When the strain amplitude is 3.5%, the residual strain of the specimen increases from 0.29% to 0.43%, with an increase of 48.3%; the energy consumption of the specimen decreases from 107.42 KN·mm to 93.16 KN·mm, with a decrease of 13.3%. When the strain amplitude is 3.75%, the residual strain of the specimen increases from 0.22% to 0.33%, with an increase of 50%; the energy consumption of the specimen decreases from 124.27 KN·mm to 99.65 KN·mm, with a decrease of 19.8%.

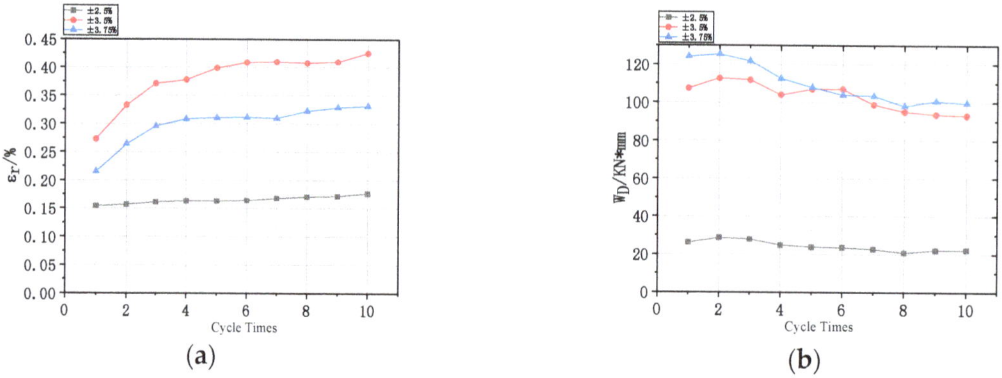

Figure 7. Effect of strain amplitude on residual strain and energy dissipation: (**a**) Effect of strain amplitude on residual strain; (**b**) effect of strain amplitude on energy consumption.

The residual strain of SMA bars at the strain amplitude of ±3.5% lies between it at a strain amplitude of ±2.5% and ±3.75%; then, the test results are in accordance with the basic law. However, the test results deviate near the sixth cycle when the strain amplitude is ±3.5%. The reasons for this are as follows: firstly, the specimens used in this test are slightly different in size from the low-circumference fatigue standard specimen, so there is a size effect. The second is that the processing quality of SMA material is slightly unstable, and the internal lattice microstructure of the material itself has an influence.

The test results show that the SMA bar with a diameter of 14 mm has small residual strain, good recovery performance and certain energy dissipation capacity under a low-cyclic reciprocating load.

4.2. Influence of Cyclic Loading Times on the Superelastic and Energy Dissipation Properties of SMA Bars

The SMA bar (specimen 3) which underwent a heat treatment condition of being held at 350 °C for 30 min was cycled for 5, 10, 20 and 30 turns with a ±3.75% strain amplitude, and its stress–strain hysteresis curves were plotted (Figure 8a–d).

It can be seen from Figure 8 that (1) when the strain amplitude is ±3.75%, the negative stress of the specimen has an obvious degradation process during cyclic loading, but the degradation of the positive stress is not obvious. (2) The area of the hysteresis curve gradually decreases with the increase in cycle times, which means that the energy consumption of the specimen gradually decreases with the increase in cycle loading times.

Figure 9 shows the variation trend of residual stress and the single-cycle energy consumption of specimen 3 with the increase in cycle times during loading: with the increase in cyclic loading times, the residual strain of the specimen gradually rises until it is stable, and the maximum residual strain in the cyclic process is about 0.4%. The energy consumption of a single cycle gradually decreases until it is stable, and the energy consumption of a single cycle is at least 75 KN·mm. According to the test results, the SMA bar with the diameter of 14 mm has a certain energy dissipation capacity and recovery performance under a reciprocating load.

Figure 10 shows the trend of the peak stress of specimen 3 with the increase in cycling period in the process of cyclic loading: when the strain amplitude is 3.75%, the maximum tensile stress of specimen remains stable, and the maximum compressive stress has a decreasing trend in the early stage of loading and gradually stabilizes after a certain cycle. It can be seen that during the loading process, the specimen first presents a certain degree of cyclic softening and then enters the cyclic stable state.

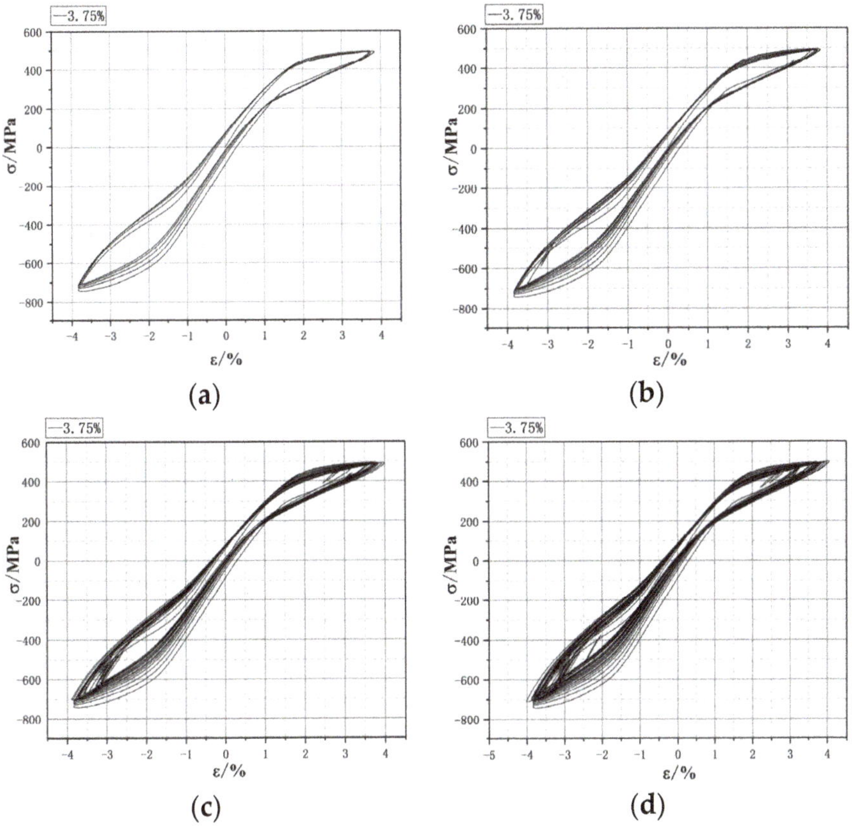

Figure 8. Influence of cyclic loading times on the stress–strain relationship: (**a**) 5-week cycle at ±3.75% of amplitude; (**b**) 10-week cycle at ±3.75% of amplitude; (**c**) 20-week cycle at ±3.75% of amplitude; (**d**) 30-week cycle at ±3.75% of amplitude.

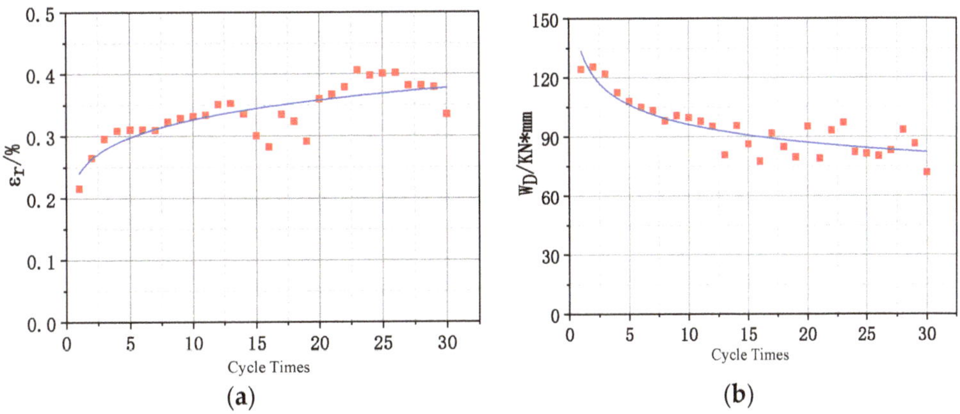

Figure 9. Trend of residual strain and energy dissipation of the specimen with increasing number of cycles; (**a**) effect of cycle number on residual strain; (**b**) effect of cycle number on energy consumption.

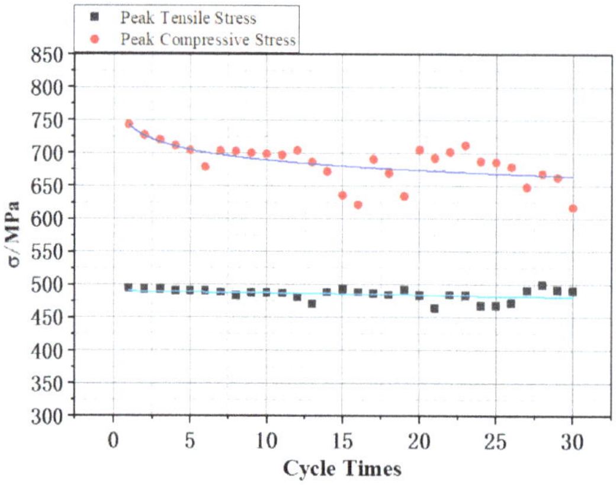

Figure 10. Comparison of peak stress variation at different numbers of cycles.

4.3. *Influence of Heat Treatment Parameters on Superelasticity and Energy Consumption of SMA Bar*

The SMA bars, after three different heat treatments of being held at 350 °C for 30 min, held at 400 °C for 15 min and held at 400 °C for 30 min, were subjected to a low-circumferential fatigue test with a strain amplitude of ±2.5%, and their stress–strain hysteresis curves are shown in Figure 11. From the test data, the residual strains of the three specimens with different heat treatments are 0.15%, 0.12% and 0.05%; the energy dissipation values are 26.03 kN·mm, 47.15 kN·mm and 41.77 kN·mm. In summary, the SMA bars with the heat treatment of being held at 400 °C for 15 min have the best performance.

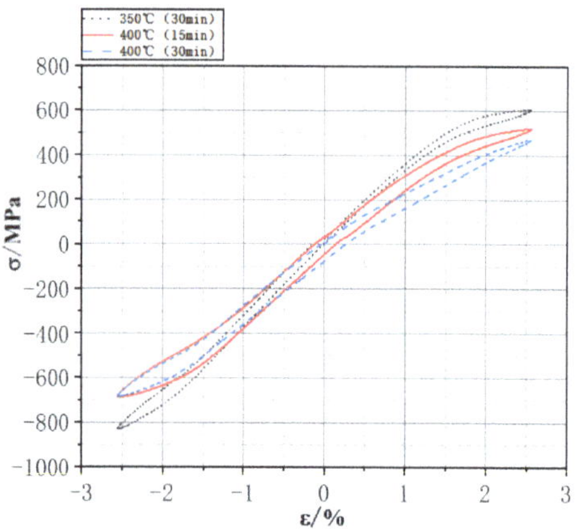

Figure 11. Effect of heat treatment parameters on stress–strain relationship.

5. FEA

According to the experimental data, the Auricchio multi-linear simplified model in ABAQUS [22,23] was used to simulate the low-cycle fatigue performance of SMA bar with the heat treatment parameters of being held at 350 °C for 30 min, and the comparison of the stress–strain curve and nephogram between the finite element and experiments are shown in Figures 12 and 13.

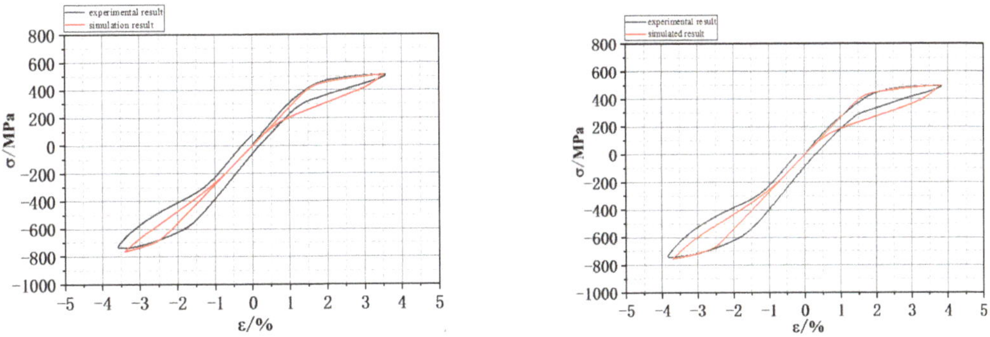

(**a**) (**b**)

Figure 12. Comparison of experimental results and simulation results of stress–strain curve: (**a**) Strain amplitude of ±3.5%; (**b**) strain amplitude of ±3.75%.

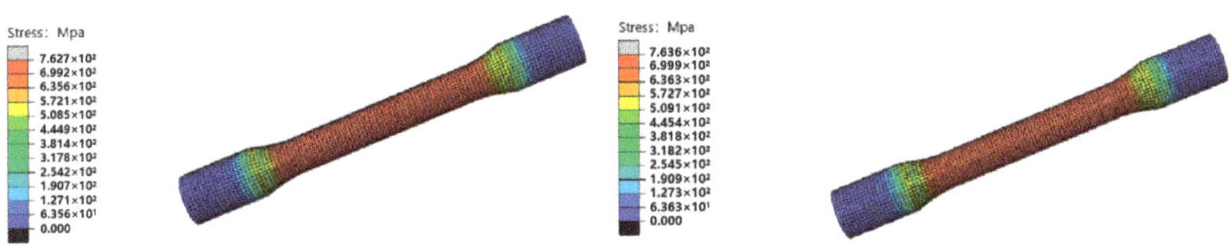

(**a**) (**b**)

Figure 13. SMA bar axial strain nephogram: (**a**) Stress amplitude of ±3.5%; (**b**) stress amplitude of ±3.75%.

The comparative analysis of the mechanical parameters in Table 4 shows that the error of the stress peak between the experimental results and the simulated analysis is within 5%, which means that the finite element simulation results are in better agreement with the experimental results. However, the energy consumption error is relatively large because the fatigue caused by the small deformation observed in the test is not considered in the numerical simulations. Moreover, a larger hysteresis loop area and increased energy consumption result from the residual strain in the specimen. Generally, the stress–strain hysteresis curves obtained from the finite element simulation are in good agreement with the stress–strain curves obtained from the tests, so it is feasible to use ABAQUS2021 to simulate the low-circumference fatigue of the shape-memory alloy bars.

Table 4. Comparative analysis of mechanical parameters.

Specimen Type	Type of Parameter	Single-Week Energy Consumption (kN·mm)	Tensile Stress Peak (MPa)	Compressive Stress Peak (MPa)
Strain Amplitude of ±3.5%	Experiment	107.42	507.1	738.1
	Simulation	56.29	505.2	752.7
	Error	47%	0.37%	1.98%
Strain Amplitude of ±3.75%	Experiment	124.27	491.1	743.2
	Simulation	76.3	500.7	752.4
	Error	38.60%	1.95%	1.23%

6. Conclusions

In this paper, the influences of strain amplitude, cycle times and heat treatment conditions on the fatigue properties of shape-memory alloy bars with the diameters of

14 mm were studied by using a low-cycle fatigue test method. The results show the energy dissipation capacity and small residual strain of the SMA bars under low cyclic fatigue loading, implying that SMA bars can be used as self-centering energy-dissipating elements in seismic design. The following conclusions are obtained:

(1). With the increase of strain amplitude, the energy dissipation performance of SMA bars shows an overall upward trend, and the residual strain is relatively small at the strain amplitude of 3.5% and 3.75%.

(2). In the test, with the increase of loading times, the peak compressive stress of SMA bars gradually tends to be stable after obvious degradation, the energy consumption of a single cycle decreases and the residual strain increases.

(3). The heat treatment parameters affect the energy dissipation and residual strain of shape-memory alloys. Larger energy dissipation and smaller residual strain of the SMA bar are obtained at 400 °C for 15 min.

Author Contributions: L.L.: Supervision, Conceptualization; Investigation, Formal analysis, Writing-original draft. X.Z.: Conceptualization, Formal analysis, Writing—Review & Editing. J.C.: Supervision, Writing—Review & Editing. All authors have read and agreed to the published version of the manuscript.

Funding: Science Foundation in the Henan province (Grant No.192102210030 and 182102210439) and the National Science Foundation in China (NSFC) (Grant No.U1404524). Tackle-Key-Program of S&T Committee of Henan Province in China (Grant No. 202102210209). Key-Research-Projects of Higher Education Institutions of Henan Province in China (Grant No. 23A590002). Key R&D Special Projects of Henan Province in China (Grant No. 221111230700).

Data Availability Statement: This paper is new. Neither the data nor any part of its content have been published or have been accepted elsewhere.

Conflicts of Interest: The authors declare no conflict of interest.

References

1. Graesser, E.J.; Cozzarelli, F.A. Shape-Memory Alloys as New Materials for Aseismic Isolation. *J. Eng. Mech.* **1991**, *117*, 2590–2608. [CrossRef]
2. Piedboeuf, M.C.; Gauvin, R.; Thomas, M. Damping behaviour of shape memory alloys: Strain amplitude, frequency and temperature effects. *J. Sound. Vib.* **1998**, *214*, 885–901. [CrossRef]
3. Dolce, M.; Cardone, D. Mechanical behaviour of shape memory alloys for seismic applications 2. Austenite NiTi wires subjected to tension. *Int. J. Mech. Sci.* **2001**, *43*, 2657–2677. [CrossRef]
4. Gong, J.; Tobushi, H. Superelastic deformation behavior of TiNi shape memory alloy subjected to various cyclic loading. *J. Funct. Mater.* **2002**, *33*, 391–393+397.
5. Gong, J.; Tobushi, H.K.; Takada, K.; Okumura, K. Analysis and simulation on cyclic superelastical deformation behavior of TiNi shape memory alloy subjected to loading and unloading. *J. Aeronaut. Mater.* **2002**, *22*, 6–12.
6. Sadiq, H.; Wong, M.B.; Al-Mahaidi, R.; Zhao, X.L. The effects of heat treatment on the recovery stresses of shape memory alloys. *SMArt Mater. Struct.* **2010**, *19*, 035021. [CrossRef]
7. Tamai, H.; Kitagawa, Y. Pseudoelastic behavior of shape memory alloy wire and its application to seismic resistance member for building. *Comp. Mater. Sci.* **2002**, *25*, 218–227. [CrossRef]
8. Zuo, X.; Li, A.; Ni, L.; Chen, Q. An Experimental Study On The Mechanical Behavior Of Superelastic Niti Shape Memory alloy Wires. *J. China Civ. Eng.* **2004**, *37*, 10–16.
9. Qian, H.; Li, J.; Li, H.; Chen, H. Mechanical behavior tests of NiTi wires with different diameters for structural vibration control. *J. Vib. Shock.* **2013**, *32*, 89–95.
10. Li, G.; Cui, D.; Hong, S. Experimental Investigation on Mechanical Properties of New Form Super-elastic Shape Memory AlloyWires. *J. Dalian Univ.* **2008**, *29*, 129–133.
11. Yan, S.; Wang, Q.; Wang, W. Experimental Research on Mechanical Performance for Pseudo-Elasticity of Shape Memory Alloy. *J. Shenyang Jianzhu Univ. (Nat. Sci.)* **2010**, *26*, 458–463.
12. Chen, W.W.; Wu, Q.P.; Kang, J.H.; Winfree, N.A. Compressive superelastic behavior of a NiTi shape memory alloy at strain rates of 0.001-750 s-1. *Int. J. Solids Struct.* **2001**, *38*, 8989–8998. [CrossRef]
13. Dolce, M.; Cardone, D. Mechanical behaviour of shape memory alloys for seismic applications 1. Martensite and austenite NiTi bars subjected to torsion. *Int. J. Mech. Sci.* **2001**, *43*, 2631–2656. [CrossRef]
14. Desroches, R.; Mccormick, J.; Liu, S.C. Properties of large-diameter shape memory alloys under cyclical loading. *Proc. Spie.* **2003**, *5057*, 187–196.

15. Wang, W.; Fang, C.; Liu, J. Large size superelastic SMA bars: Heat treatment strategy, mechanical property and seismic application. *Smart Mater. Struct.* **2016**, *25*, 075001. [CrossRef]
16. Zheng, B.-Y.; Shang, Z.-J.; Wang, Z.-M. Effects of Loading Rates on the Mechanical Behavior of Superelastic TiNi Shape Memory Alloy Bars. *Mech. Sci. Technol. Aerosp. Eng.* **2008**, *27*, 1236–1238+1242.
17. Ren, W.; Wang, L.; Jia, J.; Jia, R. Experimental study on mechanical behavior of superelastic shape memory alloy bar. *J. Funct. Mater.* **2013**, *44*, 258–261.
18. Dong, J. Experimental Research on the Self-centering Device Based on Large Diameter SMA. *Struct. Eng.* **2018**, *34*, 101–108.
19. Kang, L.; Qian, H.; Guo, Y.; Li, Z. Experimental investigation on mechanical properties of large size shape memory alloy bars underdifferent heat treatments. *J. Funct. Mater.* **2021**, *52*, 1185–1191.
20. Wang, W.; Shao, H. Experimental Investigation on Mechanical Properties of Shape Memory Alloy Bars in Different Sizes. *Struct. Eng.* **2014**, *30*, 168–174.
21. Zhuang, P.; Xue, S.; Wei, J.; Liu, Y. Research on Mechanical Performance of Superelastic NiTi Shape Memory Alloy Bars. *J. Archit. Civ. Eng.* **2015**, *32*, 96–103.
22. Tian, L.M.; Li, M.H.; Li, L.; Li, D.Y.; Bai, C. Novel joint for improving the collapse resistance of steel frame structures in column-loss scenarios. *Thin-Walled Struct.* **2023**, *182*, 110219. [CrossRef]
23. Tian, L.M.; Wei, J.P.; Huang, Q.X.; Woody, J. Collapse-Resistant Performance of Long-Span Single-Layer Spatial Grid Structures Subjected to Equivalent Sudden Joint Loads. *J. Struct. Eng.* **2021**, *147*, 04020309. [CrossRef]

buildings

MDPI

Article

Experimental and Theoretical Investigation of Viscoelastic Damper by Applying Fractional Derivative Method and Internal Variable Theory

Yeshou Xu [1,*], Qi He [1], Ying-Qing Guo [2,3], Xing-Huai Huang [1], Yao-Rong Dong [4], Zhong-Wei Hu [1] and Jinkoo Kim [5]

1 China-Pakistan Belt and Road Joint Laboratory on Smart Disaster Prevention of Major Infrastructures, Southeast University, Nanjing 211189, China
2 Mechanical and Electronic Engineering School, Nanjing Forestry University, Nanjing 210037, China
3 Nanjing Dongrui Damping Control Technology Co., Ltd., Nanjing 210033, China
4 School of Civil Engineering, Xi'an University of Architecture and Technology, Xi'an 710055, China
5 Department of Civil and Architectural Engineering, Sungkyunkwan University, Suwon 030603, Republic of Korea
* Correspondence: xuyeshou@163.com

Abstract: Viscoelastic dampers are conventional passive vibration control devices with excellent energy dissipation performance. The fractional derivative has a simple form and high accuracy in the modelling of viscoelastic materials/dampers. The internal variables reflect the internal state evolution of materials, and are often used to analyze the deformation and thermal process of materials. In the present work, the mechanical properties of a plate-shear-type viscoelastic damper at room temperature are tested under sinusoidal displacement excitations. The impacts of frequency and displacement amplitude on the dynamic properties of the viscoelastic damper in a wide frequency domain (0.1–25 Hz) are investigated. The higher-order fractional derivative model and the temperature–frequency equivalent principle are employed to characterize the frequency and temperature influence, and the internal variable theory considering the internal/microscale structure evolutions is introduced to capture the displacement affection. The higher-order fractional derivative model modified with the internal variable theory and temperature–frequency equivalent principle (ITHF) is accurate enough in describing the dynamic behaviors of viscoelastic dampers with varying frequencies and displacement amplitudes.

Keywords: viscoelastic damper; mechanical properties test; higher-order fractional derivative model; temperature–frequency equivalent principle; internal variable theory

1. Introduction

Viscoelastic dampers are perfect energy dissipation devices and have been widely used in seismic/wind vibration control, micro vibration suppression, and platform vibration isolation, etc. [1–4]. The damping performance and energy dissipation capacity of viscoelastic dampers mainly depend on the mechanical properties of viscoelastic materials [5]. Scholars have accomplished a lot of achievements in the research of viscoelastic materials and dampers [6–9].

In 1992, Zhang et al. [10] proposed a sequential design procedure for optimal placement of viscoelastic dampers in structures, and experimentally verified it with a five-story steel model structure. Aprile et al. [11] studied the influence of loading frequency and displacement amplitude on the dynamic modulus of viscoelastic dampers. Tsai et al. [12] investigated the temperature impacts on the seismic control effect of viscoelastically dampered building structures. Viscoelastic dampers can effectively reduce seismic responses of structures, and the damping effect decreases with the increment in temperature. Cazenove et al. [13] conducted numerical and theoretical observations on the self-heating effect of viscoelastic dampers. The results show that the self-heating phenomenon of viscoelastic dampers is very important and should not be neglected in vibration control analysis and structural design. Xu et al. studied the damping properties of viscoelastic dampers with different viscoelastic materials [5,14], and utilized the shaking table test to study the seismic response of structure models retrofitted with viscoelastic dampers [15–18]. Sato et al. [19] proposed an evaluation method for the practical application of viscoelastic dampers in wind vibration control by using equivalent sinusoidal waveforms of long-duration random excitations in along- and across-wind directions. Xu et al. [20] proposed a micro-vibration isolation and mitigation platform with four viscoelastic damper components to reduce disturbance generated by flywheels onboard spacecraft. He et al. [21] designed a new type of viscoelastic damper to control the translational vibration and the rotational vibration of offshore platforms simultaneously, and shaking table tests were conducted to verify the capability of dampers in mitigating the multi-dimensional seismic responses of platforms.

The dynamic characteristics and energy dissipation performance of viscoelastic dampers show a significantly nonlinear variation trend with varying temperatures, frequencies, and loading displacement amplitudes. The appropriate mathematical model is conducive to more efficient characterization of mechanical properties of dampers and response analysis of viscoelastic damper-retrofitted structures. Traditional constitutive models, the Kelvin model, Maxwell model, Zerner model, and fractional derivative models, etc. [22], of viscoelastic materials are mature in characterizing frequency dependence and rheological properties, but the temperature and frequency reliance of materials cannot be reflected well. Tsai [23] established a finite element model for viscoelastic dampers by using a fractional derivative operator and empirical formula to consider the dependence of temperature, frequency, and displacement amplitude at the same time. However, it is rarely utilized by researchers due to its complex form. Payne et al. [24] assumed that the displacement dependence of the material was mainly due to the influence of the micro filler network system, and used the Kraus model to explain the correlation characteristics between dynamic properties of the damper and displacement amplitudes. Liang et al. [25] built a constitutive model of viscoelastic materials based on a complex combination of multiple relaxation viscoelastic and viscoplastic models, which can predict the remarkable temperature- and rate-dependent deformation behaviors of materials well. Bagley et al. [26] and Lewandowski [27] investigated the mechanical behavior of viscoelastic materials and conducted a structural seismic analysis of damping structures based on fractional derivative theory, which has a more concise form and higher accuracy than integral derivative models. Xu et al. [28,29] formulated the temperature–frequency equivalent principle based on the W-L-F equation, and theoretically translated the effects of temperature into frequency impacts. Conti et al. [30] proposed a new mathematical model to describe the aging effects of viscoelastic materials. The convolution kernel of the integro-differential equation for describing the viscoelasticity was redefined as a function of time. Wang et al. [31] theoretically studied the macro properties of viscoelastic materials with the assumption that the multi-layer molecular networks interpenetrated together with different chain lengths. Lu et al. [32] investigated the constitutive relation of viscoelastic materials considering the friction effect of single molecular chains from surrounding environments at the microscale.

Most of the aforementioned research considers the constitutive relationship and mechanical properties of viscoelastic materials from the macroperspective, and there are few examples of literature from the microperspective. In traditional thermodynamic theories, the state of materials at an arbitrary moment determined by the macro variables, deformation, temperature, etc., and the internal variables, which could be one or a set of state variables, such as the damage accumulation, nonelastic deformation, phase transformation, free volume variation, and changes in material particle size, etc. [33,34]. The internal variables reflect the evolution process of the internal structure state of materials, and affect the macro mechanical properties and damping performance of viscoelastic materials significantly. The purpose of the present research is to study the dynamic performance of viscoelastic dampers with different frequencies and displacements at room temperature, and theoretically propose new mathematical models with fractional derivative and internal variable theory. The mechanical properties of a plate-shear type-viscoelastic damper at room temperature are tested under sinusoidal displacement excitations. The impacts of frequency and displacement amplitude on the dynamic properties of a viscoelastic damper in a wide frequency domain (0.1–25 Hz) are investigated. The higher-order fractional derivative model and the temperature–frequency equivalent principle are employed to predict the dynamic property variation of the damper with frequency and temperature changes, and the internal variable theory considering the internal/microscale structure evolutions is introduced to capture the displacement affection. The higher-order fractional derivative model modified with the internal variable theory and temperature–frequency equivalent principle (ITHF) is accurate enough and more appropriate than the higher-order fractional derivate model modified with the Kraus model and temperature–frequency equivalent principle (KTHF) in describing the dynamic performance of viscoelastic dampers with varying temperatures, frequencies, and displacement amplitudes.

2. Performance Test

To study the dynamic properties and vibration reduction efficiency of the viscoelastic damper with a wide frequency range (0.1~25 Hz), the hydraulic servo fatigue tests were conducted under external exactions with different loading frequencies and displacement amplitudes. The results show that the viscoelastic damper has great efficiency in energy dissipation. The properties parameters storage modulus, loss modulus, loss factor, energy dissipation, equivalent stiffness, and equivalent damping are importantly affected by the excitation frequency, and the impact of displacement amplitude on properties parameters of the viscoelastic damper is relatively slight.

2.1. Test Situation

The viscoelastic damper is fabricated based on the nitrile rubber matrix, and composed of three steel plates and two viscoelastic layers in parallel, as shown in Figure 1b. The dimensions of the viscoelastic layers are 60 mm × 50 mm × 10 mm. During the performance tests, the viscoelastic layers undergo simple shear deformation along the length direction of the damper and the external mechanical energy can be dissipated through the shear deformation. The dynamic performance tests of the damper are conducted with a 100 kN hydraulic servo fatigue test machine manufactured by w+b Company, Switzerland, as seen in Figure 1a. The viscoelastic damper is fixed tightly on the machine by steel joints and bolts (see Figure 1b). A temperature controlling box is utilized to adjust the environmental temperature to 18 °C, and the viscoelastic damper is put inside an incubator to make the temperature stable. The external displacement excitation $u_d = u_0 \sin(\omega t)$ is applied to the viscoelastic damper, where u_d is the displacement of the damper and u_0 denotes the displacement amplitude. ω is angular frequency and $\omega = 2\pi f$, where f denotes the loading frequency and t is time. All the test conditions of the viscoelastic damper are given in Table 1, and the force and displacement at each test condition are recorded by the data record system.

Table 1. Conditions of the dynamic performance tests.

Frequencies f (Hz)	Amplitudes d (mm)	Temperature T (°C)	Cycle Number
0.1, 0.2, 0.3, 0.5, 1.0, 1.5, 2.0, 3.0, 4.0, 5.0, 7.0, 10.0, 15.0, 20.0, 25.0	0.5, 1.0, 1.5, 2.5	18	5

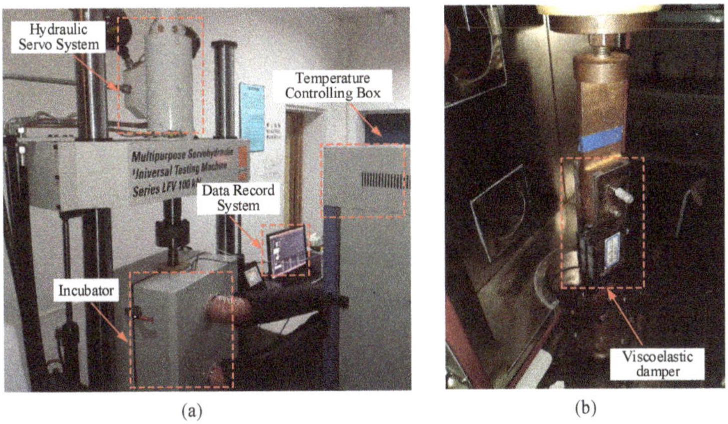

Figure 1. Dynamic properties tests of the viscoelastic damper. (**a**) Schematic diagram of the dynamic test. (**b**) Viscoelastic damper inside the incubator.

2.2. Results Analysis

To better analyze the mechanical properties and dynamic damping characteristics of the viscoelastic damper, the force–displacement hysteresis curves of the damper with changing external excitation frequencies and loading displacement amplitudes are shown in Figures 2 and 3, respectively.

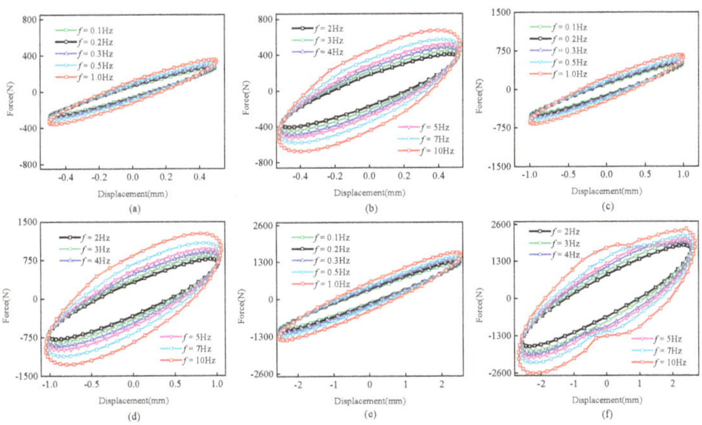

Figure 2. Hysteretic curves of the viscoelastic damper with changing frequencies. (**a**) d = 0.5 mm, f = 0.1~1 Hz. (**b**) d = 0.5 mm, f = 2~10 Hz. (**c**) d = 1 mm, f = 0.1~1 Hz. (**d**) d = 1 mm, f = 2~10 Hz. (**e**) d = 2.5 mm, f = 0.1~1 Hz. (**f**) d = 2.5 mm, f = 2~10 Hz.

The force–displacement curves of the viscoelastic damper with different loading frequencies are presented in Figure 2. It reveals that the viscoelastic damper has great energy dissipation properties, especially at high frequencies. The maximum damping force, major axis slope, smoothness, and plumpness of the hysteresis curve increase remarkably with increasing frequencies. Taking the maximum damping forces and areas at 0.1 Hz and 10 Hz as an example, the maximum damping forces and areas at frequency 10 Hz are almost twice of those at frequency 0.1 Hz.

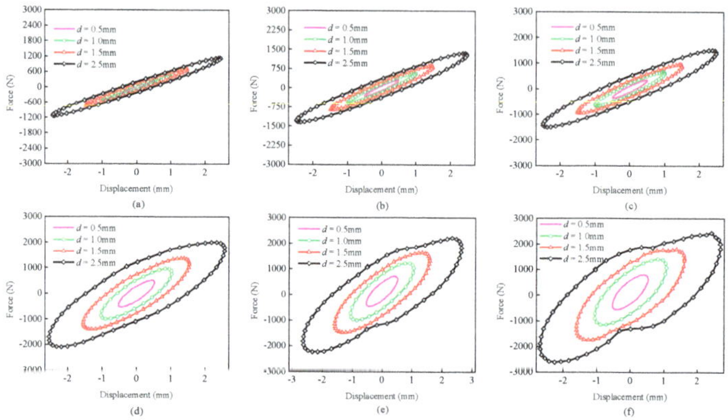

Figure 3. Hysteretic curves of the viscoelastic damper with changing displacement amplitudes. (**a**) f = 0.1 Hz. (**b**) f = 0.5 Hz. (**c**) f = 1 Hz. (**d**) f = 5 Hz. (**e**) f = 7 Hz. (**f**) f = 10 Hz.

The force–displacement curves of the viscoelastic damper with different displacement amplitudes are given in Figure 3. The damping performance of the viscoelastic damper are significantly influenced by the displacement amplitudes. The maximum damping force, smoothness, and plumpness of the force–displacement curves are increased rapidly with increasing displacement amplitudes. However, the impacts of the displacement amplitudes on the major axis slopes of the force–displacement curves are not obvious, as the equivalent stiffness changes slightly when the displacement amplitude varies from 0.5 mm to 2.5 mm.

To further analyze the mechanical properties and dynamic damping performance of the viscoelastic damper, the characteristic parameters of the viscoelastic damper storage modulus, loss modulus, loss factor, energy dissipation, equivalent stiffness, and equivalent damping at each test condition are calculated based on the force–displacement curves. According to the classical theories of viscoelastic dampers [35,36], the hysteresis curve of the viscoelastic damper can be considered as a standard ellipse as seen in Figure 4. The analytic equation of the ellipse has the form

$$\left(\frac{F_d - K_e u_d}{\eta K_e u_0}\right)^2 + \left(\frac{u_d}{u_0}\right)^2 = 1 \quad (1)$$

where u_d is the displacement and has been given as $u_d = u_0 \sin(\omega t)$ in the dynamic fatigue tests. F_d is the corresponding damping force of the viscoelastic damper at u_d. u_0 represents the displacement amplitude of the damper, and F_1 is the corresponding damping force at u_0. K_e means the equivalent stiffness, and $K_e = F_1/u_0$. F_2 denotes the damping force when $u_d = 0$ and $\eta = F_2/F_1$. F means the biggest damping force at each hysteresis curve.

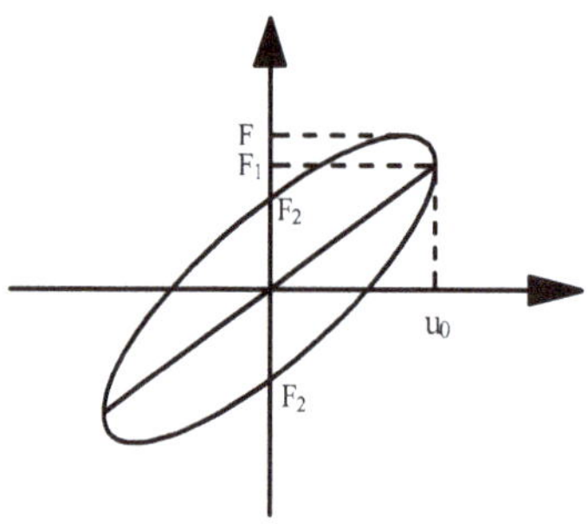

Figure 4. Force–displacement hysteresis curve.

The experimental results of the storage modulus, loss modulus, loss factor, energy dissipation, equivalent stiffness, and equivalent damping can be obtained with following equations:

$$G_1 = \frac{F_1 h_v}{n_v A_v u_0} \tag{2}$$

$$G_2 = \frac{F_2 h_v}{n_v A_v u_0} \tag{3}$$

$$\eta = \frac{F_2}{F_1} \tag{4}$$

$$E_d = \pi F_2 u_0 \tag{5}$$

$$K_e = \frac{F_1}{u_0} \tag{6}$$

$$C_e = \frac{F_2}{\omega u_0} \tag{7}$$

where G_1 denotes the storage modulus, h_v denotes the thickness of the viscoelastic layers, n_v represents the number of viscoelastic layers, and A_v means the shear area of the viscoelastic layers. For the viscoelastic damper utilized in present work, h_v = 10 mm, n_v = 2, and A_v = 60 mm × 50 mm = 3000 mm^2. G_2 means the loss modulus and η is the loss factor. E_d is the energy dissipation at each hysteresis curve, which is calculated as the area of the ellipse. K_e is the equivalent stiffness and C_e is the equivalent damping.

The characteristic parameters of the viscoelastic damper G_1, G_2, η, E_d, K_e, and C_e with changing frequencies and different amplitudes have been obtained in Tables 2 and 3, and vividly pictured in Figures 5 and 6.

Table 2. Properties parameters G_1, G_2, and η of the viscoelastic damper under different conditions.

Frequency f (Hz)	Displacement Amplitude d (mm)	Storage Modulus G_1 (MPa)	Loss Modulus G_2 (MPa)	Loss Factor η
0.1	0.5	0.8774	0.1891	0.2155
	1.0	0.8309	0.168	0.2022
	1.5	0.8123	0.1627	0.2002
	2.5	0.758	0.1472	0.1941
0.5	0.5	1.0082	0.3025	0.3
	1.0	0.9582	0.2836	0.296
	1.5	0.936	0.2711	0.2896
	2.5	0.8934	0.2501	0.28
1.0	0.5	1.1385	0.4071	0.3576
	1.0	1.1018	0.4627	0.42
	1.5	0.9952	0.3648	0.3666
	2.5	0.9638	0.3379	0.3506
5.0	0.5	1.4372	0.9166	0.6378
	1.0	1.3529	0.8668	0.6407
	1.5	1.2226	0.8076	0.6605
	2.5	1.0845	0.741	0.6833
10.0	0.5	1.6264	1.4151	0.8701
	1.0	1.4936	1.328	0.8891
	1.5	1.3255	1.2319	0.9294
	2.5	1.1564	0.9978	0.8628
20.0	0.5	1.7128	2.1143	1.2344
	1.0	1.4687	1.8828	1.2819
	1.5	1.2394	1.5164	1.2235
	2.5	1.1948	1.0873	0.91

Table 3. Properties parameters E_d, K_e, and C_e of the viscoelastic damper under different conditions.

Frequency f (Hz)	Displacement Amplitude d (mm)	Energy Dissipation E_d (N·m)	Equivalent Stiffness K_e (KN/m)	Equivalent Damping C_e (KN·s/m)
0.1	0.5	0.0859	526.4681	180.5752
	1.0	0.3053	498.5189	160.4234
	1.5	0.664	487.4013	155.3158
	2.5	1.6705	454.776	140.5169
0.5	0.5	0.1385	604.9157	57.7726
	1.0	0.5195	574.9053	54.1656
	1.5	1.1186	561.5877	51.7731
	2.5	2.861	536.03	47.7705
1.0	0.5	0.1906	683.0747	38.8732
	1.0	0.8881	661.0809	44.1857
	1.5	1.5401	597.1266	34.8371
	2.5	3.9516	578.3022	32.2703
5.0	0.5	0.4665	862.3496	17.5064
	1.0	1.7457	811.7594	16.5545
	1.5	3.6414	733.5588	15.4236
	2.5	9.2452	650.7179	14.1523
10.0	0.5	0.7364	975.8111	13.5127
	1.0	2.7693	896.144	12.681
	1.5	5.7994	795.3137	11.7641
	2.5	13.1152	693.8486	9.5278
20.0	0.5	1.3034	1027.6901	10.0951
	1.0	4.7438	881.2364	8.9897
	1.5	8.7057	743.6433	7.2403
	2.5	18.4049	716.8831	5.1913

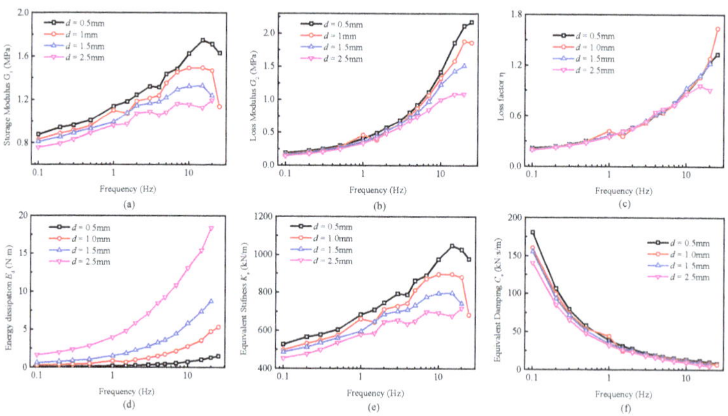

Figure 5. Dynamic properties parameters with varying frequencies. (**a**) Storage modulus. (**b**) Loss modulus. (**c**) Loss factor. (**d**) Energy dissipation. (**e**) Equivalent stiffness. (**f**) Equivalent damping.

The variations of the characteristic parameters G_1, G_2, η, E_d, K_e, and C_e with increasing frequencies are shown in Figure 5. It is shown that all the dynamic parameters increase rapidly when the loading frequency increases except the equivalent damping C_e, which is reduced rapidly with frequency increment. The changing rates of the properties parameters at low frequencies (0.1~1 Hz) are much larger than those at high frequencies (1~25 Hz). Taking the test conditions with d = 0.5 mm as an example, the characteristic parameters' changing rates with increasing frequencies are given in Table 4. These phenomena can be explained together with the micro structures of viscoelastic materials. The macro mechanical properties and energy dissipation performance are closely related to the micro molecular chain structures. When the frequency initially increases, the excitation time of external loading is gradually shortened, becoming closer to the molecular chain relaxation time. Therefore, the dynamic modulus and stiffness increase with increasing frequencies in 0.1~1 Hz. The excitation time of external loading is further shortened, becoming far less than the relaxation time of the molecular chains when the frequency further increases in the range of 1~25 Hz. The molecular chains cannot keep up with the movement of external excitations, and the increasing rates of dynamic modulus and stiffness are gradually decreased.

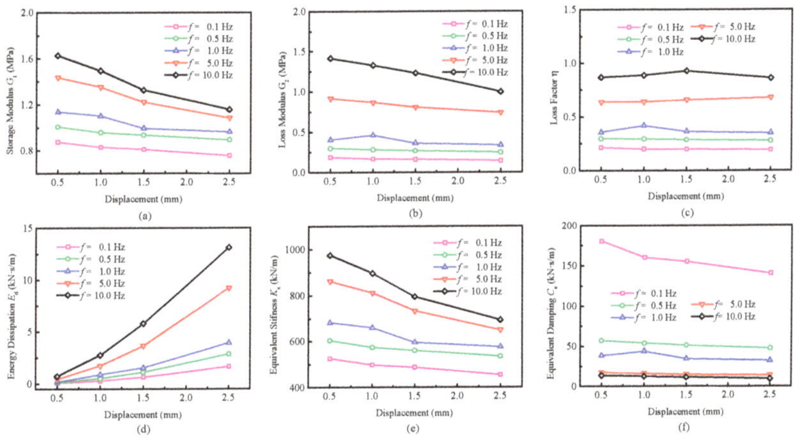

Figure 6. Dynamic properties parameters with varying displacement amplitudes. (**a**) Storage modulus. (**b**) Loss modulus. (**c**) Loss factor. (**d**) Energy dissipation. (**e**) Equivalent stiffness. (**f**) Equivalent damping.

Table 4. Parameters' changing rate with different frequencies when d = 0.5 mm.

Frequency f (Hz)	0.1~0.5	0.5~1	1~5	5~10	10~20
Storage modulus G_1	+14.91%	+12.92%	+26.24%	+13.16%	+5.31%
Loss modulus G_2	+59.97%	+34.58%	+125.15%	+54.39%	+49.41%
Loss factor η	+39.21%	+19.2%	+78.36%	+36.42%	+41.87%
Energy dissipation E_d	+61.23%	+37.62%	+144.75%	+57.86%	+77%
Equivalent stiffness K_e	+14.9%	+12.92%	+26.25%	+13.16%	+5.32%
Equivalent damping C_e	−68.01%	−32.71%	−54.97%	−22.81%	−25.29%

The equivalent damping C_e decreases with the increment of frequency, which can be explained by Equation (7). Furthermore, the characteristic parameters G_1 and K_e, and G_2 and E_d have the same change ratio with increasing frequencies, because there is a positive correlation between G_1 and K_e, and G_2 and E_d, as formulated in Equations (2)–(6). In conclusion, the dynamic performance of the viscoelastic damper is significantly affected by the excitation frequency.

Figure 6 shows the dynamic parameters' variation when displacement amplitude changes from 0.5 mm to 2.5 mm. The dynamic properties and damping parameters decrease slightly with displacement amplitude increment. This trend is more obvious at high frequencies (5 Hz and 10 Hz). The energy dissipation increases notably with the increment of excitation amplitudes. Taking the test conditions with f = 5 Hz as an example, the characteristic parameters' changing rates with varying displacement amplitudes are shown in Table 5.

Table 5. Parameters' changing rate with varying displacement amplitudes when f = 5.0 Hz.

Displacement d (mm)	0.5~1.0	1.0~1.5	1.5~2.5
Storage modulus G_1	−5.87%	−9.63%	−11.3%
Loss modulus G_2	−5.43%	−6.83%	−8.25%
Loss factor η	+0.45%	+3.09%	+3.45%
Energy dissipation E_d	+274.21%	+108.59%	+153.89%
Equivalent stiffness K_e	−5.87%	−9.63%	−11.29%
Equivalent damping C_e	−5.44%	−6.83%	−8.24%

The intermolecular force, van der Waals force, between filler particles, carbon black, silicon, etc., and the micro molecular chains adsorption on filler particle surfaces significantly enhance the modulus and strength of viscoelastic materials. With the increment of displacement amplitudes, the van der Waals force between filler particles and the molecular chain adsorption will be gradually weakened, which leads to the reduction in characteristic parameters G_1, G_2, η, K_e, and C_e. The energy dissipation E_d clearly increases when the displacement amplitude increases, which can be explained with the positive proportional relationship shown in Equation (5). The displacement amplitudes possess slight influence on mechanical behaviors but have great significance for the damping properties of viscoelastic dampers.

3. Theoretical Modeling

From the abovementioned experimental results, it can be seen that the dynamic loading frequency and displacement amplitude are two factors that significantly affect the mechanical and energy dissipation performance of viscoelastic dampers. Accurate and reasonable mathematical models are essential to describe the dynamic behaviors of viscoelastic dampers. In this part, the higher-order fractional derivative model is modified with the internal variable theory and temperature–frequency equivalent principle to characterize the performance of viscoelastic dampers with changing frequencies, displacement amplitudes, and temperatures.

3.1. Higher-Order Fractional Derivative Model

The higher-order fractional derivative model consists of a fractional Kelvin model and a fractional Maxwell model in parallel, as shown in Figure 7, which has three fractional derivative parameters and is more accurate in describing the static creep and dynamic damping characteristics of viscoelastic materials.

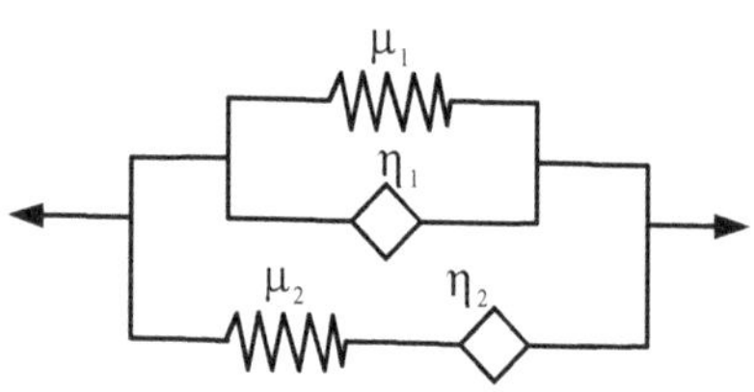

Figure 7. The higher-order fractional derivative model.

The stress–strain expressions of the higher-order fractional derivative model has the form

$$\frac{c_2}{\eta_1}\sigma(t) + \frac{1}{\eta_1}D^{\alpha}\sigma(t) = D^{\alpha+\gamma}\varepsilon(t) + c_1 D^{\alpha}\varepsilon(t) + \frac{\mu_2}{\eta_1}D^{\beta}\varepsilon(t) + c_2 D^{\gamma}\varepsilon(t) + c_1 c_2 \varepsilon(t) \quad (8)$$

where μ_1 and μ_2 are elastic modules of the spring elements. η_1 and η_2 are viscous parameters of the fractional dashpots. $c_1 = \mu_1/\eta_1$ and $c_2 = \mu_2/\eta_2$. α, β, and γ are the α-order, β-order, and γ-order fractional derivatives, respectively.

By applying the Fourier transform to Equation (8), the complex form modulus of the higher-order fractional derivative model can be obtained as

$$G^*(\omega) = \frac{\sigma^*(\omega)}{\varepsilon^*(\omega)} = \mu_1 + \eta_1(i\omega)^\gamma + \frac{\mu_2(i\omega)^\beta}{c_2 + (i\omega)^\alpha} \tag{9}$$

The dynamic modulus can be gained by decomposing Equation (9) into two parts; the real part is the storage modulus, the imaginary part is the loss modulus, and the loss factor is the ratio of the imaginary part to the real part:

$$G_1 = \mu_1 + \eta_1\omega^\gamma \cos(\gamma\pi/2) + \frac{\mu_2 c_2 \omega^\beta \cos(\beta\pi/2) + \mu_2\omega^{\alpha+\beta}\cos[(\alpha-\beta)\pi/2]}{c_2^2 + 2c_2\omega^\alpha \cos(\beta\pi/2) + \omega^{2\alpha}} \tag{10}$$

$$G_2 = \eta_1\omega^\gamma \sin(\gamma\pi/2) + \frac{\mu_2 c_2 \omega^\beta \sin(\beta\pi/2) + \mu_2\omega^{\alpha+\beta}\sin[(\beta-\alpha)\pi/2]}{c_2^2 + 2c_2\omega^\alpha \cos(\beta\pi/2) + \omega^{2\alpha}} \tag{11}$$

$$\eta = \frac{G_2}{G_1} \tag{12}$$

3.2. Internal Variable Theory

The dynamic modulus of carbon black-filled viscoelastic materials not only has temperature and frequency dependence, but also has displacement/strain amplitude reliance. Therefore, it is of crucial significance to describe the temperature, frequency, and strain amplitude dependence of the dynamic modulus of carbon black-filled viscoelastic materials with an appropriate model. Payne et al. [24] believed that the dynamic modulus of materials decreased with increments of strain amplitudes, which was due to the destruction of the filler network structures. The energy dissipation is mainly caused by the destruction and rebuilding of the filler network structures, and the Kraus model is introduced to describe the strain amplitude influence.

Both classical irreversible thermodynamics and rational thermodynamics use internal variables to describe the thermomechanical state of materials. These internal variables are microscale state variables such as damage accumulation, phase transformation, free volume variation, grain size change, inelastic strain, stress, etc., which are impossible to be observed at the macroscale. Their changes reflect the internal state evolution of materials, and have an important impact on the deformation and thermal process of materials. The deformation history and strain amplitude have a critical influence on the viscosity of viscoelastic materials. Lion et al. [37] introduced the intrinsic time with internal variables to characterize this correlation and describe the evolution process of materials:

$$u(t) = \psi(\omega, \varepsilon_0)t \tag{13}$$

where $u(t)$ is the intrinsic time, and ε_0 denotes the excitation strain amplitude. $\psi(\omega, \varepsilon_0)$ is the reduction factor, which has the form

$$\psi(\omega, \varepsilon_0) = 1 + \frac{2}{\pi}\chi\omega^\zeta\varepsilon_0 \tag{14}$$

where χ and ζ are material parameters, and $\omega = 2\pi f$.

Based on the Riemann–Liouville-type definition, the ith order fractional derivative of stress has the form

$$D^i[\sigma(t)] = \frac{\mathrm{d}^i}{\mathrm{d}t^i}\sigma(t) = \frac{1}{\Gamma(1-i)}\int_0^t \frac{\dot{\sigma}(s)}{(t-s)^i}\mathrm{d}s \tag{15}$$

According to Equation (13), by defining $z = \psi(\omega, \varepsilon_0)s$, the ith order fractional derivative of stress with intrinsic time has the form

$$D^i[\sigma(u)] = \frac{d^i}{dt^i}\sigma(u) = \frac{1}{\Gamma(1-i)}\int_0^u \frac{\dot{\sigma}(z)}{(u-z)^i}dz = \frac{1}{\psi^i(\omega,\varepsilon_0)\Gamma(1-i)}\int_0^t \frac{\dot{\sigma}(s)}{(t-s)^i}ds \tag{16}$$

Then, we obtain

$$D^i[\sigma(u)] = \frac{1}{\psi^i(\omega,\varepsilon_0)}D^i[\sigma(t)] \tag{17}$$

It also can be obtained that

$$D^i[\varepsilon(u)] = \frac{1}{\psi^i(\omega,\varepsilon_0)}D^i[\varepsilon(t)] \tag{18}$$

By replacing the physical time t in the constitutive equation of viscoelastic materials in Equation (8) with the intrinsic time $u(t)$, the viscoelastic constitutive equation based on the internal variable theory can be obtained as

$$\begin{aligned} &\frac{c_2}{\eta_1}\sigma(t) + \frac{1}{\eta_1\psi^\alpha}D^\alpha\sigma(t) \\ &= \frac{1}{\psi^{\alpha+\gamma}}D^{\alpha+\gamma}\varepsilon(t) + \frac{c_1}{\psi^\alpha}D^\alpha\varepsilon(t) + \frac{\mu_2}{\eta_1\psi^\beta}D^\beta\varepsilon(t) + \frac{c_2}{\psi^\gamma}D^\gamma\varepsilon(t) + c_1c_2\varepsilon(t) \end{aligned} \tag{19}$$

Then, Equations (10) and (11) can be rewritten as

$$\begin{aligned} G_1 &= \mu_1 + \eta_1(\omega/\psi)^\gamma\cos(\gamma\pi/2) \\ &+\frac{\mu_2c_2(\omega/\psi)^\beta\cos(\beta\pi/2) + \mu_2(\omega/\psi)^{\alpha+\beta}\cos[(\alpha-\beta)\pi/2]}{c_2^2 + 2c_2(\omega/\psi)^\alpha\cos(\alpha\pi/2) + (\omega/\psi)^{2\alpha}} \end{aligned} \tag{20}$$

$$\begin{aligned} G_2 &= \eta_1(\omega/\psi)^\gamma\sin(\gamma\pi/2) \\ &+\frac{\mu_2c_2(\omega/\psi)^\beta\sin(\beta\pi/2) + \mu_2(\omega/\psi)^{\alpha+\beta}\sin[(\beta-\alpha)\pi/2]}{c_2^2 + 2c_2(\omega/\psi)^\alpha\cos(\alpha\pi/2) + (\omega/\psi)^{2\alpha}} \end{aligned} \tag{21}$$

The performance of viscoelastic dampers is significantly affected by ambient temperature and excitation frequency, and temperature and frequency impacts are closely related. Especially for the temperature regions of T_g to T_g +100 °C, the storage modulus and loss modulus at high frequency are equal to those at low temperature, and the dynamic modules at low frequency are also equivalent to those at high temperature, which has been defined as the temperature–frequency equivalent principle [38]:

$$\begin{aligned} G_1(\omega,T) &= G_1(a_T\omega,T_0) \\ \eta(\omega,T) &= \eta(a_T\omega,T_0) \end{aligned} \tag{22}$$

where T_g is the glass transaction temperature, a_T is the shaft factor, and $a_T = 10^{-12(T-T_0)/[525+(T-T_0)]}$. T represents the environmental temperature and T_0 is the reference temperature. By applying Equation (22) to Equations (20) and (21), we have

$$\begin{aligned} G_1 &= \mu_1 + \eta_1(\alpha_T\omega/\psi)^\gamma\cos(\gamma\pi/2) \\ &+\frac{\mu_2c_2(\alpha_T\omega/\psi)^\beta\cos(\beta\pi/2) + \mu_2(\alpha_T\omega/\psi)^{\alpha+\beta}\cos[(\alpha-\beta)\pi/2]}{c_2^2 + 2c_2(\alpha_T\omega/\psi)^\alpha\cos(\alpha\pi/2) + (\alpha_T\omega/\psi)^{2\alpha}} \end{aligned} \tag{23}$$

$$G_2 = \eta_1(\alpha_T\omega/\psi)^\gamma \sin(\gamma\pi/2) + \frac{\mu_2 c_2(\alpha_T\omega/\psi)^\beta \sin(\beta\pi/2) + \mu_2(\alpha_T\omega/\psi)^{\alpha+\beta}\sin[(\beta-\alpha)\pi/2]}{c_2^2 + 2c_2(\alpha_T\omega/\psi)^\alpha \cos(\alpha\pi/2) + (\alpha_T\omega/\psi)^{2\alpha}} \tag{24}$$

Equations (22)–(24) are the expressions of the higher-order fractional derivative model modified with the internal variable theory and temperature–frequency equivalent principle (ITHF), which can reflect the impacts of surrounding temperature, excitation frequency, and loading displacement/strain amplitude on the dynamic behaviors of viscoelastic dampers well.

4. Experiments Verification

To verify the effectiveness and progressiveness of the ITHF model in describing the dynamic behaviors of viscoelastic dampers with changing displacement amplitudes and frequencies, the Kraus model [24] is introduced to modify the higher-order fractional derivative model and the comparisons of the experimental and numerical results of the viscoelastic damper are conducted. According to the Kraus model, the storage modulus $G_1(\varepsilon)$ is proportional to $\frac{1}{1+(\varepsilon/\varepsilon_c)^{2m}}$, and the loss modulus $G_2(\varepsilon)$ is proportional to $\frac{2(\varepsilon/\varepsilon_c)^m}{1+(\varepsilon/\varepsilon_c)^{2m}}$, as seen below.

$$\begin{cases} G_1(\varepsilon) = G_1(0) - \overline{G}_1 + \dfrac{\overline{G}_1}{1+(\varepsilon/\varepsilon_c)^{2m}}; \\ G_2(\varepsilon) = G_2^\infty + \dfrac{2(G_2^{\max} - G_2^\infty)(\varepsilon/\varepsilon_c)^m}{1+(\varepsilon/\varepsilon_c)^{2m}}. \end{cases} \tag{25}$$

where $G_1(0)$, $\overline{G}_1$, G_2^∞, $G_2^{\max}$, ε_c, and m are constants. By applying Equations (22) and (25) to Equations (10) and (11), the higher-order fractional derivative model modified with the Kraus model and temperature–frequency equivalent principle (KTHF) can be obtained as

$$G_1 = k_1\left\{\mu_1 + \eta_1(\alpha_T\omega)^\gamma\cos(\gamma\pi/2) + \frac{\mu_2 c_2(\alpha_T\omega)^\beta\cos(\beta\pi/2) + \mu_2(\alpha_T\omega)^{\alpha+\beta}\cos[(\alpha-\beta)\pi/2]}{c_2^2 + 2c_2(\alpha_T\omega)^\alpha\cos(\alpha\pi/2) + (\alpha_T\omega)^{2\alpha}}\right\}\frac{1}{1+(\varepsilon/\varepsilon_c)^{2m}} + b_1 \tag{26}$$

$$G_2 = k_2\left\{\eta_1(\alpha_T\omega)^\gamma\sin(\gamma\pi/2) + \frac{\mu_2 c_2(\alpha_T\omega)^\beta\sin(\beta\pi/2) + \mu_2(\alpha_T\omega)^{\alpha+\beta}\sin[(\beta-\alpha)\pi/2]}{c_2^2 + 2c_2(\alpha_T\omega)^\alpha\cos(\alpha\pi/2) + (\alpha_T\omega)^{2\alpha}}\right\}\frac{2(\varepsilon/\varepsilon_c)^m}{1+(\varepsilon/\varepsilon_c)^{2m}} + b_2 \tag{27}$$

where k_1, b_1, k_2, and b_2 are material parameters.

The least squares method is utilized for parameter identification of the models. By optimizing min $F(\omega, \varepsilon, T)$ in Equation (28) with test data of the storage modulus G_1 and the loss factor η, the parameters of the ITHF model and KTHF models can be achieved.

$$\begin{aligned} F(\omega, \varepsilon, T) &= p_1\{[G_1(\omega, \varepsilon, T) - G_{10}(\omega, \varepsilon, T)]/G_{10}(\omega, \varepsilon, T)\}^2 \\ &+ p_2\{[\eta(\omega, \varepsilon, T) - \eta_0(\omega, \varepsilon, T)]/\eta_0(\omega, \varepsilon, T)\}^2 \end{aligned} \tag{28}$$

where p_1 and p_2 are weighting parameters, $p_1 + p_2 = 1.0$, $G_1(\omega, \varepsilon, T)$ and $\eta(\omega, \varepsilon, T)$ are numerical results, and $G_{10}(\omega, \varepsilon, T)$ and $\eta_0(\omega, \varepsilon, T)$ are experimental results.

Some experimental data from the viscoelastic damper with randomly selected frequencies and displacement amplitudes are used for data fitting, and the parameters of the ITHF model are determined as $\mu_1 = 3.695 \times 10^5$, $\eta_1 = 2.0993 \times 10^{-9}$, $\mu_2 = 6.3021 \times 10^5$, $c_2 = 1.2507 \times 10^3$, $\chi = 0.6368$, $\zeta = 0.7114$, $\alpha = 0.0853$, $\beta = 0.2187$, $\gamma = 0.9654$, and $T_0 = 570.2816$ K. Similarly, the parameters of the KTHF model are also determined as $\mu_1 = 1.6086 \times 10^{10}$, $\eta_1 = 3.8768 \times 10^8$, $\mu_2 = 7.0468 \times 10^8$, $c_2 = 3.4705 \times 10^8$, $\alpha = 0.1766$, $\beta = 0.5678$, $\gamma = 0.2032$, $k_1 = 2.2937 \times 10^{-8}$, $b_1 = 8.0575 \times 10^5$, $k_2 = 1.1537 \times 10^{-8}$, $b_2 = 5.2489 \times 10^4$, $\varepsilon_c = 0.0125$,

$m = 0.3685$, and $T_0 = 634.3185$ K. The comparisons of the numerical results of the ITHF model, KTHF model, and experimental results with frequency 1.0 Hz and displacement 1.0 mm are presented in Figure 8a,b, respectively.

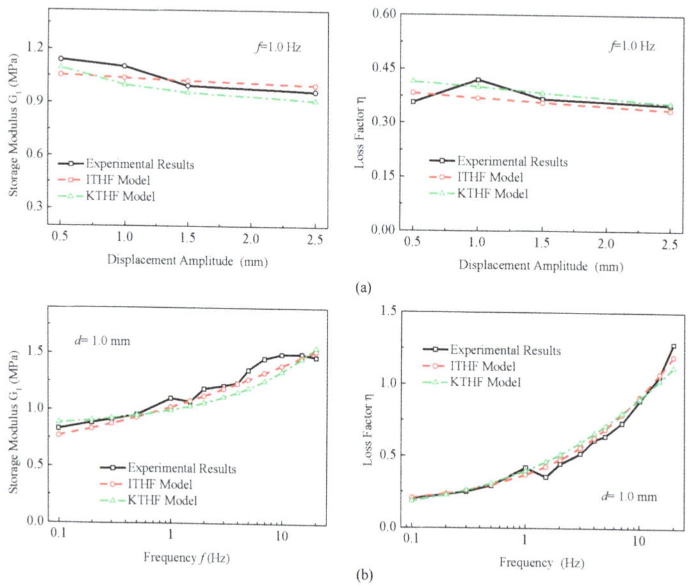

Figure 8. Comparisons of test data and numerical results for ITHF model and KTHF model. (**a**) f = 1.0 Hz. (**b**) d = 1.0 mm.

Figure 8a shows the experimental and numerical results of the storage modulus and loss factor with varying displacement amplitudes when f = 1.0 Hz. It reveals that the deviations of numerical results of both the ITHF model and the KTHF model from the experimental results are very small. The experimental and numerical results of the viscoelastic damper with changing displacement amplitudes are given in Table 6. For the storage modulus with frequency 1.0 Hz, the maximum error and root mean square error of the ITHF model in different displacement amplitudes are 7.44% and 5.28%, respectively, and the maximum error and root mean square error of the KTHF model in different displacement amplitudes are 8.59% and 6.03%, respectively. For the loss factor, the maximum error and root mean square error of the ITHF model with different displacement amplitudes are 12.04% and 7.38%, respectively, and the maximum error and root mean square error of the KTHF model in different displacement amplitudes are 16.25% and 8.78%, respectively. The ITHF model is more accurate in reflecting the dynamic behaviors of viscoelastic dampers with changing displacement amplitudes.

The experimental and numerical result comparisons with varying frequencies at displacement amplitude 1 mm are shown in Figure 8b, and the test and numerical data with the ITHF model and KTHF model are listed in Table 7. The maximum error and root mean square error of the ITHF model for the storage modulus in different frequencies are 8.53% and 4.98%, respectively, and the maximum error and root mean square error of the KTHF model for the storage modulus in different frequencies are 15.08% and 6.34%, respectively. The maximum error and root mean square error of the ITHF model for the loss factor in different frequencies are 20.98% and 8.13%, respectively, and the maximum error and root mean square error of the KTHF model for the loss factor in different frequencies are 30.34% and 11.79%, respectively.

Table 6. The comparisons of the ITHF model and KTHF model when f = 1.0 Hz.

Displacement Amplitude d (mm)	Characteristic Parameters	Experimental Results	Numerical Results			
			ITHF Model	Error	KTHF Model	Error
0.5	G_1 (MPa)	1.1385	1.0538	7.44%	1.0953	3.94%
	η	0.3576	0.3836	7.23%	0.4157	16.25%
1	G_1 (MPa)	1.1018	1.0379	5.8%	1	3.64%
	η	0.42	0.3694	12.04%	0.4015	4.41%
1.5	G_1 (MPa)	0.9952	1.0238	2.87%	0.96	6.54%
	η	0.3666	0.3573	2.53%	0.3841	4.78%
2.5	G_1 (MPa)	0.9638	1	3.76%	0.91	8.59%
	η	0.3506	0.3377	3.68%	0.3559	1.5%

Table 7. The comparisons of the ITHF model and KTHF model when d = 1.0 mm.

Frequency f (Hz)	Characteristic Parameters	Experimental Results	Numerical Results			
			ITHF Model	Error	KTHF Model	Error
0.1	G_1 (MPa)	0.8309	0.7729	6.98%	0.8895	15.08%
	η	0.2022	0.2093	3.51%	0.1902	5.96%
0.2	G_1 (MPa)	0.8874	0.8376	5.62%	0.9117	8.85%
	η	0.2342	0.2405	2.67%	0.233	0.51%
0.3	G_1 (MPa)	0.9158	0.8798	3.93%	0.9282	5.5%
	η	0.2533	0.2641	4.25%	0.2651	4.65%
0.5	G_1 (MPa)	0.9582	0.938	2.11%	0.9538	1.69%
	η	0.296	0.302	2.02%	0.3146	6.28%
1	G_1 (MPa)	1.1018	1.0263	6.85%	1	2.56%
	η	0.42	0.3741	10.93%	0.4015	4.41%
1.5	G_1 (MPa)	1.0734	1.0831	0.9%	1.0352	4.42%
	η	0.3565	0.4313	20.98%	0.4647	30.35%
2	G_1 (MPa)	1.1868	1.1256	5.15%	1.0648	5.4%
	η	0.4482	0.4803	7.15%	0.5157	15.06%
3	G_1 (MPa)	1.2141	1.1887	2.09%	1.1145	6.25%
	η	0.5197	0.5629	8.31%	0.597	14.86%
4	G_1 (MPa)	1.2393	1.2356	0.3%	1.1565	6.4%
	η	0.614	0.6322	2.97%	0.6614	7.72%
5	G_1 (MPa)	1.3529	1.273	5.91%	1.1935	6.24%
	η	0.6407	0.6925	8.08	0.7153	11.65%
7	G_1 (MPa)	1.4552	1.3311	8.53%	1.2582	5.47%
	η	0.7325	0.7944	8.45%	0.803	9.62%
10	G_1 (MPa)	1.4936	1.3945	6.64%	1.3404	3.87%
	η	0.8891	0.9171	3.14%	0.9038	1.66%
15	G_1 (MPa)	1.4949	1.4684	1.78%	1.4547	0.93%
	η	1.0592	1.0739	1.38%	1.0272	3.02%
20	G_1 (MPa)	1.4687	1.5217	3.61%	1.5521	2%
	η	1.2819	1.1954	6.75%	1.1194	12.67%

The ITHF model possesses enough precision in characterizing the dynamic and damping properties of the viscoelastic damper with changing displacement amplitudes and frequencies, and is more accurate than the KTHF model because it introduces the internal variable theory at the microscale to consider the displacement amplitude impacts. The

ITHF model has fewer parameters than the KTHF model and is more appropriate to be utilized in describing the dynamic properties of viscoelastic dampers.

5. Conclusions

Viscoelastic dampers are well-known passive vibration control devices and have been widely used in seismic/wind vibration control, micro vibration suppression, and platform vibration isolation, etc. In the present work, the dynamic properties of a viscoelastic damper at room temperature (18 °C) are tested under sinusoidal displacement excitations with a wide frequency band (0.1–25 Hz). The impacts of frequency and displacement amplitude on the dynamic properties of the viscoelastic damper are discussed. The higher-order fractional derivate model and the temperature–frequency equivalent principle are employed to characterize the frequency and temperature influence, and the internal variable theory considering the microscale structure influence is introduced to reflect the displacement amplitude affection. The ITHF model is proposed and verified with experimental results. Some notable conclusions can be obtained, such as:

(1) The viscoelastic damper has great energy dissipation properties at room temperature with frequencies 0.1 Hz~25 Hz, especially at high frequencies (1 Hz~25 Hz). The damping performance and stiffness of the damper are crucially affected by the excitation frequency, while the damping performance is greatly influenced by the displacement amplitude, and the stiffness is slightly affected.
(2) The characteristic parameters of viscoelastic dampers are significantly dependent on the external excitation frequencies and loading amplitudes. The parameters G_1, G_2, η, E_d, and K_e increase remarkably with increasing frequency, while C_e decreases. The energy dissipation E_d rises prominently with displacement amplitude, while and other parameters reduce mildly.
(3) The ITHF model possesses enough precision in characterizing the dynamic and damping properties of viscoelastic dampers with changing displacement amplitudes and frequencies, and has higher accuracy and fewer parameters than the KTHF model.
(4) The ITHF model introduces the internal variable theory to reflect the displacement amplitude impacts, which considers the internal structures' evolution process at the microscale and is of great significance for material design and damping property improvement.

Author Contributions: Conceptualization, Y.X.; data curation, X.-H.H. and Y.-Q.G.; formal analysis, Y.X., Q.H.; funding acquisition, Y.X. and Y.-Q.G.; investigation, Q.H. and X.-H.H.; methodology, Y.X., J.K.; resources, Y.X.; software, Y.-R.D. and Z.-W.H.; supervision, Y.-Q.G.; validation, Y.X., X.-H.H., Y.-R.D. and Q.H.; writing—original draft, Y.X.; writing—review and editing, Q.H. and Y.-Q.G. and J.K. All authors have read and agreed to the published version of the manuscript.

Funding: This study was financially supported by the National Natural Science Foundation of China with Grant No. 52108443, 52278505, the Jiangsu Province International Cooperation Project with Grant No. SBZ2022000169, National Key Research and Development Plans with Grant No. 2019YFE0121900, the Zhi Shan Scholarship of Southeast University with Grant No. 2242022R40041, the Innovation and Entrepreneurship Program (Innovation and entrepreneurship Doctor) of Jiangsu Province with Grant No. JSSCBS20210132, the Fundamental Research Funds for the Central Universities, the National Natural Science Foundation of China with Grant No. 52208503, and the Key Project of Collaborative Innovation Center of Shaanxi Provincial Department of Education with Grant No. 22JY029.

Informed Consent Statement: Not applicable.

Data Availability Statement: The data presented in this study are available on request from the corresponding author. The data are not publicly available due to the confidentiality requirements of the lab.

Acknowledgments: The authors are thankful for the help of Yonggang Huang's research group in Northwestern University (Evanston, IL 60201, U.S.).

Conflicts of Interest: The authors declare no conflict of interest. The funders had no role in the design of the study, in the collection, analyses, or interpretation of data, in the writing of the manuscript, or in the decision to publish the results.

Nomenclature

u_d	Displacement of the viscoelastic damper
u_0	Displacement amplitude of the damper
f	Loading frequency
ω	Angular frequency of the sinusoidal displacement loading.
d	Excitation displacement amplitude
F_d	Damping force of the viscoelastic damper at u_d
F_1	Damping force of the viscoelastic damper at u_0
K_e	Equivalent stiffness of the viscoelastic damper
F_2	Damping force when $u_d = 0$
F	Biggest damping force at each hysteresis curve
G_1	Storage modulus of the viscoelastic material/damper
G_2	Loss modulus of the viscoelastic material/damper
η	Loss factor of the viscoelastic material/damper
E_d	Energy dissipation of the viscoelastic damper at each hysteresis curve
K_e	Equivalent stiffness of the viscoelastic damper
C_e	Equivalent damping of the viscoelastic damper
h_v	Thickness of the viscoelastic layers of the viscoelastic damper
n_v	Number of viscoelastic layers of the viscoelastic damper
A_v	Shear area of viscoelastic layers of the viscoelastic damper
μ_1, μ_2	Elastic modules of the spring elements
η_1, η_2	Viscous parameters of the fractional dashpots
D^i	ith order fractional derivative, $i = \alpha, \beta, \gamma$, etc.
t	Time
$u(t)$	Intrinsic time
ε_0	Excitation strain amplitude.
$\psi(\omega, \varepsilon_0)$	Reduction factor
χ, ζ	Material parameters considering the internal variable theory
T_g	Glass transaction temperature
T	Temperature
T_0	Reference temperature
a_T	Shaft factor
$G_1(0), \overline{G}_1, G_2^{\infty}, G_2^{max}$	Storage modulus with initial strain, difference between the storage modulus of an arbitrary strain and that of an infinite strain, loss modulus of an infinite strain, and the maximum loss modulus
ε_c	Characteristic strain
m	Fractal dimension of filler structures
k_1, b_1, k_2, b_2	Linear regression parameters
$F(\omega, \varepsilon, T), p_1, p_2$	Objective optimization function and weighting parameters

References

1. Xu, Z.D. Earthquake mitigation study on viscoelastic dampers for reinforced concrete structures. *J. Vib. Control.* **2007**, *13*, 29–45. [CrossRef]
2. Mazza, F.; Vulcano, A. Control of the earthquake and wind dynamic response of steel-framed buildings by using additional braces and/or viscoelastic dampers. *Earthq. Eng. Struct. Dyn.* **2011**, *40*, 155–174. [CrossRef]
3. Lee, D.O.; Park, G.; Han, J.H. Hybrid isolation of micro vibrations induced by reaction wheels. *J. Sound Vib.* **2015**, *363*, 1–17. [CrossRef]
4. Xu, Z.D.; Huang, X.H.; Xu, F.H.; Yuan, J. Parameters optimization of vibration isolation and mitigation system for precision platforms using non-dominated sorting genetic algorithm. *Mech. Syst. Signal Process.* **2019**, *128*, 191–201. [CrossRef]
5. Xu, Z.D.; Xu, C.; Hu, J. Equivalent fractional Kelvin model and experimental study on viscoelastic damper. *J. Vib. Control.* **2015**, *21*, 2536–2552. [CrossRef]
6. Dai, J.; Xu, Z.D.; Gai, P.P.; Hu, Z.W. Optimal design of tuned mass damper inerter with a Maxwell element for mitigating the vortex-induced vibration in bridges. *Mech. Syst. Signal.* **2021**, *148*, 107180. [CrossRef]

7. Zhou, X.Q.; Yu, D.Y.; Shao, X.Y.; Zhang, S.Q.; Wang, S. Research and applications of viscoelastic vibration damping materials: A review. *Compos. Struct.* **2016**, *136*, 460–480. [CrossRef]
8. Wang, S.J.; Chiu, I.C.; Yu, C.H.; Zhang, Q.Y.; Chang, K.C. Experimental beyond design and residual performances of full-scale viscoelastic dampers and their empirical modeling. *Earthq. Engng. Struct. Dyn.* **2019**, *48*, 1093–1111. [CrossRef]
9. Ge, T.; Xu, Z.D.; Yuan, F.G. Development of viscoelastic damper based on NBR and organic small-molecule composites. *J. Mater. Civil Eng. ASCE* **2022**, *34*, 04022192. [CrossRef]
10. Zhang, R.H.; Soong, T.T. Seismic design of viscoelastic dampers for structural applications. *J. Struct. Eng.* **1992**, *118*, 1375–1392. [CrossRef]
11. Aprile, A.; Jose, A.I.; James, M.K. Evolutionary model of viscoelastic dampers for structural applications. *J. Eng. Mech. ASCE* **1997**, *123*, 551–560. [CrossRef]
12. Tsai, C.S. Temperature effect of viscoelastic dampers during earthquakes. *J. Struct. Eng. ASCE* **1994**, *120*, 394–409. [CrossRef]
13. Cazenove, J.; Rade, D.A.; Lima, A.M.G.D.; Araújo, C.A. A numerical and experimental investigation on self-heating effects in viscoelastic dampers. *Mech. Syst. Signal Process.* **2012**, *27*, 433–445. [CrossRef]
14. Xu, Z.D.; Liao, Y.X.; Ge, T.; Xu, C. Experimental and theoretical study on viscoelastic dampers with different matrix rubbers. *J. Eng. Mech.ASCE* **2016**, *142*, 04016051. [CrossRef]
15. Xu, Z.D.; Shen, Y.P.; Zhao, H.T. A synthetic optimization analysis method on structures with viscoelastic dampers. *Soil Dyn. Earthq. Eng.* **2003**, *23*, 683–689. [CrossRef]
16. Xu, Z.D.; Zhao, H.T.; Li, A.Q. Optimal analysis and experimental study on structures with viscoelastic dampers. *J. Sound Vib.* **2004**, *273*, 607–618. [CrossRef]
17. Xu, Z.D. Horizontal shaking table tests on structures using innovative earthquake mitigation devices. *J. Sound Vib.* **2009**, *325*, 34–48. [CrossRef]
18. Xu, Z.D.; Huang, X.H.; Lu, L.H. Experimental study on horizontal performance of multi-dimensional earthquake isolation and mitigation devices for long-span reticulated structures. *J. Vib. Control.* **2012**, *18*, 941–952. [CrossRef]
19. Sato, D.; Osabel, D.M.; Kasai, K. Evaluation method for practical application of viscoelastic damper using equivalent sinusoidal waveforms of long-duration random excitations in along- and across-wind directions. *Eng. Struct.* **2022**, *254*, 113735. [CrossRef]
20. Xu, C.; Xu, Z.D.; Huang, X.H.; Xu, Y.S.; Ge, T. Modeling and analysis of a viscoelastic micro-vibration isolation and mitigation platform for spacecraft. *J. Vib. Control.* **2018**, *24*, 4337–4352. [CrossRef]
21. He, X.Y.; Zhao, T.W.; Li, H.N.; Zhang, J. Multi-dimensional seismic response control of offshore platform structures with viscoelastic dampers (II-Experimental study). *Struct. Monit. Maint.* **2016**, *3*, 175–194. [CrossRef]
22. Lakes, R. *Viscoelastic Materials*; Cambridge University Press: New York, NY, USA, 2009. [CrossRef]
23. Tsai, C.S.; Lee, H.H. Applications of viscoelastic dampers to high-rise buildings. *J. Struct. Eng. ASCE* **1993**, *120*, 1222–1233. [CrossRef]
24. Payne, A.R. The dynamic properties of carbon black-loaded natural rubber vulcanizates. Part I. *J. Appl. Polym. Sci.* **1962**, *6*, 57–63. [CrossRef]
25. Liang, Z.; Li, J.; Zhang, X.; Kan, Q. A viscoelastic-viscoplastic constitutive model and its finite element implementation of amorphous polymers. *Polym. Test.* **2023**, *117*, 107831. [CrossRef]
26. Bagley, R.L.; Torvik, P.J. Fractional calculus in the transient analysis of viscoelastically damped structures. *AIAA J.* **1985**, *23*, 918–925. [CrossRef]
27. Lewandowski, R. Influence of temperature on the dynamic characteristics of structures with viscoelastic dampers. *J. Struct. Eng. ASCE* **2018**, *145*, 04018245. [CrossRef]
28. Xu, Z.D.; Wang, D.X.; Shi, C.F. Model, tests and application design for viscoelastic dampers. *J. Vib. Control.* **2011**, *17*, 1359–1370. [CrossRef]
29. Xu, Z.D.; Ge, T.; Liu, J. Experimental and theoretical study of high-energy dissipation-viscoelastic dampers based on acrylate-rubber matrix. *J. Eng. Mech. ASCE* **2020**, *146*, 04020057. [CrossRef]
30. Conti, M.; Danese, V.; Pata, V. Aging of viscoelastic materials: A mathematical model. In *Mathematical Modeling in Cultural Heritage*; Springer: Cham, Switzerland, 2021; pp. 135–146. [CrossRef]
31. Wang, Q.; Gossweiler, G.R.; Craig, S.L.; Zhao, X.H. Mechanics of mechanochemically responsive elastomers. *J. Mech. Phys. Solids* **2015**, *82*, 320–344. [CrossRef]
32. Lu, D.; Xue, B.; Cao, Y.; Chen, B. Constitutive theory for direct coupling of molecular frictions and the viscoelasticity of soft materials. *J. Appl. Mech. ASME* **2022**, *89*, 051007. [CrossRef]
33. Biot, M.A. Theory of stress-strain relations in anisotropic viscoelasticity and relaxation phenomena. *J. Appl. Phys.* **1954**, *25*, 1385–1391. [CrossRef]
34. Eringen, A.C.; Paslay, P.R. Mechanics of continua. *Crop Pasture Sci.* **1968**, *52*, 397–413. [CrossRef]
35. Chang, K.C.; Lai, M.L.; Soong, T.T.; Hao, D.S.; Yeh, Y.C. *Seismic Behavior and Design Guidelines for Steel Frame Structures with Added Viscoelastic Dampers, Technical Report NCEER-93-0009*; Department of Civil Engineering, State University of New York: Buffalo, NY, USA, 1993; Available online: https://hdl.handle.net/10477/748 (accessed on 10 October 2020).
36. Xu, Y.S.; Xu, Z.D.; Guo, Y.Q.; Ge, T.; Xu, C.; Huang, X.H. Theoretical and experimental study of viscoelastic damper based on fractional derivative approach and micromolecular structures. *J. Vib. Acoust. ASME* **2019**, *141*, 031010. [CrossRef]

37. Lion, A.; Kardelky, C. The Payne effect in finite viscoelasticity: Constitutive modelling based on fractional derivatives and intrinsic time scales. *Int. J. Plast.* **2004**, *20*, 1313–1345. [CrossRef]
38. Xu, Y.S.; Xu, Z.D.; Guo, Y.Q.; Jia, H.B.; Huang, X.H.; Wen, Y.W. Mathematical modeling and test verification of viscoelastic materials considering microstructures and ambient temperature influence. *Mech. Adv. Mater. Struct.* **2022**, *29*, 7063–7074. [CrossRef]

buildings

MDPI

Article

In Situ Concrete Bridge Strengthening Using Ductile Activated NSMR CFRP System

Jacob Wittrup Schmidt *, John Dalsgaard Sørensen and Christian Overgaard Christensen

Department of Built Environment, University of Aalborg, 9220 Aalborg, Denmark
* Correspondence: jws@build.aau.dk; Tel.: +45-9940-2904

Abstract: Novel and complex structural solutions are often challenging to introduce in the building industry since they may provide unconventional and less verified behavior combined with several novel and unique failure modes. A desired safety level may be difficult to verify due to the lack of knowledge related to failure modes and their variation. This study introduces a method that may provide a first step toward addressing such challenges. Five laboratory tests were done on a novel ductile response-controlled anchor system used for prestressed CFRP NSMR strengthening. These results were used as a basis for further implementation in a pilot project where an in situ cast concrete bridge was strengthened with the developed system. A particular focus was dedicated to the capability of the designed system response to provide a consistent load/deformation curve, yielding threshold, and yielding regime. The novel system enabled response tailoring to the specific pilot project. Installation and prestressing procedure worked as desired, where losses in the fully prestressed system, after three weeks, were approximately 0.2 to 1.6 kN (0.2–1.3 mm). In situ proof loading of all strengthening systems (20 anchor systems) was done to eliminate any possible failure modes below the proof-loading level. It was performed with no signs of distress.

Keywords: CFRP; strengthening; ductility; post-tensioning; in situ concrete structures

1. Introduction

Carbon fiber reinforced polymers (CFRP) used as near surface mounted reinforcement (NSMR) have been used as a mean to strengthen structures for several decades [1–10]. CFRP has excellent corrosion resistance, a weight that is 20% of steel, and tensile strength, which is approximately five times higher than for steel reinforcement. The material thus seems excellent for exterior structural reinforcing of existing concrete structures. Epoxy agents are typically used for the adhesive bond between the concrete structure to be strengthened and the CFRP; this may result in utilization from 0.6–0.9 of the ultimate CFRP rupture strain [11]. The reduced utilization is dedicated to premature failure modes which initiate in the concrete, such as intermediate crack (IC) de-bonding [12–17] and concrete cover separation [9,18,19]. To utilize the material further, pre-stressed CFRP plates, sheet, and wrap systems were used and researched during recent decades [20–25]. However, only a few systems are commercially available, and effective pre-stressed NSMR CFRP strengthening systems seems non-existent. Some reasons for the relatively limited use may be due to challenges related to the anchoring of the anisotropic and brittle material, as well as reluctance toward fiber reinforced polymer (FRP) creep rupture [26–28]. CFRP has been shown to be the most robust FRP material against creep rupture, whereas other materials such as Glass-FRP, Aramid-FRP, and Basalt-FRP seem more susceptible in this regard [29]. One of the keys to providing a cost-efficient system may be a well-defined and controlled pre-stressing of CFRP materials in combination with an efficient mounting procedure.

During the last decade, mechanical anchoring and pre-stressing of NSMR CFRP materials has been researched to address some of the presented challenges [30–33]. The mechanical anchorage is an essential and complex element of the CFRP strengthening system, since it has to account for the anisotropic and brittle properties of the CFRP material [34–39]. Some of the most well-known premature brittle failure modes of mechanical anchor systems are CFRP crushing, power and soft slip, cutting of fibers, and bending of fibers [40]. The highest tensile utilization of the CFRP is typically experienced at fiber failure and power slip. It is often seen that only small changes in the controlled stress transfer in the anchor can cause a change of failure mode. As also seen from conventional CFRP strengthening systems, novel pre-stressed CFRP strengthening solutions present an increased number of potentially new failure modes, and when strengthening concrete structures, these must be added to the known failure modes of such structures. This is a typical side effect when developing novel structural systems, where multiple new failure modes may initiate under laboratory and in situ conditions. The outcome is thus an increase of complexity in the capacity analysis, as well as in the safety level assessment.

In the study of [41], where concrete T-section beams were strengthened with an earlier version of the novel CFRP NSMR strengthening system applied in this paper, significant utilization of the CFRP material was achieved with tensile capacities up to 3300 MPa ($\varepsilon_{fu} \approx 0.02$). However, open discussions are still ongoing regarding an acceptable failure strain magnitude, failure modes, and related failure warning. Because of the CFRP material's reduced ductility compared to steel, which is even more amplified when inducing a pre-stressing strain, it seemed desirable to develop a CFRP strengthening system that can provide a high ductility threshold combined with complete response control. The experimental studies on T-section beams were supported through numerical studies [42].

To reduce the number of failure modes and to enable increased control of the anchor behavior, a ductile response-controlled enclosure wedge (EW) CFRP anchor system was developed [43]. Ductility was introduced by coupling the anchor with a component (the ductile mechanism) which has a tailored ductile behavior. Figure 1 shows the separate response curves of the CFRP anchor and the ductile mechanism, along with the curves of a series of combined tests (EW CFRP anchor and ductile mechanism). The ductile unit (mechanism) enables control of the CFRP anchor and thus the strengthening system response. In addition, it can be tailored to provide a desired response combined with a significant ductility fitting to the existing structure.

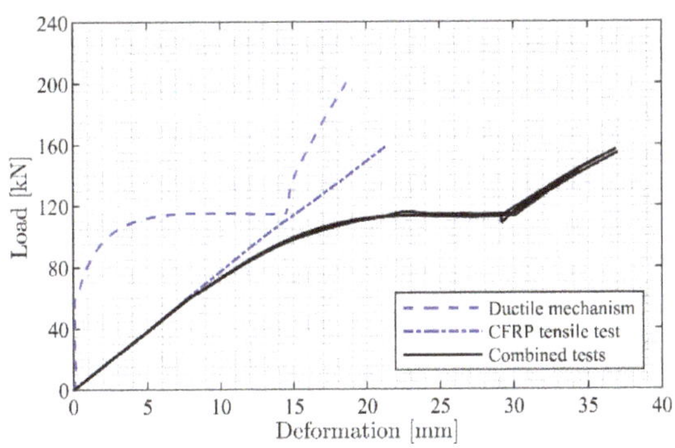

Figure 1. Increased ductility in the combined system response [43].

The effectiveness of the system for strengthening purposes was demonstrated through strengthening of concrete T-section beams [44,45]. It is seen from Figure 2 that the system seemed to work as desired, and that significant ductility was gained. "REF" represents the un-strengthened concrete beam with full ductile behavior. "PCF" represents a beam strengthened with the CFRP anchor system bonded with epoxy resin and not including the ductile mechanism; this beam shows increased capacity but reduced ductility. "PU" represents beams strengthened with the ductile response-controlled EW CFRP anchor system and bonded, as NSMR, with a flexible adhesive. It is seen that these beams exhibit ductile behavior near to the un-strengthened reference beam. In addition, the critical failure mode was also shown to be concrete crushing, not system failure, thus preventing additional novel failure modes from occurring.

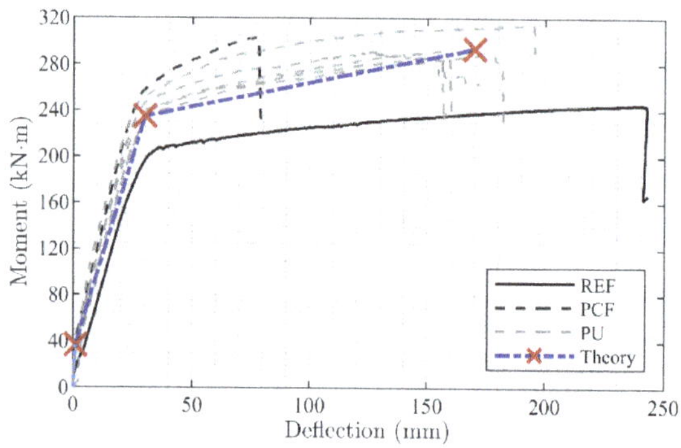

Figure 2. Strengthening effect of using the ductile CFRP strengthening system [44].

The basic reliability requirements for such a novel strengthening system are generally that the resistance/load-bearing capacity should not be smaller than the relevant load effect with a probability of failure corresponding to the reliability level required by the design codes and that the failure mode should be ductile thus providing a failure with warning. If the Design Value Format method is used as the basis for the partial factors, then the reliability requirement can be established by only considering the resistance [46]. If several failure modes may be triggered, then the reliability requirements should be satisfied for all of these. However, ensuring this requirement requires evaluation of all unique failure modes of a novel structural element or joint to be mounted in situ and may thus be very costly. In addition, it may often be uncertain if a failure mode is governing when applied to a real structure with additional failure modes unless the response is very controlled.

Scope and Research Questions

This paper provides research concerning a novel ductile response-controlled NSMR CFRP strengthening system. The system was tested as an assembled and complete anchor system under laboratory conditions and was finally applied for strengthening of an in situ bridge. The laboratory test procedure was combined with a probabilistic analysis to provide a safety assessment, which is rarely seen in such studies. In this regard, reliability relates to the load-bearing capacity and obtaining a ductile failure mode (with warning), which is the main objective. The study also includes a probabilistic model for use with the laboratory tests and in situ proof-loading of anchors to verify that a sufficient reliability level is present, concerning the load-bearing capacity. The requirement that a ductile failure mode is obtained is in this paper demonstrated by laboratory tests. A reliability-based assessment may thus be performed by system reliability modeling, which will be presented in more detail in a forthcoming paper. Consequently, the reliability assessment in this paper is based on the reliability principles in the Eurocodes, especially EN1990, Annex C and D [46], ISO 2394 [47], and the JCSS Probabilistic Model Code [48]. With a basis in the ductile behavior, it is hypothesized that:

- The developed ductile strengthening system is applicable for fast and efficient in situ installation.
- Laboratory testing of the assembled and complete CFRP anchorage system will provide high accuracy and thus a low safety factor due to a consistent response and yielding thresholds.
- Proof loading can be used to validate that all other ultimate failure modes from a theoretical safety assessment can be disregarded.

It is thus the aim to answer the following research questions:

- Does the novel activated CFRP system provide a controlled response with warning and consistent yielding thresholds?
- Can the obtained yielding threshold be used as a capacity threshold?
- Can laboratory testing ensure results that provide a basis for validating the required reliability level?
- How can an in situ proof-loading procedure support further validation of the system resistance and reliability?

2. Bridge Pilot Project

The specific configuration of the CFRP strengthening system presented in this paper was designed to strengthen a bridge in Jutland, Denmark. The bridge is depicted in Figure 3, where Figure 3a,b show a side view and a plane view, respectively. In addition, Figure 4 shows a section cut of the bridge.

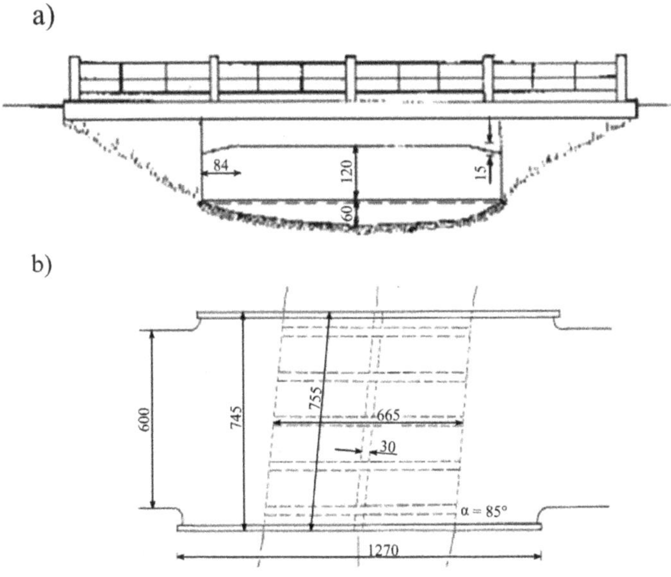

Figure 3. Bridge (**a**) side view and (**b**) plan view (all measures in cm).

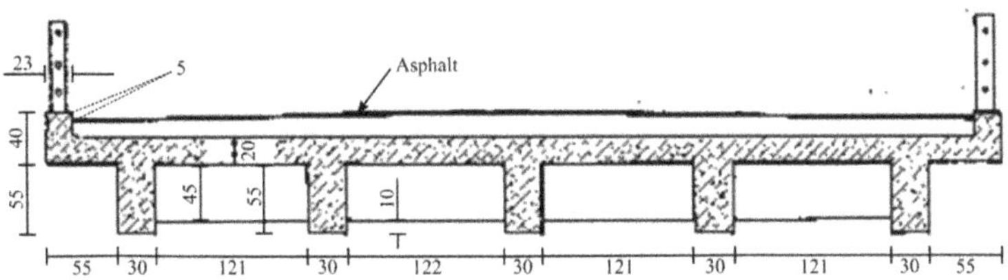

Figure 4. Section view drawing and reinforcement location and type (all measurements in cm).

The bridge was constructed as an in situ cast concrete bridge and consisted of one transverse beam and five longitudinal beams that were cast together with the bridge deck plate. The web height of the centrally placed transverse beam was 45 cm, whereas the five longitudinal beams had a height of 55 cm with a horizontal and vertical edge chamfer distance at the ends of approximately 80 cm and 15 cm, respectively. The centrally placed transverse beam height of 45 cm thus leaves 10 cm distance to the bottom face of the longitudinal beams.

The concrete was reported to have a characteristic compression strength of 22.5 MPa. The deformed reinforcement bars had a characteristic yielding strength of 225 MPa and were placed at the beam section bottom in two horizontal rows of five, with a diameter of 20 mm, see Figure 5. Stirrup reinforcement (ø10 mm) with identical yielding strength was placed at a distance of approximately 250 mm throughout the beam length. The bridge classification was Class 20 (20-tonne vehicle) for standard passages and Class 50 (50-tonne vehicle) for conditional passage (type 3), with axle configurations according to the Danish bridge classification system [49,50]. The strengthening aimed to upgrade the bridge capacity to withstand a standard passage Class 50 vehicle and a conditional passage type 3—Class 100 vehicle.

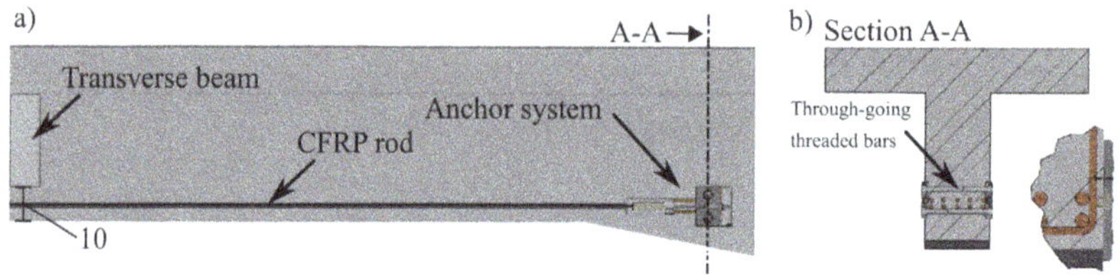

Figure 5. Side view (**a**) and section view (**b**) of the planned strengthening method (measures in cm).

Positioning of the Pre-Stressed Ductile CFRP NSMR System

When designing an anchor block, edge and mutual distances of the adhesively bonded concrete steel anchors are often the dimensioning factor. In addition, adhesively bonded threaded bar concrete anchors have to be mounted in holes drilled between the existing steel reinforcement.

For these reasons, it was impossible to place two strengthening systems on the bottom surface of the longitudinal beams of the case study bridge. However, the chamfered area at the ends of the longitudinal beams provided a potential basis for anchor placing into the concrete web side-cover. A drawing of the CFRP strengthening solution is seen in Figure 5a for this location. The anchor systems were placed on each side of the beams and clamped together by going through adhesively bonded threaded bars, Figure 5b.

In addition, this solution allowed the anchored CFRP ø8 mm rod to be glued into a groove in the concrete web-side cover. The free area underneath the transverse beam of 100 mm enabled the CFRP rod to pass the transverse beam and thus be placed in the entire length of the longitudinal concrete beams.

3. CFRP Strengthening System and Mounting Preparation

The components of the CFRP anchor system are depicted in Figure 6: (i) nut, (ii) steering washer, (iii) ductile mechanism, (iv) anchor block, (v) threaded bar, (vi) counteracting torque pin, (vii) coupler, (viii) aluminum wedge, (ix) threaded barrel. Table 1 presents the related material properties. The in situ assembling procedure of the CFRP anchors is described in Figure 7.

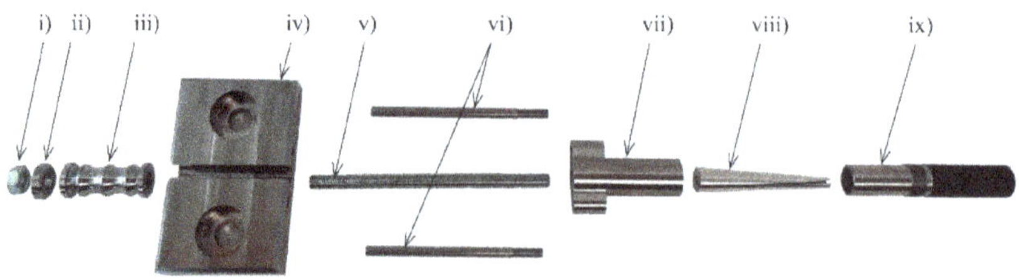

Figure 6. Anchor system parts.

Table 1. Material properties of the anchor system.

Parameter	Value [MPa]
Recommended ult. Stress of the CFRP rod	2200 [1]
E-modulus CFRP tendon, E_{cf}	160.000–170.000 [1]
Stainless steel yield/tension strength of the anchor block parts, f_y/f_u	235/700 [1]
Steel yield/tension strength of the barrel, f_y/f_u	235/340 [1]
Aluminum (6082-T6, EN 573-3) yield/tension strength	260/310 [1]
Steel yield/tension strength of the threaded activation bar, $f_{y,bar}/f_{u,bar}$	900/1000 [1]

[1] Data provided by the manufacturer.

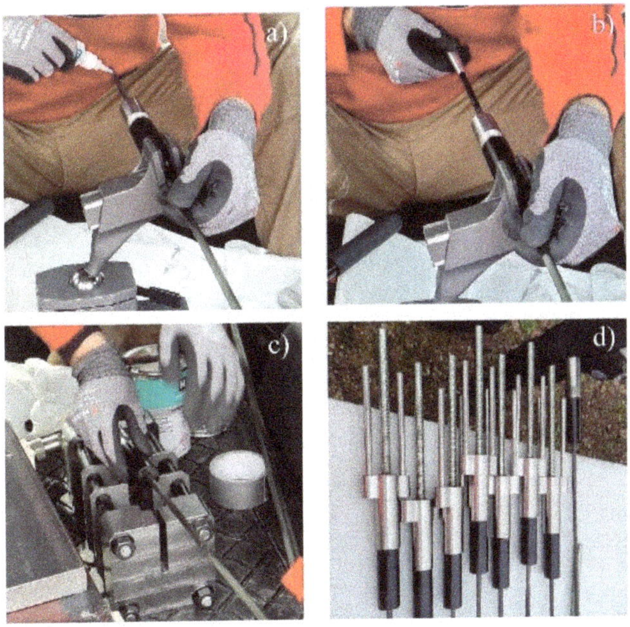

Figure 7. Assembling of the CFRP anchors. Photographs by the authors. (**a**) Grinding the CFRP surface and lubricating the surface with thin, liquid, and fast-curing glue. (**b**) Mounting of the wedge (viii) on the 8 mm CFRP rod in the lubricated area, which is then assembled with the barrel (ix). (**c**) Pre-setting of the wedge into the barrel using approximately 120 kN. (**d**) The coupler with inner threads (vii) was assembled with the threaded bar (v) as well as the end-threaded counteracting torque pins (vi) and screwed onto the threaded barrel (ix). Photographs by S&P.

The assembled parts in Figure 7d were mounted through the anchor block (iv), ductile mechanism (iii), and steering washer (ii). Pre-stressing was done by applying torque to the nut (i) mounted on the threaded bar with the steering washer (ii) applied, see Figure 8.

The counteracting torque pins (vi) prevent critical rotation in the barrel (ix), which could potentially cause premature failure of the CFRP rod. The pins are screwed into the coupler "ears" and can slide into the anchor block. Consequently, an undisturbed direct tension of the CFRP rod was deemed possible when pre-stressing was applied through the nut at the back of the anchor system.

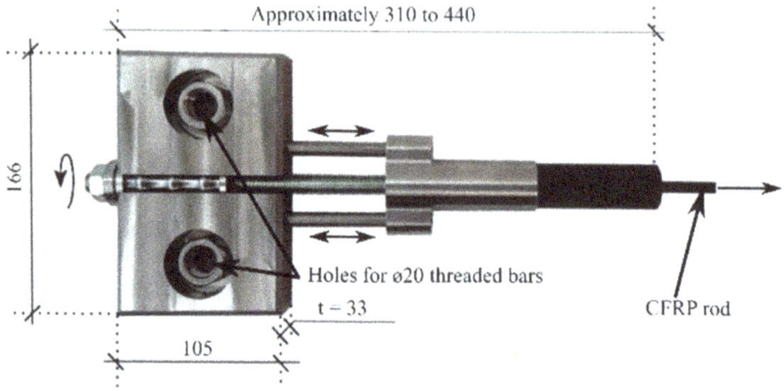

Figure 8. Pre-stressing applied by torque to the end nut on the assembled CFRP anchor system.

The dimension of the anchor block was designed to fit the structural geometry and reinforcement of the case study bridge. Figure 9 shows the cut-out area and groove in the planned location of Figure 5, in which the anchor block was mounted and the CFRP rod placed, respectively. The depth of the groove ensured that the 8 mm CFRP rod could be covered by the applied flexible adhesive throughout the beam length. The adhesive connected the otherwise free CFRP rod to the structure. The properties of the elastic adhesive are depicted in Table 2.

Figure 9. Cut grooves used for the strengthening system mounting. Photograph by S&P.

Table 2. Properties of the flexible adhesive.

Parameter	Values
Hardness	60, Shore A (ISO 868) [1]
Failure stress	3.0 N/mm^2 (ISO 37) [1]
E-modulus	3.0 MPa [1]
Extension at failure	200% (ISO 37) [1]
Elasticity	+/−20% [1]
Curing	2 mm/day (Environment dependent) [1]

[1] Data provided by manufacturer.

After the grooves were cut, holes were drilled through the beam web for the 20 mm threaded bars used for the anchor block mounting. The threaded bars were adhesively bonded to the concrete and were used to clamp the anchor block to the beam web at each side of the beam, see Figure 10.

Figure 10. Installation of the through going threaded bars. Photograph by S&P.

4. Testing of CFRP Anchor System

Before the assembled anchor system could be mounted on the structure, it had to undergo evaluations that ensured the desired response and, at the same time, provided a sufficient safety level. It was hypothesized that the designed yielding threshold provided a consistent failure mode initiation which ensures that unique ultimate failure mode resistances above this threshold can be disregarded, with sufficient reliability. This was deemed a sound approach if combined with in situ proof-loading of the system. Proof-loading of the system eliminates doubts concerning undesirable failure modes that may occur below the proof-load level.

4.1. Positioning of the Pre-Stressed Ductile CFRP NSMR System

The ductile mechanism is what defines the yielding threshold of the system, and thus, this had to be designed to provide a consistent yield strength. The yielding threshold would then function as the system design yield resistance. The desired yielding threshold, assigned in the project, for each anchor was 50 kN (as ULS design resistance). The initial design and verification of the ductile mechanism response were thus performed to comply with this demand. The initial test results are shown in Figure 11, along with the finite element (FE) simulation. It is seen that there is a slight discrepancy between the stiffness in the linear elastic regime, which is due to the test setup and machine stiffness. The commercial program ABAQUS [51] was used for the finite element evaluation, where 3D stress, quadratic (geometric order), and tetrahedron mesh were used to simulate the first part of the ductile mechanism response curve and, thus, the yielding threshold magnitude. A young's modulus of 200 GPa and Poisson's ratio of 0.3 was used as input to the analysis. Linear regression was used for the yielding regime, where yield stresses of 400 MPa and 2000 MPa were used in conjunction with plastic strains of 0 and 1, respectively. Similar yielding thresholds were obtained when comparing the two tested ductile mechanisms with theoretical simulations in Figure 11. Approximately 70 kN was obtained in the first evaluation, but due to the inclined yielding regime, this was subsequently reduced to approximately 60 kN to enable extended ductile behavior until ultimate system failure occurred.

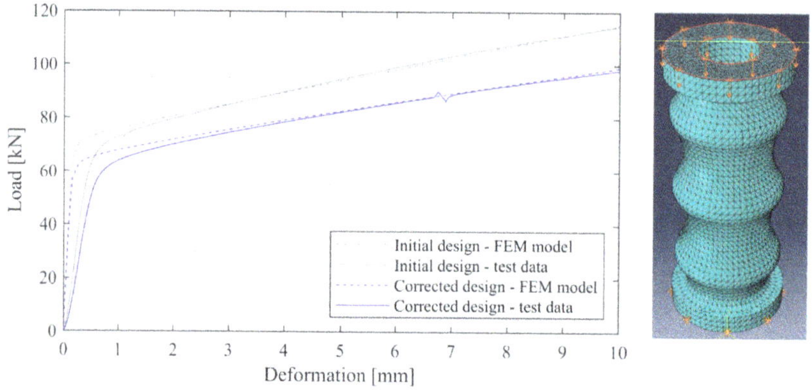

Figure 11. FE-simulation of the ductile unit.

Only minor variations were seen when comparing the force development and FE simulations. Consequently, given the correlation of the responses depicted in Figure 11 combined with earlier studies [43,44] provided background for using the 60 kN design for inclusion in the anchor system.

4.2. Anchor System Testing

Figure 12 shows the anchor system test specimen used for the pre-stressed strengthening approach, mounted in the MTS 810 material test machine. A 1 m ø8 CFRP rod test specimen was anchored by the system in the top jaw and a dead-end anchor in the bottom jaw. A deformation-controlled load rate of 2 mm/min was used for the test procedure.

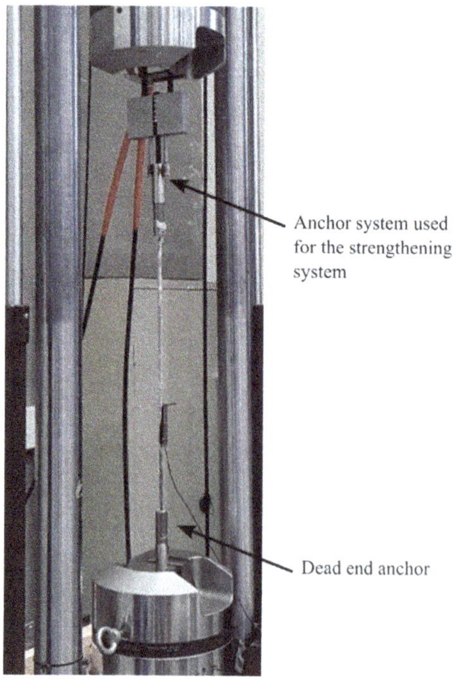

Figure 12. Tension test performed on the anchor system. Photograph by the authors.

Five system tests were performed on the anchor system. The response curves of these tests are seen in Figure 13, including a breakdown of the respective failure modes. It is seen that all ultimate failure mods, except the designed yielding threshold, are brittle. System tests 1 and 2 were CFRP anchor-related failure modes. It should be noted that the slip failure of system test 1 at 92 kN was due to the use of the lowest diameter of the produced CFRP rod batches (approximately 7.85 mm), which was deemed to provide a lower bound ultimate CFRP anchor capacity value. The failure of system test 2 at 105 kN was a frontal overload crushing of the CFRP rod. The remaining failures occurred at higher capacities in the nut and threaded components of the anchor system, which were designed with minimal dimensions for optimal design. The three tests that reached 30 mm deformation were unloaded and afterwards reloaded until failure occurred. The curves from loading and reloading are combined at the load termination point in the first cycle and yield initiation point of the second cycle (system tests 3, 4, and 5).

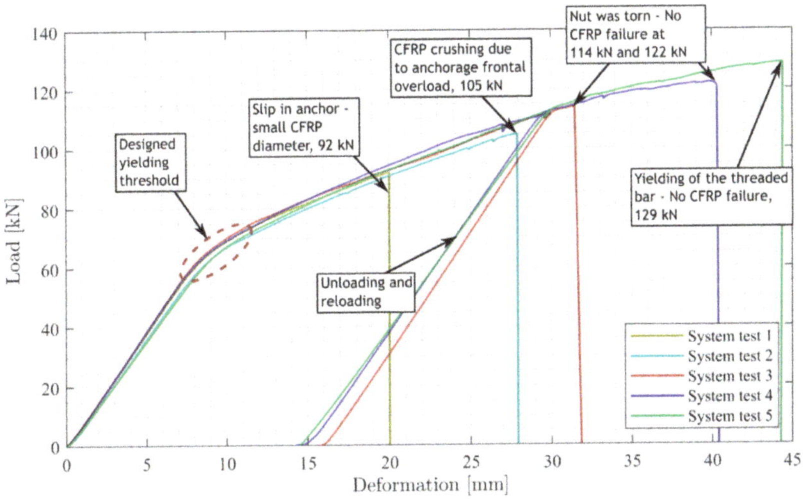

Figure 13. Load deformation test curves.

The tests provide an example of several deviating ultimate failure modes, which also may occur in any strengthening system or joint, including brittle and ductile behavior. The response curves are, however, almost identical until each of the respective ultimate failures occurs. As such, the response curves include a linear regime, a yielding threshold, and an inclined yielding regime that terminates when the failure load of a component is reached. Thus, the ductile mechanism seems to control the behavior well, with minor deviations between the curves and a pronounced yielding regime in all tests. In addition, the controlled system behavior of the tests correlates well when translating the linear regime of each test to the design "yielding" threshold using a yield strain of 0.002, which corresponds to a translation of 0.14 mm. The test curves and yielding translation line are depicted in Figure 14, where the intersection between the lines provides the yielding thresholds. The yield thresholds show only minor variation, and the results serve as input for determining the design yield threshold.

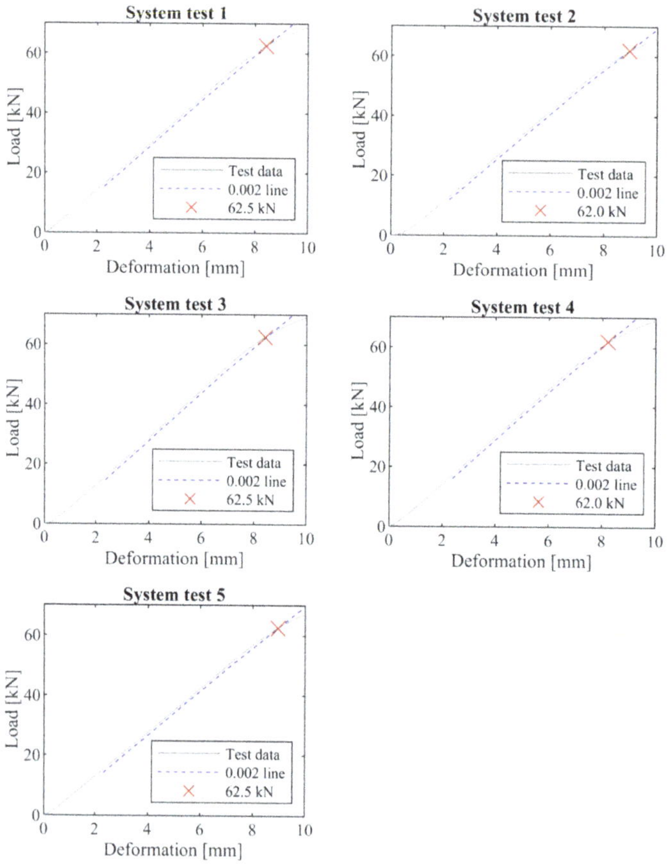

Figure 14. Translation of the linear regime (0.14 mm) and identification of the yielding threshold.

4.3. Design Yield Resistance

To the authors' best knowledge, a novel approach (apart from presenting the novel system) is to perform a system-related evaluation of the characteristic and design strength values, which is an extension of the component-based provisions in EN1990:2002 Annex D [46]. Such a system-based approach is used for other materials and structural components within the application area for the Eurocodes, but not for anchor systems as considered in this paper. The evaluation of the values is based on the Design Value Format method with parameters in EN1990:2002 Annex D, see Tables 2 and 4 in [46]. In the following analysis, characteristic values are defined as 5% quantiles, and design values were obtained using the Design Value Format method with a target 50-year reliability index equal to 3.8 and a sensitivity factor equal to 0.8. Subsequently, the characteristic and design value are determined assuming no prior knowledge (unknown coefficient of variation), including statistical uncertainty due to a limited number of tests. The characteristic resistance value is obtained by assuming a lognormal distribution of the resistance:

$$R_{k,n} = \exp(m_y - k_n s_y)$$

where:

$m_y = \frac{1}{n}\sum_{i=1}^{n} \ln(R_i)$ is the expected value of log of test results;

$s_y = \sqrt{\frac{1}{n-1} \sum_{i=1}^{n} \left(\ln(R_i) - m_y\right)^2}$ is the standard deviation of log of test results;
n is the number of data;
$R_i, i = 1, 2, \ldots, n$ are the data from tests;
$k_n = t_{n-1,p}\sqrt{1 + 1/n}$ is a factor used to obtain the characteristic value with $t_{n-1,p}$ being the p = 0.05 quantile of the Student-t distribution with $n - 1$ degrees of freedom.

The design value is obtained from:

$$R_{d,n} = \exp\left(m_y - k_{d,n} s_y\right)$$

where:
$k_{d,n} = t_{n-1,\Phi(-0.8\times 3.8)}\sqrt{1 + 1/n}$ is a factor based on the recommended lifetime reliability index = 3.8 and a sensitivity factor 0.8 implying an exceedance probability $\Phi(-0.8 \times 3.8)$ = 0.0012, with $\Phi(\,)$ being the standard Normal distribution function.

The partial material factor is calculated from:

$$\gamma_R = \frac{R_{k,n}}{R_{d,n}}$$

Probabilistic results based on the five tests are seen in Table 3.

Table 3. Probabilistic results, including design resistance and partial factor.

Mean	COV	Characteristic Resistance, ($R_{k,n}$)	ULS Design Resistance, ($R_{d,n}$)	Partial Factor, (γ_R)
62.3 kN	0.01	61.7 kN (k_n = 2.34)	60.3 kN ($k_{d,n}$ = 7.48)	1.02

Following the Eurocode recommendations, the design value in the table provides confidence in the system response and may ensure sufficient safety. However, it can be probabilistically argued that the ultimate failure modes (Figure 13), which occur above the yielding threshold, could provide a lower realization of the resistance. This could theoretically be realized through a potential large scatter (standard deviation) of one or multiple failure modes. Still, it is seen from the tests that the obtained ultimate capacities for each of these failure modes seem significantly larger than the determined ULS design yield resistance of 60.3 kN. As such, even with a significant standard deviation of these failure modes, the probability of obtaining a lower resistance value is small. This small probability of getting a lower resistance is, however, one of the most challenging aspects when introducing novel systems with many potential failure modes. The issue thus deserves significant attention, not only for novel systems but also for existing construction parts and joints, where similar realizations may be obtained. For the presented case, it still needs to be verified probabilistically that there is a low or no probability that an ultimate capacity value will be lower than the lower boundary provided through the designed and tested yield resistance threshold. This cannot be verified through the current methods. A detailed probabilistic modeling and safety verification of this hypothesis will be investigated further in a follow-up paper. A more direct approach was applied in the presented research pilot project to verify further that no systems with lower resistance arise when applying these. This approach relates to the installation procedure, which includes proof loading of each system used in this project, thus verifying the resistance. It should, however, be noted that this procedure was for detailed investigations of the novel system and is thus unique to the particular project.

5. Mounting, Prestressing, and Verification

Following the experimental testing of the anchor system, the system could be mounted on the case-study bridge, as shown in Figure 10, where the pre-stressing procedure was performed in two phases. The described proof loading, Phase 1, consisted of a step-wise pre-stressing to the desired proof-loading magnitude. Phase 2 related to the time-dependent effects, where the system was unloaded after three weeks and pre-stressed again to evaluate and account for potential pre-stress losses.

5.1. Pre-Stressing, Phase 1

Figure 15 shows the numbering of the beams on which the anchors were step-wise pre-stressed to 10 kN, 25 kN, and 55 kN in the beam sequence; 3, 2, 4, 1, and 5. The pre-stressing level (proof loading level) of 55 kN was chosen in relation to the required design value of 50 kN and the ULS design resistance estimated to 60.3 kN. As such, the proof-loading eliminates any possible failures lower than 55 kN, and thus significantly reduces the risk of an undesired brittle failure without warning. It should be noted that the reliability assessment of the proof load level depends on the level of knowledge on the resistance, and since the resistance of the strengthening system can be modeled with relatively small uncertainty, the selected proof load level was considered sufficient. Pre-stressing at each step was finalized for all anchors before the next phase was initiated.

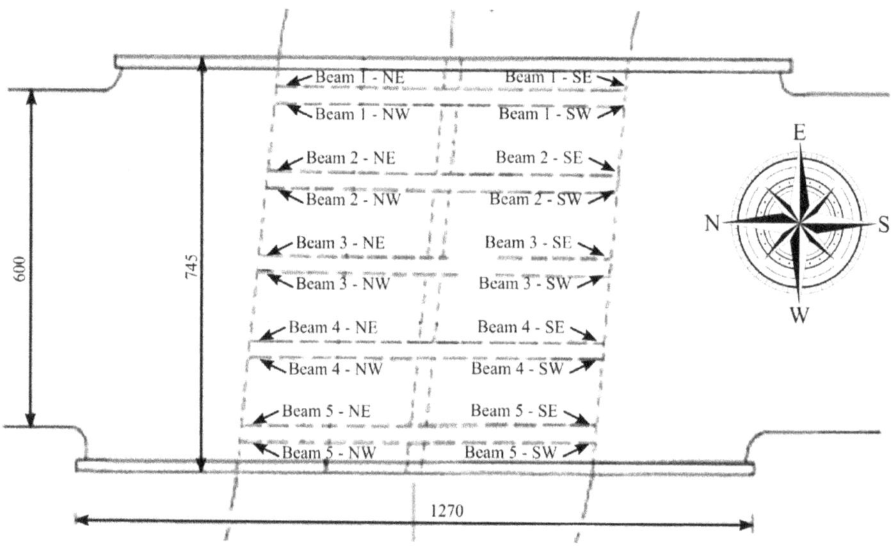

Figure 15. Beam numbering used for the pre-stressing procedure.

Table 4 depicts the actual pre-stressing magnitudes of the proof-loadings, με/(kN), measured by a strain gauge of type HBM 6/350VE LY41 mounted on each CFRP rod. An ø8 mm CFRP rod and an elastic modulus of 160 GPa were used as background for the reference values in each loading step of 1250 με (10 kN), 3150 με (25 kN), and 6875 με (55 kN). After three hours, the tensile levels were measured again to evaluate the system seating, ε_3 (kN_3). It was seen that the strain values were acceptable after three hours, where only an insignificant reduction was seen. Accordingly, the achieved pre-stressing values fulfilled the request of 50 kN ULS design resistance. No additional post-tensioning was thus needed to adjust tensioning magnitudes. After the values were approved, the strain gauge wires were disconnected and later reconnected for phase 2, which was initiated after three weeks.

Table 4. Measured pre-stressing values during Phase 1, measured in microstrain with kilonewtons in parenthesis. In the final pre-stressing level, initial values are given along with measurements after three hours.

	Location	10 kN, Measures με/(kN)	25 kN, Measures με/(kN)	55 kN, Measures με/με$_3$; (kN/kN$_3$)
Beam 1	Ø (A)	1269 (10.2)	3189 (25.6)	6867/6785; (55.2/54.6)
	V (B)	1185 (9.5)	3196 (25.7)	6852/6743; (55.1/54.2)
Beam 2	Ø (A)	1211 (9.7)	3102 (24.9)	6877/6821; (55.3/54.9)
	V (B)	1187 (9.5)	3069 (24.7)	6841/6696; (55.0/53.9)
Beam 3	Ø (A)	1281 (10.3)	3161 (25.4)	6870/6762; (55.3/54.4)
	V (B)	1248 (10.0)	3200 (25.7)	6882/6795; (55.3/54.6)
Beam 4	Ø (A)	1226 (9.9)	3091 (24.9)	6877/6780; (55.3/54.5)
	V (B)	1215 (9.8)	3142 (25.3)	6883/6804; (55.4/54.7)
Beam 5	Ø (A)	1354 (10.9)	3116 (25.1)	6847/6748; (55.1/54.3)
	V (B)	1278 (10.3)	3141 (25.3)	6844/6780; (55.0/54.5)

Figure 16 depicts the site installation of the strengthening system where pre-stressing was performed with a single-head wrench, applying torque to the system nut.

Figure 16. Site installation of the throughgoing threaded bar and pre-stressing of the system. Photograph by S&P.

The pre-stressing procedure worked as intended up to the desired pre-stressing levels at which no damage was identified for the full strengthening systems and associated anchor locations. Proof-loading by pre-stressing to 55 kN furthermore demonstrated the desired resistance for all 20 unique anchor systems and ensured that no failure modes were initiated at lower load levels. The CFRP rod of the strengthening system was left unbonded until the initiation of phase 2.

5.2. Pre-Stressing, Phase 2

The second phase was initiated after three weeks, where acquisition was performed after re-connecting the strain gauge wires. In this phase, the system was unloaded and pre-stressed again to 55 kN using identical loading steps as for phase 1. The unloading was done to account for possible relaxation effects of the unbonded strengthening system.

Table 5 depicts the initial three-hour strain value of phase 1 ($\mu\varepsilon_3$) compared to the measured value after three weeks ($\mu\varepsilon_w$). It is seen that the pre-stressing magnitudes are reduced by values ranging from approximately 0.2 to 1.6 kN, corresponding to a full system deformation of 0.2–1.3 mm, respectively (using a strengthening system length of 6500 mm).

Table 5. Initial three-hour strain value of phase 1 ($\mu\varepsilon_3$) compared to measured value after three weeks.

	Location	55kN, Measures $\mu\varepsilon_3/\mu\varepsilon_w$ (kN_3/kN_w)	Difference $\mu\varepsilon_3$-$\mu\varepsilon_w$/(kN_3-kN_w)
Beam 1	Ø (A)	6785/6666 (54.6/53.6)	119/(1.0)
	V (B)	6743/6582 (54.2/52.9)	161/(1.3)
Beam 2	Ø (A)	6821/6758 (54.9/54.3)	63/(0.5)
	V (B)	6696/6545 (53.9/52.6)	151/(1.2)
Beam 3	Ø (A)	6762/6568 (54.4/52.8)	194/(1.6)
	V (B)	6795/6627 (54.6/53.3)	168/(1.4)
Beam 4	Ø (A)	6780/6630 (54.5/53.3)	150/(1.2)
	V (B)	6804/6749 (54.7/54.3)	55/(0.4)
Beam 5	Ø (A)	6748/6648 (54.3/53.5)	100/(0.8)
	V (B)	6780/6750 (54.5/54.3)	30/(0.2)

The values depicted in Table 4. describe the full system behavior. Separating these values between the anchor systems, a deformation ranging from 0.1–0.8 mm was thus obtained. Table 6 shows the final pre-stressing magnitudes approximately at the desired 55 kN (1094 MPa).

Table 6. Final pre-stressing magnitudes just before adhesive bonding.

	Location	55kN, Measures $\mu\varepsilon_3/\mu\varepsilon_w$ (kN_3/kN_w)
Beam 1	Ø (A)	6899/55.5/1104
	V (B)	6938/55.8/1110
Beam 2	Ø (A)	6861/55.2/1098
	V (B)	6869/55.2/1099
Beam 3	Ø (A)	6846/55.1/1095
	V (B)	6963/55.9/1114
Beam 4	Ø (A)	6938/55.8/1110
	V (B)	6888/55.4/1102
Beam 5	Ø (A)	6863/55.2/1098
	V (B)	6924/55.7/1108

After verifying the desired pre-stressing magnitudes, adhesive bonding of the CFRP rod was done by using the flexible adhesive in pre-cut grooves. Figure 17 shows the final result of the applied strengthening with the adhesive applied into the groove, thus bonding it to the structure along the beam length and how it terminates at the anchor system location.

Figure 17. The final result of the applied strengthening. Photograph by S&P.

6. Conclusions

The paper presents research that demonstrates the response of a novel response-controlled pre-stressing anchor system used for CFRP NSMR strengthening. An initial assessment of an in situ pre-cast concrete bridge was performed to provide a basis for the most desirable anchor system placement. Five anchor systems were tested in the laboratory to verify the anchor system's controlled response and consistency with a special focus on the designed yielding threshold and yielding regime. The ultimate capacities occurred at significantly higher magnitudes than the yielding threshold for all tests. A concern was, however, that a large standard deviation of a unique ultimate failure mode could lead to a realization of a lower resistance value than the yielding threshold. The probabilistic anchor system assessment based on laboratory tests in combination with in situ proof loading was deemed a sound approach to disregard the probability of obtaining the hypothesized lower value realization. Based on the research hypotheses and questions, the following conclusions can be drawn:

- The laboratory test showed excellent consistency when comparing the load/deformation curves of the anchor system.
- The response can be tailored, primarily controlled by the ductile mechanism.
- Controlled yielding was obtained for all anchor systems tested in the laboratory, thus ensuring a low-yielding threshold safety factor (γ_R = 1.02).
- The ULS design (yielding threshold) resistance of 60.3 kN fulfilled the required demand for the ULS design resistance of 50 kN, required from the presented unique pilot project.
- In situ proof loading of the strengthening systems (20 anchor systems) showed no signs of distress.
- Proof loading ensured that variations in the unique ultimate failure modes of the mounted anchor systems did not compromise the desired resistance (yielding threshold).
- A good basis is provided for upcoming systems, which can resist significantly higher tailored load magnitudes combined with a desired response.
- A more detailed probabilistic verification seems necessary to ensure that the yielding threshold provides a minimum capacity level not compromised by ultimate capacity variations and to avoid proof-loading entirely.
- The developed NSMR CFRP pre-stressed strengthening system worked as desired and ensured fast mounting and effective controlled prestressing
- After three weeks, acceptable prestress losses ranging from approximately 0.2 to 1.6 kN were observed, corresponding to a full strengthening system (including anchors) and a deformation of 0.2–1.3 mm, respectively.

- The applied strengthening resulted in a bridge class upgrading from standard passage class 20 to class 50 and conditional passage (type 3) class 50 to class 100, as required.

It is thus seen that the ductile response controlled strengthening system works as desired, but a probabilistic assessment may provide uncertainty in relation to the reliability and probability of failure. It is still an open question as to whether laboratory testing for such systems verifies all possible failure modes and whether these are representative if forced to occur. From an economical perspective, such evaluations can cause a significant burden to a research project. In situ proof loading can ensure validation of a system, but such a procedure would often be considered an undesirable and cumbersome process. In addition, proof loading does not verify that the yielding regime will be activated. The designed ductile unit response may solve this challenge since it can be constructed to meet a desired capacity demand and response.

Author Contributions: Conceptualization, J.W.S.; methodology, J.W.S. and J.D.S.; software, J.W.S., J.D.S. and C.O.C.; validation, J.W.S. and J.D.S.; formal analysis, J.W.S. and J.D.S.; investigation, J.W.S.; resources, J.W.S.; data curation, J.W.S. and J.D.S.; writing—original draft preparation, J.W.S.; writing—review and editing, J.W.S., J.D.S. and C.O.C.; visualization, J.W.S. and C.O.C.; supervision, J.W.S. and J.D.S.; project administration, J.W.S.; funding acquisition, J.W.S. All authors have read and agreed to the published version of the manuscript.

Funding: The funding source do not wish to be revealed as stated in this section.

Data Availability Statement: All data and information needed for the evaluations discussed are provided in the paper.

Acknowledgments: A sincere gratitude is addressed to S&P Clever Reinforcement Company AG and Business Manager Morten Frost Kamphøvener from S&P Reinforcement Nordic for supporting the presented pilot project and ongoing research at Aalborg University (AAU). Thank you very much to Vejen municipality and COWI A/S, for giving us the opportunity to initiate and carry out the pilot project. Additionally, a great appreciation is dedicated to craftsmen from CG Jensen and technical staff from AAU, for their professionalism and excellent work.

Conflicts of Interest: The authors declare no conflict of interest.

References

1. El-Hacha, R.; Rizkalla, S.H. Near-Surface-Mounted Fiber-Reinforced Polymer Reinforcements for Flexural Strengthening of Concrete Structures. *ACI Struct. J.* **2004**, *101*, 717–726.
2. Mohamed Ali, M.S.; Oehlers, D.J.; Griffith, M.C.; Seracino, R. Interfacial Stress Transfer of near Surface-Mounted FRP-to-Concrete Joints. *Eng. Struct.* **2008**, *30*, 1861–1868. [CrossRef]
3. Rashid, R.; Oehlers, D.J.; Seracino, R. IC Debonding of FRP NSM and EB Retrofitted Concrete: Plate and Cover Interaction Tests. *J. Compos. Constr.* **2008**, *12*, 160–167. [CrossRef]
4. Seracino, R.; Jones, N.M.; Ali, M.S.; Page, M.W.; Oehlers, D.J. Bond Strength of Near-Surface Mounted FRP Strip-to-Concrete Joints. *J. Compos. Constr.* **2007**, *11*, 401–409. [CrossRef]
5. Sena-Cruz, J.M.; Barros, J.A.O.; Coelho, M.R.F.; Silva, L.F.F.T. Efficiency of Different Techniques in Flexural Strengthening of RC Beams under Monotonic and Fatigue Loading. *Constr. Build. Mater.* **2012**, *29*, 175–182. [CrossRef]
6. Al-Mahmoud, F.; Castel, A.; Franois, R. Failure Modes and Failure Mechanisms of RC Members Strengthened by NSM CFRP Composites—Analysis of Pull-out Failure Mode. *Compos. Part B Eng.* **2012**, *43*, 1893–1901. [CrossRef]
7. Schmidt, J.W.; Krabbe, J.; Sørensen, N.O.; Hertz, K.D.; Goltermann, P.; Sas, G. CFRP Strengthening of RC Beams Using a Ductile Anchorage System. In Proceedings of the 8th International Conference on Fibre-Reinforced Polymer (FRP) Composites in Civil Engineering, CICE 2016, Hong Kong, China, 14–16 December 2016.
8. Schmidt, J.W.; Hertz, K.D.; Goltermann, P. NSMR Strengthening of Short RC Beams Using Activated Anchorage. In Proceedings of the 9th International Conference on Fibre-Reinforced Polymer (FRP) Composites in Civil Engineering, CICE 2018, Paris, France, 17–19 July 2018.
9. Sabau, C.; Popescu, C.; Sas, G.; Schmidt, J.W.; Blanksvärd, T.; Täljsten, B. Strengthening of RC Beams Using Bottom and Side NSM Reinforcement. *Compos. Part B Eng.* **2018**, *149*, 82–91. [CrossRef]
10. Nordin, H.; Täljsten, B. Concrete Beams Strengthened with Prestressed Near Surface Mounted CFRP. *J. Compos. Constr.* **2006**, *10*, 60–68. [CrossRef]
11. American Concrete Institute. *ACI 440.2R-17: Guide for the Design and Construction of Externally Bonded FRP Systems for Strengthening Concrete Structures*; American Concrete Institute: Indianapolis, IN, USA, 2017; ISBN 9781945487590.

12. Täljsten, B. Strengthening of Beams by Plate Bonding. *J. Mater. Civ. Eng.* **1997**, *9*, 206–212. [CrossRef]
13. Triantafillou, T.C.; Antonopoulos, C.P. Design of Concrete Flexural Members Strengthened in Shear with FRP. *J. Compos. Constr.* **2000**, *4*, 198–205. [CrossRef]
14. Smith, S.T.; Gravina, R.J. Modeling Debonding Failure in FRP Flexurally Strengthened RC Members Using a Local Deformation Model. *J. Compos. Constr.* **2007**, *11*, 184–191. [CrossRef]
15. Teng, J.G.; Smith, S.T.; Yao, J.; Chen, J.F. Intermediate Crack-Induced Debonding in RC Beams and Slabs. *Constr. Build. Mater.* **2003**, *17*, 447–462. [CrossRef]
16. Said, H.; Wu, Z. Evaluating and Proposing Models of Predicting IC Debonding Failure. *J. Compos. Constr.* **2008**, *12*, 284–299. [CrossRef]
17. Tworzewski, P.; Alexy, J.K.; Barnes, R.W. Intermediate Crack Debonding of Externally Bonded FRP Reinforcement—Comparison of Methods. *Materials* **2022**, *15*, 7390.
18. Gao, B.; Leung, C.K.Y.; Kim, J.K. Prediction of Concrete Cover Separation Failure for RC Beams Strengthened with CFRP Strips. *Eng. Struct.* **2005**, *27*, 177–189. [CrossRef]
19. Corden, G.; Ibell, T.; Darby, A. Concrete Cover Separation Failure in Near-Surface Mounted CFRP Strengthened Concrete Structures. *Struct. Eng.* **2008**, *86*, 19–21.
20. Wight, R.G.; Green, M.F.; Erki, M.-A. Prestressed FRP Sheets for Poststrengthening Reinforced Concrete Beams. *J. Compos. Constr.* **2001**, *5*, 214–220. [CrossRef]
21. Diab, H.; Wu, Z.; Iwashita, K. Short and Long-Term Bond Performance of Prestressed FRP Sheet Anchorages. *Eng. Struct.* **2009**, *31*, 1241–1249. [CrossRef]
22. Yang, D.S.; Park, S.K.; Neale, K.W. Flexural Behaviour of Reinforced Concrete Beams Strengthened with Prestressed Carbon Composites. *Compos. Struct.* **2009**, *88*, 497–508. [CrossRef]
23. Chen, C.; Chen, J.; Zhou, Y.; Sui, L.; Hu, B. Design of Ductile H-Anchorage for Strengthening Reinforced Concrete Beams with Prestressed FRP. *Constr. Build. Mater.* **2021**, *307*, 124883. [CrossRef]
24. D'Amato, M.; Laterza, M.; Casamassima, V.M. Seismic Performance Evaluation of a Multi-Span Existing Masonry Arch Bridge. *Open Civ. Eng. J.* **2018**, *11*, 1191–1207. [CrossRef]
25. Modena, C.; Tecchio, G.; Pellegrino, C.; da Porto, F.; Donà, M.; Zampieri, P.; Zanini, M.A. Reinforced Concrete and Masonry Arch Bridges in Seismic Areas: Typical Deficiencies and Retrofitting Strategies. *Struct. Infrastruct. Eng.* **2015**, *11*, 415–442. [CrossRef]
26. American Concrete Institute. *ACI 440.4R-04: Guide for Prestressing Concrete Structures with FRP Tendons*; American Concrete Institute: Farmington Hills, MA, USA, 2004.
27. Nanni, A. North American Design Guidelines for Concrete Reinforcement and Strengthening Using FRP: Principles, Applications and Unresolved Issues. *Constr. Build. Mater.* **2003**, *17*, 439–446. [CrossRef]
28. Jiang, Z.; Fang, Z.; Fang, C.; Li, Q.; Wang, Z. Experimental Investigation on High-Temperature Creep Behavior of Carbon Fiber Reinforced Polymer Cable. *Compos. Struct.* **2022**, *291*, 115533. [CrossRef]
29. Schmidt, J.W.; Bennitz, A.; Täljsten, B.; Goltermann, P.; Pedersen, H. Mechanical Anchorage of FRP Tendons—A Literature Review. *Constr. Build. Mater.* **2012**, *32*, 110–121. [CrossRef]
30. Bennitz, A.; Grip, N.; Schmidt, J.W. Thick-Walled Cylinder Theory Applied on a Conical Wedge Anchorage. *Meccanica* **2011**, *46*, 959–977. [CrossRef]
31. Hansen, C.S.; Schmidt, J.W.; Stang, H. Transversely Compressed Bonded Joints. *Compos. Part B Eng.* **2012**, *43*, 691–701. [CrossRef]
32. Schmidt, J.W.; Bennitz, A.; Täljsten, B.; Pedersen, H. Development of Mechanical Anchor for CFRP Tendons Using Integrated Sleeve. *J. Compos. Constr.* **2010**, *14*, 397–405. [CrossRef]
33. Grelle, S.V.; Sneed, L.H. Review of Anchorage Systems for Externally Bonded FRP Laminates. *Int. J. Concr. Struct. Mater.* **2013**, *7*, 17–33. [CrossRef]
34. Cuntze, R.G.; Freund, A. The Predictive Capability of Failure Mode Concept-Based Strength Criteria for Multidirectional Laminates. *Compos. Sci. Technol.* **2004**, *64*, 343–377. [CrossRef]
35. Schmidt, J.W. *External Strengthening of Building Structures with Prestressed CFRP*; Technical University of Denmark: Lyngby, Denmark, 2011.
36. Bennitz, A.; Schmidt, J.W.; Täljsten, B. Failure Modes of Prestressed CFRP Rods in a Wedge Anchored Set-Up. In Proceedings of the Advanced Composites in Construction 2009, ACIC 2009—Proceedings of the 4th International Conference, Toronto, ON, Canada, 16–17 July 2009.
37. Piątek, B.; Siwowski, T. Experimental Study on Flexural Behaviour of Reinforced Concrete Beams Strengthened with Passive and Active CFRP Strips Using a Novel Anchorage System. *Arch. Civ. Mech. Eng.* **2022**, *22*, 1–17. [CrossRef]
38. Heydarinouri, H.; Motavalli, M.; Nussbaumer, A.; Ghafoori, E. Development of a Mechanical Wedge–Barrel Anchor for CFRP Rods: Static and Fatigue Behaviors. *J. Compos. Constr.* **2021**, *25*, 04021015. [CrossRef]
39. Heydarinouri, H.; Vidovic, A.; Nussbaumer, A.; Ghafoori, E. FE Analysis and Experimental Validation of Mechanical Wedge–Barrel Anchors for CFRP Rods. *Compos. Struct.* **2021**, *275*, 114509. [CrossRef]
40. Schmidt, J.W.; Smith, S.T.; Täljsten, B.; Bennitz, A.; Goltermann, P.; Pedersen, H. Numerical Simulation and Experimental Validation of an Integrated Sleeve-Wedge Anchorage for CFRP Rods. *J. Compos. Constr.* **2011**, *15*, 284–292. [CrossRef]
41. Schmidt, J.W.; Christensen, C.O.; Goltermann, P.; Hertz, K.D. Shared CFRP Activation Anchoring Method Applied to NSMR Strengthening of RC Beams. *Compos. Struct.* **2019**, *230*, 111487. [CrossRef]

42. Schmidt, J.W.; Sena-Cruz, J.; Goltermann, P.; Christensen, C.O. Experimental and Numerical Studies on the Shared Activation Anchoring of NSMR CFRP Applied to RC Beams. In Proceedings of the APFIS 2019 Proceedings—7th Asia-Pacific Conference on FRP in Structures, Gold Coast, Australia, 10–13 December 2019.
43. Schmidt, J.W.; Christensen, C.O.; Goltermann, P. Ductile Response Controlled EW CFRP Anchor System. *Compos. Part B Eng.* **2020**, *201*, 108371. [CrossRef]
44. Schmidt, J.W.; Christensen, C.O.; Goltermann, P.; Sena-Cruz, J. Activated Ductile CFRP NSMR Strengthening. *Materials* **2021**, *14*, 2821. [CrossRef]
45. Schmidt, J.W.; Christensen, C.O.; Goltermann, P.; Sena-Cruz, J. Activated CFRP NSMR Ductile Strengthening System. In *RILEM Bookseries*; Springer: Berlin/Heidelberg, Germany, 2022; Volume 34, pp. 349–361.
46. *EN 1990*; Eurocode: Basis of Structural Design. European Committee for Standardization: Bruxelles, Belgium, 2002.
47. *ISO 2394*; General Principles on Reliability for Structures: International Standard ISO; Bind 2394. International Organization for Standardization: Geneva, Zwitserland, 2015.
48. Joint Committee on Structural Safety. *Probabilistic Model Code*; Joint Committee on Structural Safety: Deift, The Netherlands, 2002.
49. Vejdirektoratet (The Danish Road Directorate). *Vejledning Til Belastnings- Og Beregningsgrundlag*; Vejdirektoratet (The Danish Road Directorate): Hedehusene, Denmark, 2010.
50. Vejdirektoratet (The Danish Road Directorate). *DS/EN 1991-2 DK NA:2017, Annex A: Lastmodeller for Klassificering Og Bæreevnevurdering (Models of Special Vehicles for Road Bridges)*; Vejdirektoratet (The Danish Road Directorate): Hedehusene, Denmark, 2017.
51. *ABAQUS, Version Abaqus/CAE*; Dassault Systèmes Simulia Corp: Johnston, RI, USA, 2021.

buildings

MDPI

Article

Assessment of Durability Indicators for Service Life Prediction of Portland Limestone Cementitious Systems Produced with Permeability-Reducing Admixtures

Athanasios Malakopoulos [1,2] and Athanasios Salifoglou [1,*]

1 Laboratory of Inorganic Chemistry and Advanced Materials, School of Chemical Engineering, Aristotle University of Thessaloniki, 54124 Thessaloniki, Greece
2 VINCI Technology Centre UK Ltd., Leighton Buzzard LU7 4QH, UK
* Correspondence: salif@auth.gr; Tel.: +30-2310-996179

Abstract: Supplementary cementing materials (SCMs) and the challenges associated with admixture compatibility and durability performance are continuous challenges for the construction industry. Utilizing SCMs has apparent benefits in reducing the carbon footprint and improving the durability performance of concrete structures. This work investigates the performance of mortars composed of Portland limestone cement, calcium carbonate, butyl stearate, and oleic acid. The effort focuses on transport properties using electrical resistivity, chloride migration, porosity, and water permeability measurements. Then, various methods based on the last parameters are compared to assess the changes in the effective chloride diffusion coefficient and the intrinsic liquid water permeability. Measurements for drying shrinkage, thermal expansion, and compressive strength are conducted to verify durability and mechanical performance. The effect of the admixture addition on the service life of a structure fully submerged in marine exposure conditions is then evaluated using Fick's second law of diffusion and the approach described by FIB model code 34. The results indicate that incorporating calcium carbonate, butyl stearate, and oleic acid in mortar mixtures provides enhanced durability compared to plain Portland limestone cement mortars.

Keywords: limestone cement; durability; butyl stearate; calcium carbonate; oleic acid; migration; service life

1. Introduction

In recent years, sustainable development and long-term durability of concrete structures have become a significant challenge. Lean construction and design, the use of recycled materials, alternative binders, new admixtures, and concrete surface treatments are some approaches to reducing the societal, economic, technical, and environmental impact [1–3]. There is a consensus in the literature stating that incorporation of cement replacements, such as limestone dust, ground granulated blast furnace slag, or fly ash in concrete, improves resistance to chloride ingress, but the resulting concrete may be less resistant to carbonation-induced corrosion [4–6]. In addition, superplasticizers and water-repellent admixtures reduce porosity and retard or inhibit chemical agent movement [7,8]. Consequently, there is an increasing demand for the current design practices and concepts of service life prediction to include performance-based and probabilistic approaches that incorporate the effect of additives and admixtures.

A concept to evaluate durability performance is through a system of classes for durability indicators (DIs) such as porosity, diffusion, and permeability [9]. The evaluation compares the measured DI to the threshold value for each class. Furthermore, the measured values can be used as input data in multi-level modelling concepts for service life predictions. However, to perform an objective assessment, reviewing all of the available data and understanding the processes leading to different material performances is necessary. Hence, it is essential to focus on the synergistic effect of the various material combinations and investigate whether the DIs can be assessed for those mixtures using conventional methods.

In this context, the present work developed several mortar combinations with Portland limestone binder, calcium carbonate filler, butyl stearate, and oleic acid. The partial replacement of Portland cement with limestone filler undoubtedly brings environmental benefits, but comparable mechanical properties and durability must be achieved compared to ordinary Portland cement mixtures. At replacement levels of up to 15%, limestone filler can improve compressive strength and reduce chloride permeability [10,11]. However, replacements above that level typically lead to lower strengths, greater porosity, and permeability.

The hydrophobic action of butyl stearate has been investigated in previous studies [12–14] and found similar to that of soaps. During cement hydration, butyl stearate hydrolyses in the alkaline environment and produces precipitates that form hydrophobic coatings and pore blockers in the hardened concrete. While this effect is beneficial for reducing agent movement in the pore structure, it appears to influence the shrinkage mechanism [12]. Furthermore, combinations of different types of stearates may activate chemical reactions that complicate further durability assessment.

This study focused on assessing the effective chloride diffusion coefficient and intrinsic liquid water permeability with various methods in a range of mixture combinations. The range of admixture contents was selected according to BS EN 934-2 [15] recommendations and the standard practice for mortar production. Oleic acid was incorporated into the mixtures to promote material dispersion and improved homogeneity. The purpose was to validate rapid assessment methods and compare them with more theoretical and sophisticated approaches. An additional goal was to contribute to existing databases for the estimation of effective chloride diffusion coefficients and intrinsic liquid water permeability by considering compositional parameters, such as the presence of SCMs and permeability-reducing admixtures.

2. Materials and Methods

2.1. Materials

The evaluated cementitious systems were produced using the following materials: Portland limestone cement (CEM II A-L 32,5R) conforming to BS EN 197-1 [16], containing between 6% and 20% by weight of limestone, with a total organic content not exceeding 0.5% by weight, and being of a strength class 32.5 MPa with an ordinary early strength; natural quartz sand; distilled water; and the additives butyl stearate, oleic acid, and calcium carbonate.

Table 1 presents the chemical composition of cement and sand, accrued with X-ray fluorescence analysis. Cement had a specific gravity equal to 3100 kg/m^3, and sand, 2650 kg/m^3. The particle size distribution of sand, measured through the sieving method according to BS EN 933-1 [17], is presented in Figure 1.

The three different additives considered in this study were acquired from Sigma-Aldrich. Butyl stearate ($C_{22}H_{44}O_2$) was of technical grade 58.2% and consisted of a mixture of butyl stearates, palmitates, and other fatty acids with a density equal to 861 kg/m^3 at 20 °C. Oleic acid ($C_{18}H_{34}O_2$) and calcium carbonate ($CaCO_3$) were of technical grade with higher purity levels corresponding to 90% and 98%, respectively. The particle size of $CaCO_3$ was smaller than 50 μm, and the specific gravity was equal to 2930 kg/m^3.

2.2. Specimen Preparation

Ten cementitious systems were produced for testing, as depicted in Table 2. Each mix was assigned an identification code O_x-B_y-C_z, in which O, B, and C denoted the $C_{18}H_{34}O_2$, the $C_{22}H_{44}O_2$, and the $CaCO_3$ component, respectively. The x, y, and z symbols indicated the content of each additive in the mix as a percentage of the total binder content. The reference mixes with no additives were assigned a code R_w, where w denoted the W/C ratio (water to cement ratio). Two conventional reference mortars were used: (a) LC-$R_{0.5}$, in the proportions 1.0:3.0:0.5 (cement:sand:water), and (b) LC-$R_{0.45}$, whose unit formulation was 1:0:3.0:0.45 (cement:sand:water).

Table 1. Chemical composition of cement and sand (wt.%).

	CEM II A-L 32,5R	Sand
Chemical components		
CaO	64.80	<0.01
SiO_2	15.66	99.72
Al_2O_3	4.03	0.07
Fe_2O_3	2.78	0.05
SO_3	3.18	-
MgO	0.98	0.02
K_2O	0.54	0.02
Na_2O	0.11	0.04
Free lime	1.24	-
Ignition loss	7.76	0.05
Alkali equivalent	0.47	-
Limestone	16.8	-
Mineralogical composition [1]		
C_3S (%)	69.4	-
C_2S (%)	0.6	-
C_3A (%)	6.0	-
C_4AF (%)	8.3	-

[1] According to the ASTM C150.

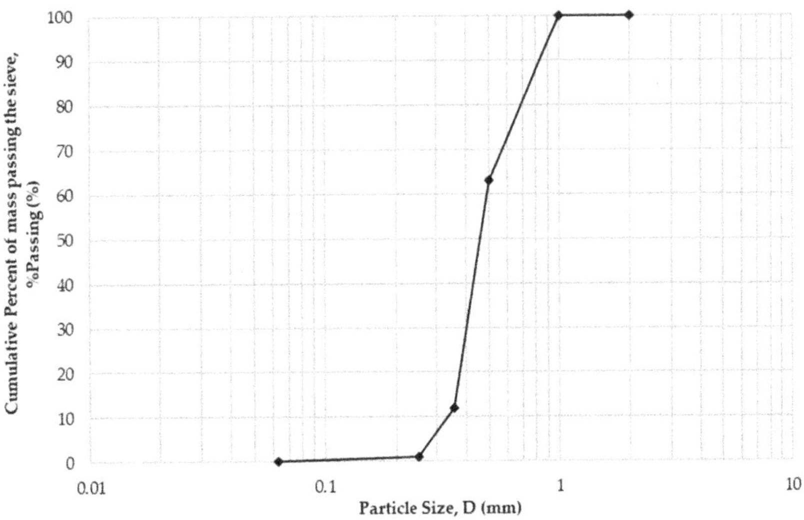

Figure 1. Sand particle size distribution.

The mortar mixes with a designation code O_x-B_y-C_z had an additive content of 0.7–6.0% by weight of Portland limestone cement. Percentages were selected according to their expected reactivity. The process involved partially replacing cement, sand, and water for volumetric purposes. The effect of these changes produced, at the most, approximately a 1.5% change in the expected density of fresh mortar.

All materials were stored under standard laboratory conditions or as prescribed by the manufacturer. Cement was sampled and homogenized according to BS EN 196-7 [18] before taking sub-samples for specimen preparation. The sand was cured at 105 °C until the total moisture content was reduced to less than 0.2% by mass of sand. $C_{22}H_{44}O_2$ was brought to the liquid state upon heating up to 30 °C prior to adding to the blend.

Table 2. Composition of cementitious systems.

Mix Designation	Water (kg/m³)	Cement (kg/m³)	Sand (kg/m³)	Water to Cement Ratio (W/C)	Admixture (wt.% of Total Binder) $C_{18}H_{34}O_2$	$C_{22}H_{44}O_2$	$CaCO_3$
LC-$R_{0.50}$	256	512	1535	0.50	-	-	-
LC-$R_{0.45}$	236	525	1575	0.45	-	-	-
LC-$O_{0.7}B_1C_3$	233	517	1551	0.45	0.7	1.0	3.0
LC-$O_{0.7}B_3C_3$	230	511	1532	0.45	0.7	3.0	3.0
LC-$O_{1.4}B_1C_3$	232	515	1545	0.45	1.4	1.0	3.0
LC-$O_{1.4}B_3C_3$	229	509	1526	0.45	1.4	3.0	3.0
LC-$O_{0.7}B_1C_6$	231	514	1543	0.45	0.7	1.0	6.0
LC-$O_{0.7}B_3C_6$	229	508	1524	0.45	0.7	3.0	6.0
LC-$O_{1.4}B_1C_6$	231	512	1536	0.45	1.4	1.0	6.0
LC-$O_{1.4}B_3C_6$	228	506	1519	0.45	1.4	3.0	6.0

Test specimens were prepared according to BS EN 196-1 [19], with admixtures added after an initial 90 s mixing period of water with sand and cement. At the end of the mixing period, three mortar increments were placed in each mold. Compaction of each increment was conducted with 25 jolts, using a jolting table. At the end of the process, the specimens were covered with impermeable polyethylene sheets for 24 h in standard laboratory conditions. Subsequently, the specimens were removed from the molds and placed in a water tank containing a lime-saturated distilled water solution at 20 ± 1 °C. Curing was continuous up to the testing age.

2.3. Experimental Procedure

Electrical resistivity measurements, non-steady-state migration tests, and water permeability under hydrostatic pressure were conducted. The following sections detail the test procedures. Compressive strength tests were performed on 40 mm nominal dimension cubes according to BS EN 196-1 [19], with the loading rate regulated using an ELE compression machine (ELE ADR-Auto V2.0 2000 kN). Drying shrinkage was determined from day 2 to day 28 according to ASTM C157 [20]. The thermal expansion coefficient experiments were carried out according to BS EN 1770 [21], with readings at −20, 0, +23, +40, and +60 ± 1 °C. In addition, the porosity accessible to water (Φ) was measured using the vacuum water absorption method [22]. All tests except for the thermal expansion measurements were carried out under standard laboratory conditions of 23 ± 2 °C and 65% RH.

2.4. Electrical Resistivity

An indirect method to assess durability performance of cementitious systems is through electrical resistivity or conductivity testing. Whittington et al. defined the electrical resistivity of any material as the resistance in ohms per unit length between the opposite faces of a unit cube of the material [23]. The resistivity of mortar (ρ') depends on the moisture content, porosity, tortuosity, and interconnectivity of the pore system. Surface conductivity of the pore walls, conductivity of the pore solution, and the nature of aggregates may further influence this parameter [24,25].

The standard method for electrical resistivity measurements is ASTM C1760 [26]. However, the standard is rarely followed, probably because authors prefer to use the available instrumentation in their labs, electrode configurations, and different geometrical dimensions of the specimen concerning the standard [27–30]. The technique used in the present study involved two cylindrical specimens of 100 mm in diameter and 50 mm in length, extracted from a Ø100 × 200 mm cast cylinder. Each specimen was saturated with a calcium hydroxide saturated solution in a vacuum container at a pressure of 5 kPa. The specimen was sealed into a rubber sleeve, exposing the two parallel surfaces. A pair of stainless steel electrodes were then placed on the parallel surfaces of the sample, and current I (A) was induced through application of a potential drop (ΔE) of 10 V. Measurements were taken of 28-d age specimens for each mortar mix designation, with readings taken at first contact and after 24 h of continuous exposure to electric current. Resistivity was then calculated according to Ohm's law:

$$\rho' = \frac{\Delta E}{I} \cdot \frac{A}{e} \quad (1)$$

where A (m^2) was the cross-sectional area of the electrodes and e (m) was the distance between electrodes.

2.5. Effective Chloride Diffusion Coefficient

2.5.1. Definition and Significance

The effective chloride diffusion coefficient, D_{eff} (m^2/s), is one of the governing parameters in the chloride-induced corrosion models, reflecting the ability of the material to resist chloride penetration. In the Nernst–Planck equation or Fick's first law, this is an intrinsic property, independent of the chloride binding capacity or chloride concentration [31,32]. This can be measured by means of a steady-state migration test in saturated conditions or calculated using analytical formulas from parameters such as the apparent chloride diffusion coefficient or the electrical resistivity [25].

Ion movement in solution arises from differences in electrical or chemical potential or from hydrodynamic transport (convection). Mass transfer to an electrode can be described by the Nernst–Planck equation, using the approach for one-dimensional mass transfer along the x axis [33,34]:

$$J_i(x) = -D_i\frac{\partial C_i(x)}{\partial x} - \frac{z_i F}{RT} D_i C_i \frac{\partial \phi(x)}{\partial x} + C_i v(x) \quad (2)$$

where x is the distance from the surface, $J_i(x)$ is the flux of species i [mol/(s·cm^2)], D_i is the diffusion coefficient (cm^2/s), $\partial\phi(x)/\partial x$ is the potential gradient, $\partial C_i(x)/\partial x$ is the concentration gradient, z_i is the charge, C_i is the concentration (mol/L) of species i, and $v(x)$ is the velocity with which a volume element in solution moves along the axis, expressed in cm/s.

In a steady-state migration test, the chloride ion content is measured against time in the downstream compartment of the migration cell. The effects of chemical activity and diffusion flow are neglected concerning the electrical migration flow. An effective chloride diffusion coefficient $D_{ss(mig)}$ can be calculated when the potential drop and chloride flux are constant, using the modified Nernst–Planck equation [35]:

$$D_{ss(mig)} = \frac{RT}{ZF'} \cdot \frac{e}{\Delta E} \cdot \frac{Q}{\gamma C_0 t} \quad (3)$$

where R is the ideal gas constant (R = 8.3144 J/mol·K), T the absolute temperature (K), Z the valence number of chloride ion (Z = 1), F' the Faraday constant (F' = 96,480 J/V·mol), e the sample thickness (m), ΔE denotes the electrical potential drop between the two sides of the sample (V), Q the cumulative quantity of chloride ions arriving in the downstream compartment during the test duration t (s), and γ and c_0 (mol/m^3) are the activity coefficient and concentration of chloride ions, respectively, in the catholyte solution.

Other techniques can also be used to assess $D_{ss(mig)}$ via conductivity measurements in the downstream compartment or titration in the upstream [31,36,37]; however, $D_{ss(mig)}$ measurements can suffer from poor accuracy and reproducibility. An easy method to assess the D_{eff} is through the non-steady-state migration test with some assumptions for chloride binding.

2.5.2. Assessment of the Effective Chloride Diffusion Coefficient from Non-Steady-State Migration Tests

A method to determine the non-steady-state migration coefficient $D_{ns(mig)}$ is presented in the standardized procedure NT Build 492 [38]. According to this method, an external electrical field is applied axially across a cylindrical concrete or mortar specimen. After a certain test period, the specimen is axially split, and the freshly split sections are sprayed with silver nitrate solution. The chloride penetration depth is measured from the visible white silver precipitation. The $D_{ns(mig)}$ is then calculated by the following solution:

$$D_{ns(mig)} = \frac{RT}{ZF'E} \cdot \frac{x_d - a\sqrt{x_d}}{t} \tag{4}$$

where x_d is the average penetration depth (m) and E and a are supplemental terms defined in [38] as functions of the test conditions.

During this test, the contact time is short (usually less than 48 h), and the velocity of the ions transported through the pore structure is remarkably high. Therefore, a non-equilibrium state is expected, and chloride binding can be neglected, according to L. Tang and L.O. Nilsson [39,40]. Furthermore, L. Tang [41] derived the following relationship between $D_{eff(mig)}$ and $D_{ns(mig)}$:

$$D_{eff(mig)} = D_{ns(mig)} \cdot \left(1 + \frac{f_a}{1000} (\gamma C_0)^{B-1} \frac{W_{gel}}{\varepsilon}\right) \tag{5}$$

where W_{gel} denotes the gel content in concrete (kg/m^3), ε is the volume fraction of pore or porosity (m^3/m^3), and f_a and B are adsorption constants. The weak interaction of chlorides with cementitious systems during non-steady-state migration tests was further observed by M. Castellote et al. [42]. According to V. Baroghel-Bouny et al. [25], by assuming no binding, $D_{eff(mig)}$ can be deduced from $D_{ns(mig)}$ according to the Equation (6):

$$D_{eff(mig)} = D_{ns(mig)}\ \Phi \tag{6}$$

where Φ represents the porosity accessible to water.

2.5.3. Assessment of the Effective Chloride Diffusion Coefficient from Electrical Resistivity Measurements

If the mortar is considered a solid electrolyte, then the D_{eff} of charged species i in mortar is related to its partial conductivity σ_i. This relationship is expressed through the Nernst–Einstein Equation (7) [43]:

$$D_{eff} = \frac{RT\sigma_i}{Z_i^2 F'^2 C_i} \tag{7}$$

A more simplified equation for $D_{eff(resistivity)}$ is presented in [25]. According to this study, the effective chloride diffusion coefficient is inversely proportional to the water-saturated material's electrical resistivity (ρ').

The pore solution in a mortar specimen has relatively high conductivity, whereas the solid phase has a low value. The solid and pore solution conductivity σ is defined according to [44] as the ratio of the pore solution electrical conductivity to the formation factor:

$$\sigma = \frac{\sigma_0}{F} = \frac{1}{\rho'} \tag{8}$$

The σ_0 of the pore solution can be determined experimentally for each mortar designation with ionic concentration analysis [45]. However, the technique may require high precision equipment to avoid errors during sampling and testing, or measurement could be impossible for specimens with low porosity. Another method to determine σ_0 is through algebraic forms, aiming at an accuracy within 10%. According to [34,46], σ_0 can be expressed as the weighted sum of the equivalent ionic conductivity λ_i of each ionic species:

$$\sigma_0 = \sum_i Z_i C_i \lambda_i \tag{9}$$

where Z_i and C_i (mol/L) are the species valence and molar concentration.

The OH^- ion is the most significant contributor to the pore solution conductivity of a well-hydrated specimen. Other ionic species, such as Na^+, K^+, Ca^{2+}, and SO_4^{2-}, should be secondary contributors due to their concentration changes resulting from cement dissolution, product precipitation, and ettringite growth after the first 24 h of cement hydration [47]. According to [46], the electrical conductivity of most pore solutions can be accurately estimated from the contribution of the Na^+, K^+, and OH^- ions. For concentrations near 1 mol/L, an approximation of the λ_i can be given by Equation (10):

$$\lambda_i = \frac{\lambda_i^o}{1+G_i I_M^{1/2}} \tag{10}$$

$$I_M = \frac{1}{2} \sum_i Z_i^2 C_i \tag{11}$$

where the equivalent conductivity of the ionic species at infinite dilution, λ_i^o, can be found in [48], the ionic strength, I_M, can be estimated using Equation (11), and the empirical conductivity coefficients, G_i for each ion, are reported in [46].

A method to describe the alkali content in cement pore solution was developed by H.F.W. Taylor [49]. The total number of Na^+ and K^+ moles per cement mass is calculated from the corresponding mass fraction and molecular weight. Alkalis in cement are partly bound in the clinker phase and partly present as readily soluble sulfates. The available moles of alkali ions in the pore solution, released through cement hydration, are predicted as [50]:

$$n_i^s = \frac{n_i^T m_{OPC}\left(f_i^{sul} + \alpha f_i^{hyd}\right)}{(V_w + b_i \alpha)} \tag{12}$$

where n_i^T is the total quantity of the alkali ion I in the anhydrous cement (mol/g), f_i^{sul} is the fraction of alkali as sulfate, f_i^{hyd} is the fraction of alkali ion i in the clinker phase, m_{OPC} is the mass of ordinary Portland cement (g), V_w is the volume of water in the pore solution (cm^3), b_i is the binding factor (cm^3), and α is the hydration degree.

S.Y. Hong and F.P. Glasser [51,52] studied the alkali binding in synthetic hydration products (C-S-H (calcium silicate hydrate) and C-A-S-H (calcium aluminate silicate hydrate)) using the concept of the distribution ratio, R_d (cm^3/g), to describe the alkali-binding capacity of the synthetic gels. The alkali-binding capacity of the C-S-H increases linearly with increasing alkali concentrations in the solution. Employing the concept of binding yields:

$$n_i^s = \frac{n_i^T m_{OPC}\left(f_i^{sul} + \alpha f_i^{hyd}\right)}{(V_w + R_d m_{C-S-H})} \tag{13}$$

in which m_{C-S-H} is the mass of C-S-H in the hydrated paste (g).

The alkali binding capacities of hydrated Portland cement pastes were investigated by Brouwers and Van Eijk [50] using a three-dimensional computer model CEMHYD3D, the H.F.W. Taylor's method and the concept of the distribution ratio. They proposed that only the hydration product C-S-H can bind alkalis instead of the relative fraction of C-S-H and AFm phases forming in the paste.

In the present study, the C-S-H content for each mixture was estimated using the analytical formula for cement hydration kinetics proposed by Mounanga [53], and the mix composition of each designation. The Na^+ and K^+ fractions in sulfate phases and clinker were taken from [50], and the distribution ratios from the literature [51,52]. Table 3 presents the numerical predictions for the alkali content based on the binder type and content and the W/C ratio [49–52]. The concentration of OH^- was deduced from the electroneutrality condition [31].

2.6. Intrinsic Liquid Water Permeability

The liquid water permeability of mortar can be measured using the Taywood Engineering permeability test rig and the flow method. The intrinsic permeability can be expressed using Darcy's law [54], with the assumption that the effective porosity is fully saturated, and no interactions occur between water and mortar:

$$Q = \frac{kA\Delta P}{\mu L} \quad (14)$$

where Q is the flow rate (m^3/s), k is the water permeability coefficient under steady-state conditions (m^2), A is the specimen's ingress area (m^2), ΔP is the difference in pressure between the two ends of the test specimen (Pa), μ is the water viscosity (Pa·s), and L is the length of the test specimen (m).

Table 3. Pore solution composition (mol/m^3) based on Taylor [49], H and G [51,52], and B and VE [50] expressions for the binder type, content, and W/C ratios of the present study.

Time (Hours)	a (%)	φ_p/φ_p^0	n^s_{Na} Taylor	n^s_{Na} H and G	n^s_{Na} B and VE	n^s_K Taylor	n^s_K H and G	n^s_K B and VE
a. Calculated pore solution based on a W/C ratio of 0.50								
0.01	0.0	1.00	24.85	24.95	24.85	126.11	126.11	126.11
1	1.4	0.97	26.25	25.96	26.25	131.33	129.88	131.33
16	19.1	0.62	54.43	43.87	54.43	235.83	191.02	235.83
27	30.4	0.39	99.35	61.85	99.36	402.43	252.97	402.45
64	34.9	0.30	135.26	71.24	135.27	535.58	285.56	535.63
158	36.9	0.26	159.32	76.07	159.34	624.82	302.38	624.89
311	38.2	0.24	180.31	79.60	180.35	702.70	314.71	702.78
615	40.1	0.20	218.50	84.83	218.55	844.34	332.98	844.47
1000	40.7	0.19	233.81	86.58	233.87	901.11	339.13	901.26
b. Calculated pore solution based on a W/C ratio of 0.45								
0.01	0.0	1.00	27.61	27.61	27.61	140.13	140.13	140.13
1	1.4	0.97	29.27	28.90	29.27	146.40	144.60	146.40
16	19.1	0.58	64.93	50.45	64.94	281.33	219.81	281.34
27	30.4	0.32	133.45	73.55	133.46	540.53	301.37	540.57
64	34.9	0.23	201.99	86.26	202.04	799.87	346.63	799.97
158	36.9	0.18	257.22	92.97	257.29	1008.81	370.66	1008.99
311	38.2	0.15	313.27	97.96	313.38	1220.92	388.58	1221.18
615	40.1	0.11	440.63	105.47	440.84	1702.83	415.61	1703.37
1000	40.7	0.10	503.95	108.02	504.24	1942.47	424.85	1943.18

An equivalent expression to Darcy's law was developed by Valenta [55] for specimens exhibiting low permeability and where the flow rate cannot be measured:

$$K = \frac{e^2 v}{2ht} \quad (15)$$

where K is the water permeability coefficient (m/s), e is the depth of water penetration (m), v is the void content which does not become filled with water except under pressure (%), h is the hydraulic head (m), and t is the exposure period under hydraulic pressure (s). The relationship between K and k is expressed in Equation (16) [55]:

$$k = \frac{K\mu}{\rho g} \quad (16)$$

where ρ is the water density (kg/m^3) and g is the acceleration due to gravity (m/s^2).

According to the literature [56–58], for normal strength ordinary Portland cement (OPC) hardened pastes, k can be estimated using the Katz–Thompson relationship (see Equation (17)), where the critical pore diameter d_c and the formation factor F of the material are known.

$$k = \frac{d_c^2}{226\,F} \quad (17)$$

The pore diameter can be measured using the mercury intrusion porosimetry method; however, the measurements are strongly influenced by parameters such as sampling, specimen preparation, drying methods, heterogeneity of the specimens, and the colloidal cement matric [59,60]. As a result, the method gives a wide range of distributions for hydrated cement materials that do not reflect the actual sizes of the present pores. In the present study, the k and d_c were obtained from water permeability measurements using Valenta's expression and Equations (16) and (17). The values for F are given in Section 3.2.

2.7. Shrinkage Strains

Cracking occurs when tensile strain arising from restrained contraction or strain differential exceeds the tensile strain capacity of mortar and concrete structures. Early thermal strains, autogenous and drying shrinkage, expansion from water absorption, and annual temperature variations contribute to early and long-term contraction [61,62]. Numerous factors, including the coefficient of thermal expansion (CTE), the restraint to movement, the strain differential within an element, and the resistance to tensile strain influence the risk of early-age cracking. According to BS EN 1922-1-1 [63], these strains should be included in the design process as excessive cracking in a concrete structure reduces its strength, durability, and service life.

Physical actions such as drying shrinkage are influenced by changes in the hardened cement paste's moisture content, the degree of hydration, and the external environment's relative humidity (RH). The binder and alkali content, the cement fineness, and the binder type are known parameters for governing the associated capillary stresses [64–67]. Furthermore, some admixtures may compensate for shrinkage by minimizing the moisture losses or by producing expansive compounds, resulting in an initial net volume increase in the mortar, which offsets subsequent shrinkage [12,68]. BS EN 1992-1-1 provides a method for calculating drying shrinkage based on the RH, the dimensions of the element, and the strength class of the concrete. In addition, this parameter can be determined experimentally according to ASTM C157 [20] using a length comparator.

The main factors affecting the thermal expansion coefficient, a_c, are the moisture content and the type and quantity of the aggregate. The influence of other factors such as cement type, mix proportions, and age on this property is of a lesser extent [69]. A value equal to 10 $\mu\varepsilon/°C$ for a_c is given by BS EN 1992-1-1 and BS EN 1992-3 with limited consideration on early-age thermal strains [63,70]. Characteristic values for a_c based on the aggregate type are given in [61]; however, it is recommended for this parameter to be determined experimentally. The test methods involve subjecting mortar specimens to a change in temperature and measuring deformation [21] or by measurement of strain during the cooling arm of the heat cycle [61].

3. Results and Discussion

3.1. Results from Electrical Resistivity Measurements

Figure 2 displays the average electrical resistivity results with their standard deviations measured upon initial contact and after 24 h of continuous exposure to electric current. Resistivity and rapid chloride penetration testing methods take the electrical resistance of cementitious systems as an indication of their resistance to chloride ion penetration. An empirical classification, given in [71], defines high chloride permeability for bulk resistivity < 50 Ω·m and moderate permeability for resistivity 50–100 Ω·m. Furthermore, estimated chloride diffusion coefficients are $>15 \times 10^{-12}$ m^2/s and 10–15 $\times 10^{-12}$ m^2/s for high and moderate chloride permeability, respectively.

The results suggested that electrical resistivity was reduced with the period of exposure to direct current (DC). Resistivity was independent of specimen length, but DC measurements were expected to be affected by polarization. The relatively high DC combined with the increased exposure period made the DC measurements susceptible to the Joule effect [71,72], thereby, resulting in a current increase and further resistivity reduction over time.

The difference in resistivity between the LC-$R_{0.5}$ and LC-$R_{0.45}$ indicated that the lower W/C ratio influenced the volume of pores facilitating ionic conduction in the mortar specimens [73,74]. The mixtures containing 3% $CaCO_3$ by weight of cement exhibited a corrosion probability similar to the reference LC-$R_{0.45}$ regardless of the presence of $C_{22}H_{44}O_2$ or $C_{18}H_{34}O_2$. A minor improvement to the resistivity of specimens containing 1.4% $C_{18}H_{34}O_2$ by weight of cement was attributed to the plasticizing effect of this polymer [75] and the better degrees of compaction and deflocculation of cement particles.

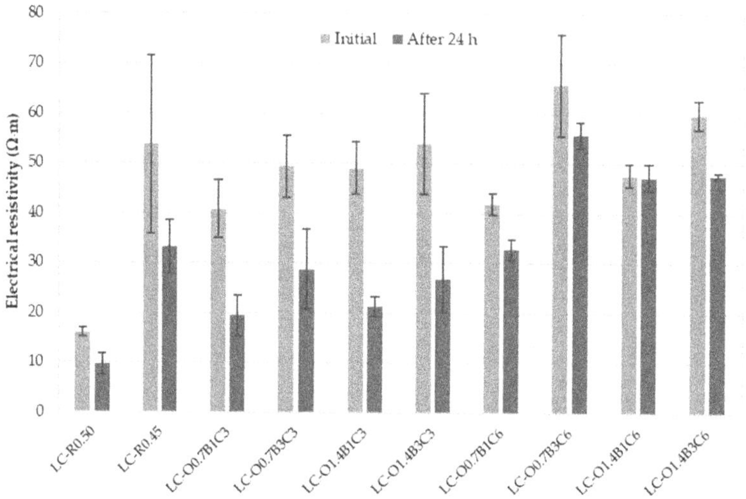

Figure 2. Experimental mean values of electrical resistivity measured on water-saturated mortar specimens initially and after 24 h of continuous exposure to electric current.

The synergistic effect of $CaCO_3$, $C_{22}H_{44}O_2$, and $C_{18}H_{34}O_2$ was evident on the specimens LC-$O_{0.7}B_3C_6$, LC-$O_{1.4}B_1C_6$, and LC-$O_{1.4}B_3C_6$. Figure 3 presents the cylindrical specimens produced for electrical resistivity and rapid chloride migration testing for each mixture. The absence of surface defects on these specimens served as evidence of improved homogeneity and compaction. In addition, the lower standard deviation of the electrical resistivity measurements of these specimens compared to the reference LC-$R_{0.45}$ provided greater confidence concerning the expected performance.

3.2. *Effective Chloride Diffusion Coefficient*

Figure 4 presents a comparison of the D_{eff} from resistivity and chloride migration measurements. Excellent agreement was observed for D_{eff} obtained through the Taylor, Brouwers, and Van Eijk models for pore solution resistivity estimation. The following order describes the correlation of effective diffusivities obtained from the non-steady-state migration and the conductivity test method:

$$D_{eff(resist)H\&G} > D_{eff(resist)Taylor/B\&VE} > D_{eff(mig)}$$

The non-steady-state (nss) migration method provided the lowest effective diffusivity, being several times lower than that of the conductivity method using the S.Y. Hong and F.P. Glasser model. It should be noted that the values of $D_{eff(resist)}$ could be questioned by the assumptions made for the distribution ratio and the Na^+ and K^+ fractions in sulfate and clinker phases. However, a specific correlation between the diffusivities may have still existed.

As shown in Figure 5a,b, the *W/C* ratio and the exposure period to DC influenced the correlation between $D_{eff(resist)}$ and $D_{eff(mig)}$. The $D_{eff(mig)}$ was reduced with the increase in the water content. This occurrence could be attributed to the lower porosity or tortuosity caused by improved homogeneity and particle distribution at the higher *W/C* ratio. A linear increase in the chloride transport with the pore size and volume was reported by

$CaCO_3$ by weight of cement. The performance further improved when $C_{22}H_{44}O_2$ content rose from 1 to 3% by cement weight, confirming that F was related to the B_y-C_z content. In the case of $C_{18}H_{34}O_2$ incorporation, the effect was not sufficiently clear. Figure 6b points out the inverse effect when $C_{18}H_{34}O_2$ content rose from 0.7 to 1.4% in mixtures with 3% $C_{22}H_{44}O_2$ by weight of cement. This observation was also consistent with the chloride migration measurements.

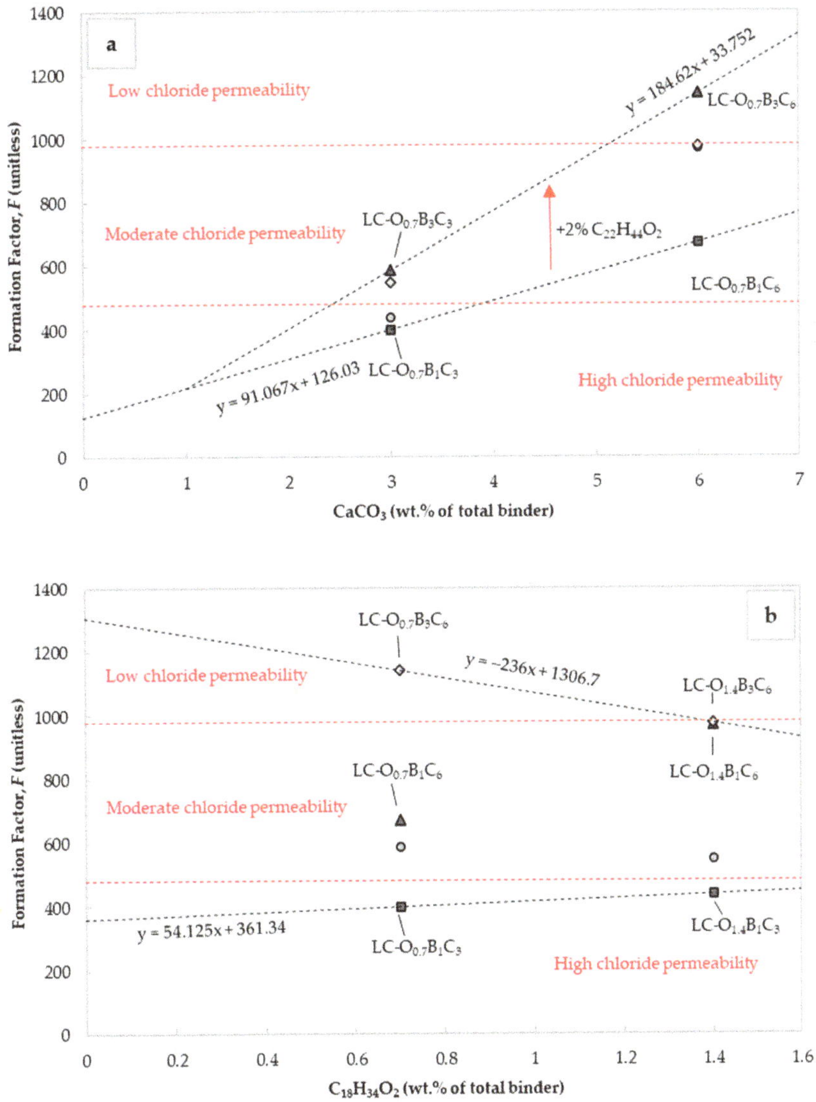

Figure 6. Influence of (**a**) $CaCO_3$ and (**b**) $C_{18}H_{34}O_2$ content on formation factor. The indicated chloride permeability classes were taken from [79].

3.3. Results from Water Permeability Measurements

Figure 7a shows the relationship between the measured k and the derived d_c for the different mixes. The increase in the W/C ratio significantly reduced the water permeability and critical pore diameter. This was consistent with observations made and discussed in the previous sections. For specimens with 6% $CaCO_3$ by weight of cement, the d_c was reduced below 0.11 μm and the k below 7.4×10^{-20} m^2. This corresponded to a water permeability reduction of 85.2% and a critical pore diameter decrease of 60.7% compared to the reference LC-$R_{0.45}$. The lower permeability exhibited by these specimens could be attributed to the void filling capacity of $CaCO_3$ and the plasticizing effect of $C_{22}H_{44}O_2$ and $C_{18}H_{34}O_2$.

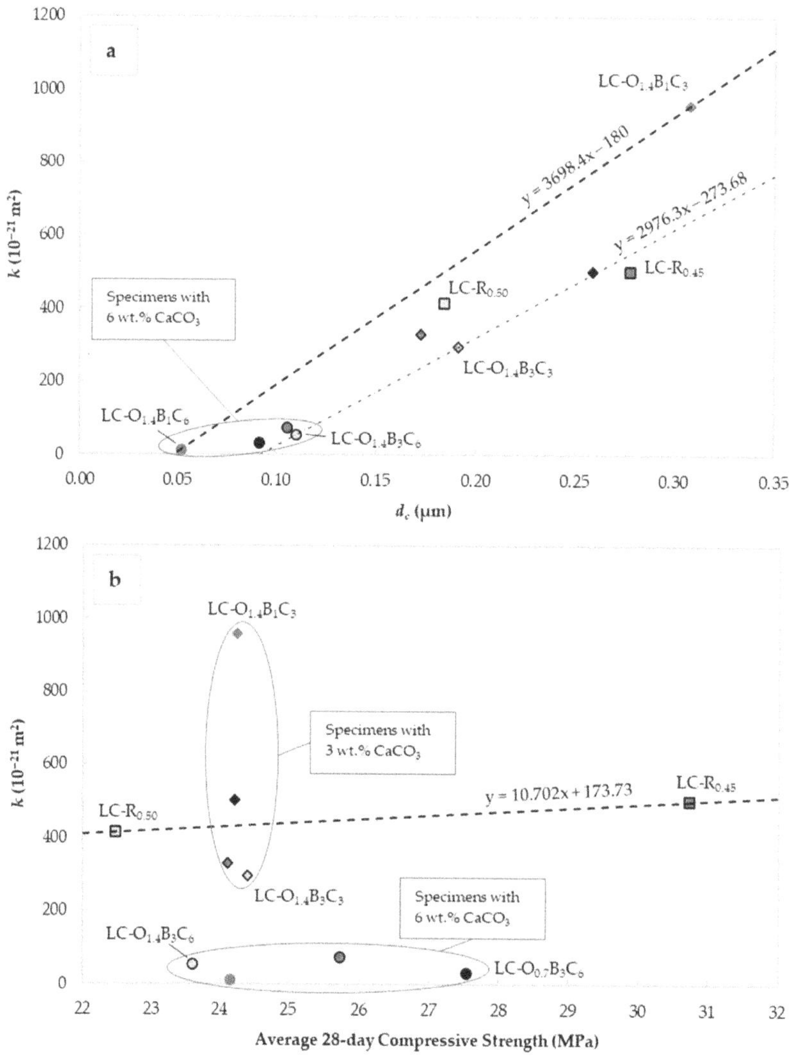

Figure 7. Mean values of "intrinsic" liquid water permeability k assessed using Valenta's expression vs. (**a**) critical pore diameter determined with the Katz–Thompson relationship and (**b**) the average 28-day compressive strength for each mix design.

The $C_{22}H_{44}O_2$ capacity to perform as an effective water repellent admixture under hydrostatic pressure seemed to depend on the $C_{18}H_{34}O_2$ content and the degree of dispersion in the specimen. While some improvement was observed between the specimens LC-$O_{1.4}B_1C_3$ and LC-$O_{1.4}B_3C_3$, this was not consistent with specimens LC-$O_{0.7}B_1C_3$ and

$LC\text{-}O_{0.7}B_3C_3$. The latter could be assigned to variations in the homogeneity of the specimens obtained with the lower $C_{18}H_{34}O_2$ content.

Figure 7b depicts the values of k vs. the average 28-day cube compressive strength for each mortar specimen. The standard deviation varied from 0.7 to 2.4 MPa, with the highest value observed in the specimens for $LC\text{-}O_{1.4}B_3C_6$ and the lowest in $LC\text{-}O_{1.4}B_1C_3$ and $LC\text{-}O_{0.7}B_1C_6$. For ease of assessment, the standard deviations are excluded from Figure 7b. The plain OPC mixtures exhibited ranges from 22.5 to 30.7 MPa, with the lower values observed in the $LC\text{-}R_{0.50}$, indicative of the lower strengths usually observed at higher W/C ratios. Incorporating $CaCO_3$, $C_{22}H_{44}O_2$, and $C_{18}H_{34}O_2$ had an inverse effect on strength, with values generally rising at higher $CaCO_3$ and $C_{22}H_{44}O_2$ contents. Nevertheless, these results confirmed that permeability and strength provided complementary information concerning durability, and justified the selection of both parameters as durability indicators.

3.4. Results from Drying Shrinkage and Thermal Expansion Measurements

Using the Controls digital length comparator 62-L0035/A, drying shrinkage measurements were undertaken at 2- and 28-day age, according to ASTM C157. The a_c parameter was determined with readings taken at −20, 0, +23, +40, and +60 ± 1 °C on 28-day age specimens, according to BS EN 1770. The specimens were initially cured for seven days in lime-saturated water at 23 ± 2 °C, followed by a twenty-one-day curing period at 65% RH. Figure 8 presents the drying shrinkage vs. compressive strength from experimental measurements and estimations using BS EN 1992-1-1 for cement class CEM 32,5R [16]. Findings showed that for the same binder type, the experimental measurements for LC-R0.50 demonstrated similar values to those predicted by BS EN 1992-1-1. The lower shrinkage strains observed in LC-R0.45 could be attributed to the lower water content and volume contraction during the different stages of curing and drying.

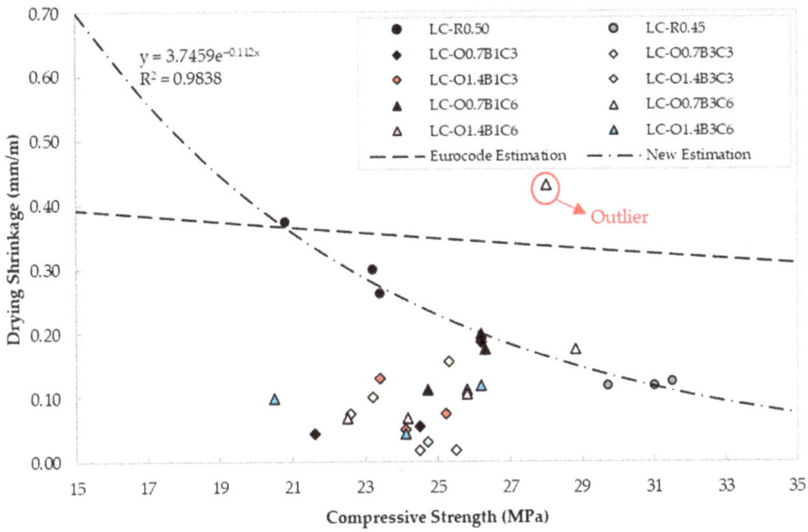

Figure 8. Drying shrinkage values vs. compressive strength for each mix design.

The specimens containing $CaCO_3$, $C_{22}H_{44}O_2$, and $C_{18}H_{34}O_2$ generally demonstrated similar to or lower shrinkage strains than the reference LC-R0.45. A specific relationship between the admixture content and the shrinkage cannot be verified from the present results of this study. $CaCO_3$ is less reactive than Portland cement, but it may provide nucleation sites for hydration products to precipitate and fill the voids between cement particles, thus, leading to improved particle size distribution, higher density and milder moisture loss compared to the reference specimens. This, in turn, can reduce the magnitude of internal strains from drying shrinkage [80]. In addition, the presence of $CaCO_3$ promotes transformation of monosulfoaluminates to ettringite, which increases the total volume of hydrate products, thus reducing the chemical shrinkage [81,82].

More recent studies showed that soluble soaps of unsaturated fatty acids function as surfactants and swell when in contact with water [83,84]. This reaction may start at any stage of cement hydration, resulting in expansion phenomena. In addition, calcium stearate is the product of butyl stearate hydrolysis. Calcium stearate may react during cement hydration and generate lighter hydrophobic compounds that fill smaller voids. This process promotes a higher BET surface area and mitigates moisture loss from evaporation. As a result, entrapped solvent water leads to different hydration rates, capillary tensions, and contraction.

The CTEs from −20 to 60 °C for all mixtures are reported in Figure 9. According to [60], concrete containing quartzite or limestone aggregate exhibits a CTE in the range of 11.7–14.6 and 4.3–10.3 $\mu\varepsilon/°C$, respectively. The values reported in Figure 9 appeared in their majority to be within the range for the characteristic CTE of quartzite mortars. Moreover, there was no distinct pattern between the admixture content and this property, thus, providing evidence that the three admixtures investigated in this study did not affect this parameter negatively.

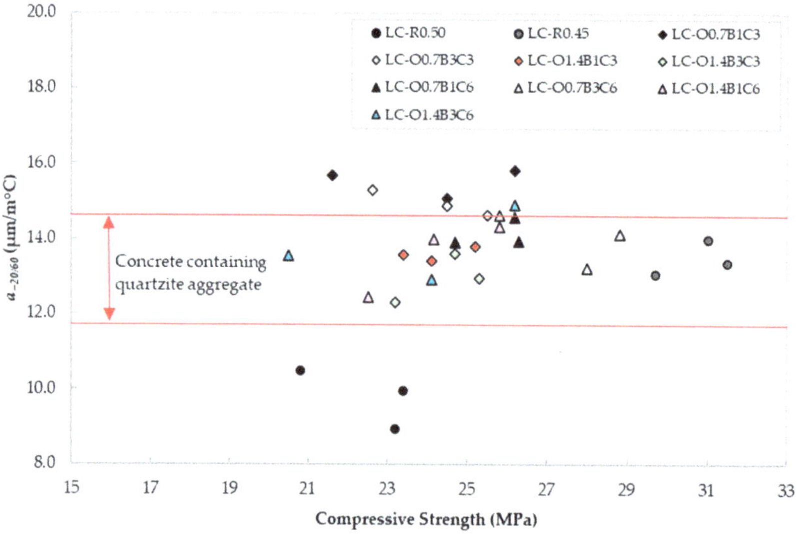

Figure 9. CTE vs. compressive strength for each mix design.

3.5. Estimated Service Life

The service life definition given by FIB model code 34 [85] is the time elapsed for the reinforcement to reach a certain chloride concentration level, where depassivation and corrosion initiation occurs. A full probabilistic design approach in [85] is based on Fick's second law of diffusion, with the condition that diffusion processes dominate the transport of chlorides. Furthermore, an increase in temperature increases the rate of diffusion and lowers the effective threshold value to initiate corrosion. The model employed in this study assumes that the diffusion coefficient and threshold values are estimated at 20 °C. The initial chloride content of mortar was calculated from the chloride content of each component of the mix.

The service life estimates for the types of mortars studied in this work are based on submerged marine exposure conditions at 1 m depth. On this basis, it is assumed that chlorides for marine conditions are 18.3 ± 2.5 g/L (North Sea) converted to chloride surface concentration using the methods developed by Lindvall [86]. Based on [85], the chloride threshold is considered approximately 0.6% by weight of binder but varies greatly depending on the concrete, steel properties, and exposure conditions. In the submerged marine zone, where little oxygen is available, the chloride threshold can increase to 2.0% [86].

Considering a 50 mm coverage with no convection zone and an aging factor between 0.17 and 0.20 [87], which was not taken into consideration beyond 40 years, an estimate of a 34–43-year service life was predicted for the reference specimen LC-$R_{0.45}$. For the same composition, with no admixtures and a *W/C* ratio equal to 0.5, the service life extended up to 78 years. As expected, incorporating $CaCO_3$, $C_{22}H_{44}O_2$, and $C_{18}H_{34}O_2$ generally extended the service life. The results summarized in Table 4 indicate that the mixtures with 6.0% $CaCO_3$, 1.4% $C_{18}H_{34}O_2$, and $C_{22}H_{44}O_2$ content greater than 0.7% by weight of cement exceeded the 100-year life expectancy. Combinations with lower contents exhibited comparable performance depending on material dispersion and compaction degree. Nonetheless, the overall extension of life expectancy by incorporating these admixtures was 76.7 to 269.8%.

Table 4. Service life estimate using two aging factors.

Mix Design	Service Life (Years)	
	a = 0.17	a = 0.20
LC-$R_{0.50}$	65	78
LC-$R_{0.45}$	34	43
LC-$O_{0.7}B_1C_3$	79	95
LC-$O_{0.7}B_3C_3$	112	135
LC-$O_{1.4}B_1C_3$	97	117
LC-$O_{1.4}B_3C_3$	110	133
LC-$O_{0.7}B_1C_6$	63	76
LC-$O_{0.7}B_3C_6$	132	159
LC-$O_{1.4}B_1C_6$	108	131
LC-$O_{1.4}B_3C_6$	115	139

4. Conclusions

Based on the experimental findings of this study, the following conclusions can be drawn:

1. Portland limestone cement mixtures with no plasticizing additives may exhibit lower strengths and higher drying shrinkage when the *W/C* ratio increases from 0.45 to 0.50.
2. Portland limestone cement mixtures with a lower *W/C* ratio require additional effort or the addition of plasticizing admixtures to achieve the required level of compaction. Otherwise, durability may be impaired, as was proven in the present study by the higher chloride migration coefficients observed in specimens with lower *W/C* ratios.
3. Incorporating $CaCO_3$ in Portland limestone cement mortars may improve durability, reduce shrinkage, and lower compressive strength marginally. The mechanical and durability properties extensively depend on the material dispersion degree and, thus, the presence of supplementary admixtures with plasticizing properties that improve rheology and compaction.
4. Good durability-related properties were pointed out for Portland limestone cement mortars containing butyl stearate and oleic acid, particularly where a 6% $CaCO_3$ by weight of cement was incorporated. The effective chloride diffusion and intrinsic liquid water permeability were significantly lower than the reference specimen LC-$R_{0.45}$ due to lower porosity. This further confirms the complementary nature of the two DIs to assess durability to potential corrosion.
5. The thermal expansion coefficients were indicative of the aggregate type, and the presence of $CaCO_3$, $C_{22}H_{44}O_2$, and $C_{18}H_{34}O_2$ in the mixtures did not alter this property. The drying shrinkage at 28 days was equal to or lower than the reference specimen LC-$R_{0.45}$. Therefore, it was concluded that material combinations of this study cannot impose risks for shrinkage strains to exceed the design codes and model recommendations.

6. The methods presented in this paper to assess DIs can be helpful to engineers in selecting material types or methodologies for durability assessment. For example, the data displayed can provide helpful references when testing or entering data in predictive models. However, some of the parameters used for the service life predictions, such as the aging factor, may require further investigation, particularly for cementitious composites containing the aforementioned admixtures.
7. The depassivation time and chloride threshold depend on the pore solution OH^- content. While some estimates were made in the present study using several approaches, these excluded the potential effect of the admixture hydrolysis. Furthermore, the potential effect of the sulfate content on the early hydration kinetics requires further assessment and validation. For future work, the pore solution chemistry is recommended to be determined experimentally for a more accurate calculation of the service life.

Author Contributions: Conceptualization, A.M. and A.S.; methodology, A.M. and A.S.; project administration, A.S.; validation, A.M. and A.S.; formal analysis, A.M.; investigation, A.M.; data curation, A.M.; writing—original draft preparation, A.M.; writing—review and editing, A.M. and A.S.; visualization, A.M.; supervision, A.S.; All authors have read and agreed to the published version of the manuscript.

Funding: This research received no external funding.

Data Availability Statement: Not applicable.

Conflicts of Interest: The authors declare no conflict of interest.

Abbreviations

CTE	Coefficient of thermal expansion
C-S-H	Calcium silicate hydrate
C-A-S-H	Calcium aluminate silicate hydrate
DIs	Durability indicators
OPC	Ordinary Portland cement
RH	Relative humidity
SCMs	Supplementary cementing materials
W/C	Water to cement ratio

References

1. Basheer, P.A.M. Durability of Concrete Structures. In Proceedings of the Sixth International Conference (ICDCS2018), University of Leeds, Leeds, UK, 18–20 July 2018; Whittles Publishing Ltd.: Dunbeath, UK, 2018.
2. Badr, A.; Fentiman, C.; Grantham, M.; Mangabhai, R. Concrete for the modern age, Developments in materials and processes. In Proceedings of the 1st International ICT Conference on Cement and Concrete Technology, Military Technology College, Muscat, Oman, 20–22 November 2017; Whittles Publishing Ltd.: Dunbeath, UK, 2017.
3. Dhir, R.K.; Hewlett, P.C.; Csetenyi, L.; Newlands, M.D. Role for concrete in global development. In Proceedings of the International Conference, University of Dundee, Scotland, UK, 10 July 2008; IHS BRE Press: London, UK, 2008.
4. Sun, J.; Chen, Z. Influences of limestone powder on the resistance of concretes to the chloride ion penetration and sulfate attack. *Powder Technol.* **2018**, *338*, 728. [CrossRef]
5. Dhir, R.K.; Jones, M.R. Development of chloride-resisting concrete using fly ash. *Fuel* **1999**, *78*, 137–142. [CrossRef]
6. Luo, R.; Cai, Y.; Wang, C.; Huang, X. Study of chloride binding and diffusion in GGBS concrete. *Cem. Concr. Res.* **2003**, *33*, 1–7. [CrossRef]
7. Dransfield, J.M.; Hewlett, P.C.; Hodgkinson, L.; Gibb, I.; Price, W.F. *A Guide to the Selection of Admixtures for Concrete*, 2nd ed.; Technical Report 18; The Concrete Society: Blackwater, UK, 1980; p. 28.
8. Neville, A.M.; Brooks, J.J. *Concrete Technology*, 2nd ed.; Pearson: Harlow, UK, 2010; p. 158.
9. Baroghel-Bouny, V. *Evaluation and Prediction of Reinforced Concrete Durability by Means of Durability Indicators, Part I: New Performance-Based Approach, ConcreteLife '06, International RILEM-JCI Seminar on Concrete Durability and Service Life Planning: Curing, Crack Control, Performance in Harsh Environments*; RILEM Publications SARL: Paris, France, 2006; pp. 259–269. [CrossRef]
10. Panesar, D.K.; Zhang, R. Performance comparison of cement replacing materials in concrete: Limestone fillers and supplementary cementing materials—A review. *Constr. Build. Mater.* **2020**, *251*, 118866. [CrossRef]

11. Dhir, R.K.; Limbachiya, M.C.; McCarthy, M.J.; Chaipanich, A. Evaluation of Portland limestone cements for use in concrete construction. *Mater. Struct.* **2007**, *40*, 459–473. [CrossRef]
12. Malakopoulos, A.; Chatzigeorgiou, M.; Boukos, N.; Salifoglou, A. Durability performance of Portland limestone cement mortar containing butyl and zinc stearate admixtures. *Mater. Struct.* **2021**, *54*, 6–8. [CrossRef]
13. Hewlett, P.C. *Lea's Chemistry of Cement and Concrete*, 4th ed.; Elsevier Ltd.: Oxford, UK, 2006; pp. 869,887–890.
14. Lagazzo, A.; Vicini, S.; Cattaneo, C.; Botter, R. Effect of fatty acid soap on microstructure of lime-cement mortar. *Constr. Build. Mater.* **2016**, *116*, 384–390. [CrossRef]
15. *BS EN 934-2:2009+A1:2012*; Admixtures for Concrete, Mortar, and Grout—Part 2: Concrete Admixtures—Definitions, Requirements, Conformity, Marking and Labelling. The British Standards Institution: London, UK, 2012.
16. *BS EN 197-1:2011*; Incorporating Corrigenda November 2011, October 2015, and February 2019; Cement—Part 1: Composition, Specifications, and Conformity Criteria for Common Cements. The British Standards Institution: London, UK, 2011.
17. *BS EN 933-1:2012*; Tests for Geometrical Properties of Aggregates—Part 1: Determination of Particle Size Distribution—Sieving Method. The British Standards Institution: London, UK, 2012.
18. *BS EN 196-7:2020*; Method of Testing Cement—Part 7: Methods of Taking and Preparing Samples of Cement. The British Standards Institution: London, UK,, 2020.
19. *BS EN 196-1:2016*; Methods of Testing Cement—Part 1: Determination of Strength. The British Standards Institution: London, UK, 2016.
20. *ASTM C157-75*; Standard Test Method for Length Change of Hardened Hydraulic-Cement Mortar and Concrete. ASTM International: West Conshohocken, PA, USA, 2017.
21. *BS EN 1770*; Products and Systems for the Protection and Repair of Concrete Structures—Test Methods, Determination of the Coefficient of Thermal Expansion. The British Standards Institution: London, UK, 1998.
22. RILEM TC. *CPC 11.3 Absorption of Water by Concrete by Immersion under Vacuum*; Recommendations for the Testing and Use of Construction Materials; E & FN SPON: London, UK, 1994; pp. 36–37. [CrossRef]
23. Wittington, H.W.; McCarter, J.; Forde, M.C. The conduction of electricity through concrete. *Mag. Concr. Res.* **1981**, *33*, 48–60. [CrossRef]
24. Bamforth, P.B.; Price, W.F. *An International Review of Chloride Ingress into Structural Concrete*; Contractor Report 359; Transport Research Laboratory: Scotland, UK, 1997; pp. 83–87.
25. Baroghel-Bouny, V.; Kinomura, K.; Thiery, M.; Moscardelli, S. Easy assessment of durability indicators for service life prediction or quality control of concretes with high volumes of supplementary cementitious materials. *Cem. Concr. Compos.* **2011**, *33*, 832–847. [CrossRef]
26. *ASTM C1760-12*; Standard Test Method for Bulk Electrical Conductivity of Hardened Concrete. ASTM International: West Conshohocken, PA, USA, 2012.
27. Zhang, L.; Ding, S.; Han, B.; Yu, X.; Ni, Y.Q. Effect of water content on the piezoresistive property of smart cement-based materials with carbon nanotube/nanocarbon black composite filler. *Compos. Part A Appl. Sci. Manuf.* **2019**, *119*, 8–20. [CrossRef]
28. Ding, S.; Ruan, Y.; Yu, X.; Han, B.; Ni, Y.Q. Self-monitoring of smart concrete column incorporating CNT/NCB composite fillers modified cementitious sensors. *Constr. Build. Mater.* **2019**, *201*, 127–137. [CrossRef]
29. D'Alessandro, A.; Meoni, A.; Ubertini, F.; Luigi Materazzi, A. Strain measurement in a reinforced concrete beam using embedded smart concrete sensors. In *Conference on Italian Concrete Days, Proceedings of the Italian Concrete Days 2018, Lecco, Italy, 14–15 June 2018*; Springer: Cham, Switzerland, 2019; pp. 289–300. [CrossRef]
30. Cosoli, G.; Mobili, A.; Tittarelli, F.; Revel, G.M.; Chiariotti, P. Electrical resistivity and electrical impedance measurement in mortar and concrete elements: A systematic review. *Appl. Sci.* **2020**, *10*, 9152. [CrossRef]
31. Nguyen, T.Q.; Baroghel-Bouny, V.; Dangla, P. Prediction of chloride ingress into saturated concrete on the basis of a multi-species model by numerical calculations. *Comput. Concr.* **2006**, *3*, 401–422. [CrossRef]
32. Baroghel-Bouny, V.; Thiery, M.; Wang, X. Performance-based assessment of durability and prediction of RC structure service life: Transport properties as input data for physical models. *Mater. Struct.* **2014**, *47*, 1669–1691. [CrossRef]
33. Bard, A.J.; Faulkner, L.R. *Electrochemical Methods: Fundamentals and Applications*, 2nd ed.; John Wiley & Sons: Hoboken, NJ, USA, 2001; pp. 28, 29, 138.
34. Bockris, J.O.; Reddy, A.K.N. *Modern Electrochemistry*, 2nd ed.; Kluwer Academic Publishers: New York, NY, USA, 2002; p. 1.
35. Andrade, C. Calculation of chloride diffusion coefficients in concrete from ionic migration measurement. *Cem. Concr. Res.* **1993**, *23*, 724–742. [CrossRef]
36. Truc, O.; Ollivier, J.P.; Carcasses, M. A new way for determining the chloride diffusion coefficient in concrete from steady state migration test. *Cem. Concr. Res.* **2000**, *30*, 217–226. [CrossRef]
37. Castellote, M.; Andrade, C.; Alonso, C. Measurement of the steady and non-steady-state chloride diffusion coefficients in a migration test by means of monitoring the conductivity in the anolyte chamber Comparison with natural diffusion tests. *Cem. Concr. Res.* **2001**, *31*, 1411–1420. [CrossRef]
38. *Nordtest Method NT Build 492*; Concrete, Mortar, and Cement-Based Repair Materials: Chloride Migration Coefficient from Non-Steady-State Migration Experiments. Nordic Council of Ministers: Copenhagen, Denmark, 1999.
39. Tang, L.; Nilsson, L.O. Rapid determination of the chloride diffusivity in concrete by applying an electrical field. *ACI Mater. J.* **1992**, *81*, 49–53.

40. Samson, E.; Marchand, J.; Snyder, K.A. Calculation of ionic diffusion coefficients on the basis of migration test results. *Mater. Struct.* **2003**, *36*, 156–165. [CrossRef]
41. Tang, L. *On Chloride Diffusion Coefficients Obtained by Using the Electrically Accelerated Methods. 1st RILEM Workshop on Chloride Penetration into Concrete*; RILEM: St Rémy lès Chevreuse, France, 1995.
42. Castellote, M.; Andrade, C.; Alonso, C. Chloride-binding isotherms in concrete submitted to non-steady-state migration experiments. *Cem. Concr. Res.* **1999**, *29*, 1799–1806. [CrossRef]
43. Lu, X. Application of the Nernst-Einstein equation to concrete. *Cem. Concr. Res.* **1997**, *27*, 293–302. [CrossRef]
44. Collins, R.E. *Flow of Fluids through Porous Materials*; PennWell Books: Tulsa, OK, USA, 1961; p. 38.
45. Page, C.L.; Vennesland, O. Pore solution composition and chloride binding capacity of silica-fume cement pastes. *Matér. Constr.* **1983**, *16*, 19–25. [CrossRef]
46. Snyder, K.A.; Feng, X.; Keen, B.D.; Mason, T.O. Estimating the electrical conductivity of cement paste pore solutions from OH^-, K^+ and Na^+ concentrations. *Cem. Concr. Res.* **2003**, *33*, 793–798. [CrossRef]
47. Taylor, H.F.W. *Cement Chemistry*, 2nd ed.; Thomas Telford: London, UK, 1997; pp. 212–218.
48. Haynes, W.M. *Handbook of Chemistry*, 97th ed.; CRC Press: Boca Raton, FL, USA; Taylor & Francis Group: Boca Raton, FL, USA, 2017; pp. 5-75–5-76.
49. Taylor, H.F.W. A method for predicting alkali ion concentrations in cement pore solutions. *Adv. Cem. Res.* **1987**, *1*, 5–16. [CrossRef]
50. Brouwers, H.J.H.; Van Eijk, R.J. Alkali concentrations of pore solution in hydrating OPC. *Cem. Concr. Res.* **2003**, *33*, 191–196. [CrossRef]
51. Hong, S.Y.; Glasser, F.P. Alkali binding in cement pastes: Part I. The C-S-H phase. *Cem. Concr. Res.* **1999**, *29*, 1893–1903. [CrossRef]
52. Hong, S.Y.; Glasser, F.P. Alkali sorption by C-S-H and C-A-S-H gels: Part II. Role of alumina. *Cem. Concr. Res.* **2002**, *32*, 1101–1111. [CrossRef]
53. Mounanga, P.; Khelidj, A.; Loukili, A.; Baroghel-Bouny, V. Predicting $Ca(OH)_2$ content and chemical shrinkage of hydrating cement pastes using analytical approach. *Cem. Concr. Res.* **2004**, *34*, 255–265. [CrossRef]
54. A Concrete Society Working Group. *Permeability Testing of Site Concrete*; Technical Report No. 31; The Concrete Society: Witney, UK; Alden Press: Witney, UK, 2008; pp. 31–41.
55. Neville, A.M. *Properties of Concrete*, 5th ed.; EYROLLES: Paris, France, 2011; pp. 486,496.
56. Tumidajski, P.J.; Lin, B. On the validity of the Katz-Thompson equation for permeabilities in concrete. *Cem. Concr. Res.* **1998**, *28*, 643–647. [CrossRef]
57. El-Dieb, A.S.; Hooton, R.D. Evaluation of the Katz-Thompson model for estimating the water permeability of cement-based materials from mercury intrusion porosimetry data. *Cem. Concr. Res.* **1994**, *24*, 443–455. [CrossRef]
58. Scherer, G.W.; Valenza, J.J., II; Simmons, G. New methods to measure liquid permeability in porous materials. *Cem. Concr. Res.* **2007**, *37*, 386–397. [CrossRef]
59. Rübner, K.; Fritz, T.; Jacobs, F. Precision of porosity measurements on cementitious mortars. *Stud. Surf. Sci. Catal.* **2002**, *144*, 459–466. [CrossRef]
60. Diamond, S. *Durability of Concrete and Cement Composites, Physical and Chemical Characteristics of Cement Composites*; Woodhead Publishing Series in Civil and Structural Engineering; Purdue University: West Lafayette, IN, USA, 2007; pp. 10–44. [CrossRef]
61. Bamforth, P.B. *Control of Cracking Caused by Restrained Deformation in Concrete*; C766; CIRIA: London, UK, 2018; pp. 47–49.
62. Bensted, J.; Barnes, P. *Structure and Performance of Cements*, 2nd ed.; CRC Press: Boca Raton, FL, USA, 2008; pp. 158,163,194,384.
63. *BS EN 1992-1-1:2004+A1:2014*; Eurocode 2: Design of Concrete Structures—Part 1-1: General Rules and Rules for Building. The British Standards Institution: London, UK, 2014.
64. O' Donoghue, V. *Shrinkage of Concrete*; Concrete Advice No. 56; The Concrete Society: Blackwater, UK, 2016.
65. Bentz, D.P.; Jensen, O.M.; Hansen, K.K.; Olesen, J.F.; Stang, H.; Haecker, C.J. Influence of cement particle size distribution on early age autogenous strains and stresses in cement-based materials. *J. Am. Ceram. Soc.* **2001**, *84*, 129–135. [CrossRef]
66. Wittmann, F.H.; Beltzung, F.; Zhao, T.J. Shrinkage mechanisms, crack formation and service life of reinforced concrete structures. *Int. J. Struct. Eng.* **2009**, *1*, 13–28. [CrossRef]
67. Beltzung, F.; Wittmann, F.H. Role of disjoining pressure in cement-based materials. *Cem. Concr. Res.* **2005**, *35*, 2364–2370. [CrossRef]
68. *ACI 364.15T*; Technical Note: Significance of the Shrinkage-Compensating and Nonshrink Labels on Packaged Repair Materials, ACI Committee 364. American Concrete Institute: Indianapolis, IN, USA, 2018.
69. *ACI 209R*; Prediction of Creep, Shrinkage, and Temperature Effects in Concrete structures, ACI Committee 209. American Concrete Institute: Farmington Hills, MI, USA, 1997.
70. *BS EN 1992-3*; Eurocode 2: Design of Concrete Structures Part 3: Liquid Retaining and Containment Structures. The British Standards Institution: London, UK, 2006.
71. Thomas, R.J.; Ariyachandra, E.; Lezama, D.; Peethamparan, S. Comparison of chloride permeability methods for Alkali-Activated concrete. *Constr. Build. Mater.* **2018**, *165*, 104–111. [CrossRef]
72. Stanish, K.D.; Hooton, R.D.; Thomas, M.D.A. *Testing the Chloride Penetration Resistance of Concrete: A Literature Review, FHWA Contract DTFH61-97-R-00022*; University of Toronto: Toronto, ON, Canada, 1997.
73. Amorim, J.N.S.; Silva, G.A.O.; Dias, C.M.R.; Ribeiro, D.V. Concrete containing recycled aggregates: Estimated lifetime using chloride migration test. *Constr. Build. Mater.* **2019**, *222*, 108–118. [CrossRef]

74. Sigh, N.; Singh, S.P. Carbonation and electrical resistance of self-compacting concrete made with recycled concrete aggregates and metakaolin. *Constr. Build. Mater.* **2016**, *121*, 400–409. [CrossRef]
75. Sander, M.M.; Nicolau, A.; Guzatto, R.; Samios, D. Plasticiser effect of oleic acid polyester on polyethylene and polypropylene. *Polym. Test.* **2012**, *31*, 1077–1082. [CrossRef]
76. Li, S.; Roy, D.M. Investigation of relations between porosity, pore structure, and Cl^- diffusion of fly ash and blended cement pastes. *Cem. Concr. Res.* **1986**, *16*, 756. [CrossRef]
77. Spragg, R.; Bu, Y.; Snyder, K.A.; Bentz, D.P.; Weiss, J. *Electrical Testing of Cement-Based Materials: Role of Testing Techniques, Sample Conditioning, and Accelerated Curing*; Report FHWA/IN/JTRP-2013/28; Joint Transportation Research Program, Indiana Department of Transportation and Purdue University: West Lafayette, IN, USA, 2013; p. 16. [CrossRef]
78. Archie, G.E. The electrical resistivity log as an aid in determining some reservoir characteristics. *Trans. AIME* **1941**, *146*, 54–62. [CrossRef]
79. *AASHTO PP 84-18*; Standard Practice for Developing Performance Engineered Concrete Pavement Mixtures. American Association of State and Highway Transportation Officials: Washington, DC, USA,, 2018.
80. Meddah, M.S.; Lmbachiya, M.C.; Dhir, R.K. Potential use of binary and composite limestone cements in concrete production. *Constr. Build. Mater.* **2014**, *58*, 193–205. [CrossRef]
81. Wang, D.; Shi, C.; Farzadnia, N.; Zhenguo, S.; Huangfei, J. A review on effects of limestone powder on the properties of concrete. *Constr. Build. Mater.* **2018**, *192*, 153–166. [CrossRef]
82. Valcuende, M.; Marco, E.; Parra, C.; Serna, P. Influence of limestone filler and viscosity-modifying admixture on the shrinkage of self-compacting concrete. *Cem. Concr. Res.* **2012**, *42*, 583–592. [CrossRef]
83. Maryoto, A.; Gan, S.B.; Hermanto, N.I.S.; Setijadi, R. Effect of calcium stearate in the mechanical and physical properties of concrete with PCC and fly ash binders. *Materials* **2020**, *13*, 1394. [CrossRef] [PubMed]
84. Stolz, H.J. Oleochemicals—Important additives for building protection. *ZKG Int.* **2008**, *61*, 78–86.
85. Task Group 5.6. *Bulletin 34—Model Code for Service Life Design*; International Federation for Structural Concrete: Lausanne, Switzerland, 2006.
86. Lindvall, A. Environmental Actions on Concrete Exposed in Marine and Road Environments and Its Response. Consequences for the Initiation of Chloride Induced Reinforcement Corrosion. PhD Thesis, Department of Building Technology, Building Materials, Chalmers University of Technology, Göteborg, Sweden, 2003.
87. Attari, A.; McNally, C.; Richardson, M.G. A probabilistic assessment of the influence of age factor on the service life of concretes with limestone cement/GGBS binders. *Constr. Build. Mater.* **2016**, *111*, 488–494. [CrossRef]

buildings

MDPI

Article

Study on Mechanical Properties and Durability of Alkali-Activated Silicomanganese Slag Concrete (AASSC)

Baifu Luo [1,*], Dong Wang [1] and Elchalakani Mohamed [2]

[1] College of Civil Engineering and Mechanics, Xiang Tan University, Yang Gu Tang Street, Xiangtan 411105, China
[2] School of Engineering, Civil, Environmental and Mining Engineering, The University of Western Australia (M051), 35 Stirling Highway, Perth 6009, Australia
* Correspondence: luobaifu@xtu.edu.cn

Abstract: Alkali-activated materials are produced by chemically polymerizing the aluminosilicate materials using alkaline activators, which can effectively lower the greenhouse-gas emissions (approximately 73%) released by ordinary Portland cement (OPC). Silicomanganese slag is a large solid waste discharged from the ferroalloy industry in China that can pollute the environment and occupy resources. In this paper, the slag in alkali-activated material was replaced with silicomanganese slag to address the disposal of silicomangaese slag. The flowability, setting times, compressive and flexural strengths, micro-structure and freeze-thaw resistance of alkali-activated silicomanganese slag concrete (AASSC) with varied substitution ratios, volume fractions of steel fibers and alkali-activated modulus (M_S) were exploited. As a results the compressive strengths at 56 days of AASSC with a 10% substitution ratio of silicomanganese slag reached over 80 MPa and over 132 MPa with the 3% steel fiber dosage. AASSC still reached 91 MPa and 45 MPa with 60% and 100% substitution ratios by cooperating 2% steel fibers, respectively. When the freeze-thaw cycle number reached 300, the compressive strengths of AASSC with the replacement ratios of 10%, 60% and 100% were 84%, 74% and 51% of their original values by cooperating 2% steel fibers, respectively; AASSC with the numerous substitutions of 60% and 100% were destroyed at 600 and 300 freeze-thaw cycles, respectively. AASSC with a 10% substitution ratio and 2% steel fiber content is suitable for excellent performance, and a 60% substitution ratio can also be applied to construction for the massive utilization of silicomanganese slag.

Keywords: mechanical properties; freeze-thaw resistance; micro-structure; alkali-activated; silicomanganese slag

1. Introduction

With further improvement and development in technology, silicomanganese alloys have been widely used in the steel industry. According to International Manganese Institute statistics, around 16 million tons of SiMn alloy was produced in 2018 worldwide, which increased by 16% from 2017, and production also continuously increased in 2019, with China sharing the majority [1]. It is about 1.3 tons of silicomanganese slag generated per ton of alloy production [2]. With the rapid development of silicon manganese alloy, the main metals used in China are silicon manganese alloy and low-carbon ferromanganese alloy, followed by medium-low ferromanganese alloy [3]. Silicomanganese slag is made of a variety of raw materials such as manganese ore, lime, silica, etc. The chemical components of silicon manganese slag mainly include silicon dioxide and calcium oxide, and the sum of both exceeds 55% with the silicomanganese slag having high potential hydration activity [4]. For the granular silicomanganese slag, the mineral compositions are mainly vitreous, have high activity and contain fewer minerals such as $AlSiO_5$ [5]. Silicomanganese slag can be better ground than other metal slag due to its internal composition. After testing the density and specific surface area, researchers also found that the specific surface area of granular silicon manganese slag was larger than that of other massive silicomanganese slags [6]. It was found that silicon manganese slag could complement cement and be used as cementitious materials in concrete. Silicomanganese slag has a good activity that could affect the performance of concrete [7].

Due to the rapid development of the steel industry, a large amount of metallurgical slag needs to be recycled. Explorations have been made into the recycling and utilization of metallurgical slag. Metallurgical slag has largely used in cement production, concrete admixture and road materials, but metallurgical slag has the characteristics of unstable internal chemical composition and low activity, which can reduce its practical application. To improve the activity and utilization of metallurgical slag, physical activation, chemical activation and thermal activation have mainly employed in previous literature.

(1) Physical activation

The application of mechanochemistry technology in environmental protection is a research hotspot [8]. The common form of physical activation is mechanical grinding. Through mechanical grinding, there is a large amount of mechanical energy in the metallurgical slag, and the surface has a high activity. Physical activation not only cause physical and chemical changes in its internal structure, but develop also obvious micro-cracks on the surface of the material. At the same time, crystal defects and lattice distortion occur in the structure, which accelerates the ability of other substances to participate in the reaction [9].

(2) Chemical activation

The activators can stimulate and improve the intrinsic activity of metallurgical slag for hydration. Common activators include acid–base activators, seed excitation [10], sulfate activators [11], chloride activators [12], etc., which can promote the disintegration of metallurgical slag and generate stable hydration products.

(3) Thermal activation

The common methods of thermal activation are high-temperature calcination and quenching [13], and the use of high-temperatures with the addition of structural adjusters [14].

The popular research concentrating on silicomanganese slag has been applied to the substitution of cementitious material in concrete. He et al. [15] investigated the feasibility of preparing belite–calcium sulphoaluminate cement by using a synergistic reaction of electrolytic manganese residue and barium slag, as well as limestone, and bauxite. The results showed that the compressive strengths of samples at 3, 7, and 28 days were 37.9, 47.3, and 60.0 MPa, respectively, which were better than those of OPC425. Wang et al. [16] utilized cement, sand, crushed stone, water reducer and silicomanganese slag powder to prepare C60 (compressive strength f_{cu} equals 60 MPa) concrete. The mechanical and durability performance of concrete was enhanced by substituting 10–20% of cement with ultra-fine silicomanganese ash, because the silicon manganese ash can react with the activator gypsum or lime and also fill the gaps between cement and particles. It was preliminarily estimated that each cubic meter of concrete could save 200 kg of cement. Therefore, the application of silicomanganese slag in high-strength concrete can achieve economic and mechanical effects. Srikavya et al. [17] replaced natural fine aggregates and coarse aggregates with silicomanganese slag by 0%, 5%, 10%, and 15%, respectively. The result indicated that 15% substitution ratios to natural fine aggregates and coarse aggregates could reduce a certain amount of aggregates in concrete with a limited strength reduction. However, these methods of substituting cement in concrete had a limited substitution ratio, and high-strength concrete-containing silicomanganese slag required the use of ultra-fine particles and had a limited ratio of 20%, which restricted its practical utilization.

Alkali-activated materials are seen as the alternative materials to cement-based concrete in recent years based on their low carbon emission and the use of amorphous aluminosilicate-based powder, it is activated by alkalis such as granulated blast furnace slag (GBFS) and fly ash. Zhang et al. [18] studied the durability of alkali-activated concrete and found out that the durability of alkali-activated concrete was better than cement-based concrete under the same strength level. A research review [19] indicated that the early and ultimate mechanical strengths of alkali-activated cementitious systems are better than that of the traditional cement-based binder. Ambily et al. [20] reported that UHPGC with the highest compressive strength of 175 MPa was successfully produced by mixing steel fibers under curing at ambient temperature. You et al. [21] also reported that ultra-high-performance alkali-activated concrete can be applied to structures with a lightweight. A research review [22] studied the properties of ultra-high-performance alkali-activated concrete by considering variables such as the steel fibers and activators, which discovered that the ultra-high performance alkali-activated concrete is an eco-friendly material with great mechanical strength and durability. However, the potential problem of alkali-activated materials is their costs, which are higher than those of ordinary cement-based concretes.

In the previous literature, Su et al. [23] utilized silicomanganese (SiMn) slag (ultra-fine powder, no more than 3% above 80 μm) as the cementitious material, activated by sodium hydroxide. The results indicated that SiMn slag can be applied to construction with high strength. Huang et al. [24] used three common activators to study the mechanical properties of alkali-activated materials with the addition of granular silica manganese slag and analyzed the effects of anhydrous sodium sulfate, sodium hydroxide and water glass on the performance of cementitious materials. The strength of the materials with sodium hydroxide addition increased rapidly in the early stage and slowly in the later stage. It was appropriate to use 4% sodium hydroxide to excite the cementitious materials, which could gain a better effect with water glass addition. Through comparative analysis of the three materials, it was found that sodium hydroxide is the best among all materials when the content of the activator is lower than 3%. Nath et al. [25] added granular silicon manganese slag to fly-ash-based alkali-activated cementitious material to replace a part of the fly ash. When the silicon manganese slag content increased, the reaction rate of the material accelerated. Navarro et al. [26] studied the varied activators on the alkali-activated SiMn slag pastes. The results showed that the main reaction product formed in the alkaline activation of SiMn slag is C-S-H with a low Ca/Si ratio, and it is higher for pastes activated with NaOH compared to those with water glass. The method of using activators for silicomanganese slag generally requires massive a ground effort for high-strength. The way to substitute the cementitious material (fly ash) uses a small substitution ratio. So, it is worthwhile to research a new way to improve the utilizable amount of silicomanganese slag with excellent strength to meet engineering needs.

This paper used silicomanganese slag powder to replace the grounded blast furnace slag to generate alkali-activated silicomanganese slag concrete (AASSC) with excellent performance. The main cementitious materials of slag or fly ash in alkali-activated concrete, generally have a greater percentage in alkali-activated concrete in comparison to the cement in cement-based concrete, which implies that silicomanganese slag can be utilized more in alkali-activated concrete than in cement-based concrete. Compared to silicomanganese slag in alkali-activated concrete in a previous study, the method in this study can activate the grounded blast furnace slag and silicomanganese slag both by sodium hydroxide, and silicomanganese slag powder can also act as a filler to fill the micro-voids inside cementitious materials, resulting in improved strength and durability. To systematically analyze the AASSC, the mechanical properties, freeze-thaw resistance and micro-structures of AASSC were studied by considering variables such as the substitution ratio of silicomanganese slag, alkali-activated modulus (M_S) and volume fraction of steel fiber. The list of abbreviations is shown in Table 1.

2. Raw Materials and Experimental Details

2.1. Raw Materials

Raw materials used in this study were fly ash (FA), silica fume (SF), grounded blast furnace slag (GBFS), fine sand and sand, water (W), steel fibers, polycarboxylic acid superplasticizer (SP), NaOH and water glass (WG). The chemical and physical properties of the main materials are shown in Tables 2–6 and Figures 1 and 2. The silicomanganese slag used in the test was provided by an alloy factory in Anyang, Henan Province. The treatment of silicomanganese slag encompassed grinding for 3 h. The silicomanganese slag mainlycontains Si, Ca, Mg, Al, Mg and other elements, and has certain hydraulic and volcanic properties.

Table 1. List of abbreviations.

Acronym	Abbreviations
FA	Fly ash
SF	Silica fume
GBFS	Grounded blast furnace slag
AAMs	Alkali-activated materials
AASSC	Alkali-activated silicomanganese slag concrete
UHPGC	Ultra-high performance alkali-activated concrete
LOI	Loss of ignition
WG	Water glass
SP	Polycarboxylic acid superplasticizer
XRD	X-ray diffractometer
SEM	Scanning Electron Microscope
C-S-H	hydrated calcium silicate
M_S	Alkali-activated modulus

Table 2. Chemical composition of raw materials (%).

Materials	SiO_2	Al_2O_3	K_2O	TiO_2	SO_3	Na_2O	MgO	CaO	Fe_2O_3	LOI
FA	53.36	29.09	-	-		-	0.81	2.27	3.87	2.48
SF	95.26	-	0.98	-	-	0.98	0.29	1.93	0.57	2.48
GBFS	34.21	14.15	1.95	-	0.12	-	10.32	39.11	0.82	0.7
Silicomanganese slag	36.27	7.82	-	0.67	0.74	0.28	11.01	26.49	0.79	3.98

Table 3. Physical properties of GBFS and SF.

Material	Specific Surface Area (m^2/kg)	Specific Density (kg/m^3)	Mass Coefficient (Kkc)	28 Days Activity Coefficient (%)
GBFS	519	2980	1.697	110
SF	2500	2700	-	121

Table 4. Particle size of GBFS and SF.

Particle Size (μm)	<2	<4	<10	<16	<20	<32	<64	<80	<100
GBFS	4.1	12.78	36.3	41.8	49.5	65.9	87.2	94.2	100.0
SF	98.21	100.0	100.0	100.0	100.0	100.0	100.0	100.0	100.0

Table 5. Properties of fine sand and sand.

Material	Particle Size (mm)	Bulk Density (kg/m^3)	SiO_2 (%)	Fe_2O_3	Water Content (%)
Fine sand	0.08–0.15	1600	90.5–92.7	0.03–0.10	1.5
Sand	0.178–0.425	1450	90.5–92.7	0.03–0.10	1.5

Table 6. Properties of steel fiber.

Shape	Length (mm)	Diameter (mm)	Density (g/cm^3)	Roughness	Tensile Strength (MPa)	Elongation (%)
Straight	6	0.16	2.56	smooth	2500	3.5

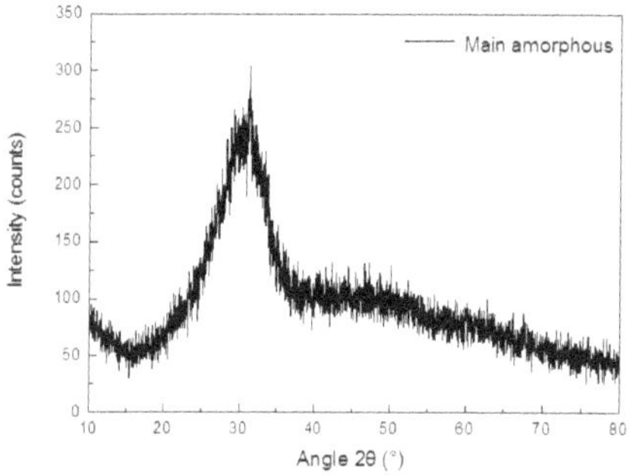

Figure 1. XRD patterns of GBFS.

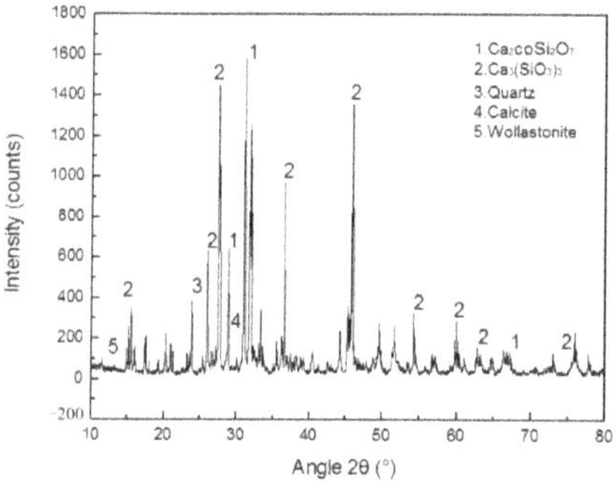

Figure 2. XRD patterns of silicomanganese slag.

As seen in Figure 1, the mineral composition of the GBFS is amorphous with high activity, which means that GBFS has high pozzolanic effects and can fill the inner voids in the matrix to improve the density of the alkali-activated concrete. As seen in Figure 2, the main mineral components of the silicomanganese slag are calcium silicate, cobalt chromite, quartz, calcite and wollastonite. The activity of the silicomanganese slag can be improved by grinding.

2.2. Proportioning and Mixing Details of Study

As shown in Table 7, the water-to-binder ratio used in this research was 0.32, the sand-to-binder ratio was 1.0, and the alkali content (calculated by Na_2O) was 7%, which was derived from previous research [27] and revised on our own raw materials. The variables such as the replacement percentage of GBFS, alkali-activated modulus (M_S) and steel fiber content of the AASSC were studied. Group S10–S30 were the factors related to the ratios of GBFS replaced with silicon manganese slag (10–100%), S10-M1 to S10-M2 were factor related to modulus M_S (1.0, 1.5 and 2.0), S10-SF1 to S10-SF3were factors related to volume ratios of the steel fiber and the sample (1%, 2% and 3%).

Table 7. Mix proportion of AASSC (kg/m³).

No.	GBFS	FA	SF	Silicon Manganese Slag	Fine Sand	Sand	Steel Fiber (%)	SP	W	NaOH	WG
S0	688	172	45	0	271.5	633.5	0	13.8	36.2	36.2	416.3
S10	619.2	172	45	68.8	271.5	633.5	0	13.8	36.2	36.2	416.3
S20	550.4	172	45	137.6	271.5	633.5	0	13.8	36.2	36.2	416.3
S30	481.6	172	45	206.4	271.5	633.5	0	13.8	36.2	36.2	416.3
S10-M1	619.2	172	45	68.8	271.5	633.5	0	13.8	162.9	54.3	208.2
S10-M1.5	619.2	172	45	68.8	271.5	633.5	0	13.8	97	45	314
S10-M2	619.2	172	45	68.8	271.5	633.5	0	13.8	36.2	36.2	416.3
S10-SF1	619.2	172	45	68.8	271.5	633.5	1	13.8	36.2	36.2	416.3
S10-SF2	619.2	172	45	68.8	271.5	633.5	2	13.8	36.2	36.2	416.3
S10-SF3	619.2	172	45	68.8	271.5	633.5	3	13.8	36.2	36.2	416.3
S60-SF2	275.2	172	45	412.8	271.5	633.5	2	13.8	36.2	36.2	416.3
S100-SF2	0	172	45	688	271.5	633.5	2	13.8	36.2	36.2	416.3

The letter 'S' represents the substitution ratio of silicomanganese slag, and the letter 'M' represents the modulus M_S. For example, S10–M1.5 indicates that the GBFS is replaced with a 10% substitution ratio of silicomanganese slag, and the modulus is 1.5. The letter 'SF' represents the volume ratio of steel fiber. For example, S10–SF1 indicates that the GBFS is replaced with 10% silicomanganese slag and the content of steel fiber is 1% of the sample volume.

Before the experiment, the alkali activator was prepared by mixing water, NaOH, and water glass for 24 h in advance. The preparation of samples was as follows. (1) Firstly, we mixed slag, fly ash, silica fume, and silicomanganese slag for 2 min at a low speed; (2) then, we added quartz sand for 1 min at a low speed; (3) finally, we slowly added alkali activator and mixed for 3 min, and the steel fibers were slowly added into the mixture for another 2 min. The fluidity test was conducted immediately after the mixing process with the conical mold, then mixtures were cast into the mold. A previous study [28] showed that 24 h of steam-curing at 80 °C could make alkali-activated slag concrete reach the strength development plateau, a method which was also applied in another study [29]. Therefore, a duration of 24 h steam-curing at 90 °C was applied in this study before the standard curing.

2.3. Experimental Methods

2.3.1. Fluidity and Setting Times

The fluidity tests of samples were conducted at room temperature, 20 ± 2 °C, and a humidity of more than 50%, and three specimens were tested for each mixture, according to the standard 'Test method for fluidity of cement mortar' (GB/T2419-2016) [30]. As seen in Figure 3, a Vicat instrument was used to measure the initial and final setting times of the samples. The tested method was as follows: (1) the inner sides of the mold were first coated with a thin layer of oil, and the round mold was placed on the glass plate; (2) we poured the slurry into the round mold, put it into the curing box, and started to record the setting time; (3) we recorded the samples as set when the pointer sank and the test needle could fall into the slurry without seeing a ring mark.

2.3.2. Mechanical Properties

The compressive strength tests of samples were prepared by pouring the slurry into the cubic mold of 40 mm, and bending strength tests were prepared by a 40 × 40 × 160 mm prism mold according to a previous study [31]. All samples were covered with plastic film and cured by steam for 1 day after demolding, and finally placed in a standard curing room at a temperature of 20 ± 5 °C with humidity of more than 95%, based on standard DL/T5144-2001 [32]. The partial specimens after steam curing are shown in Figure 4a. Three specimens were tested for each mixture at the desired age for mechanical and flexural strengths (third-point bending as shown in Figure 4b) based on the Chinese standard GB/T 17671-2020 [33].

Figure 3. Liquidity test.

(**a**) Partial specimens

(**b**) Mechanical test

Figure 4. Samples and test details.

2.3.3. Freeze-Thaw Cycle Test

Three 40 × 40 × 40 mm specimens from each mixture were applied for the freeze-thaw cycle test based on the Chinese standard GB/T50082-2009 [34]. The cycle time was 4.5 h, the temperature was set at (-18 ± 2)~(5 ± 2) °C, and the number of freeze-thaw cycles was 300. Three specimens were prepared for each mixture and firstly immersed in distilled water for more than 0.5 h. After the immersion, we dried and weighed the samples accurately before the tests. The details are shown in Figure 5.

After n freeze-thaw cycles, the mass loss rates of the samples were calculated according to Formula (1):

$$W_{\mathrm{n}} = \frac{G_0 - G_{\mathrm{n}}}{G_0} \times 100\% \tag{1}$$

where W_{n} represents the weight loss rate of the sample after n times of freezing and thawing (%), G_0 represents the weight of the sample before the freeze-thaw test (g), G_{n} represents the weight of the sample after freezing and thawing (g).

After n freeze-thaw cycles, the compressive loss rates of the samples were calculated according to Formula (2):

$$Q_{\mathrm{n}} = \frac{P_0 - P_{\mathrm{n}}}{P_0} \times 100\% \tag{2}$$

where Q_{n} represents the loss rate of compressive strength of the sample after n freeze-thaw tests (%), P_0 represents the compressive strength of the sample before the test (MPa), P_{n} represents the compressive strength of the sample after the test (MPa).

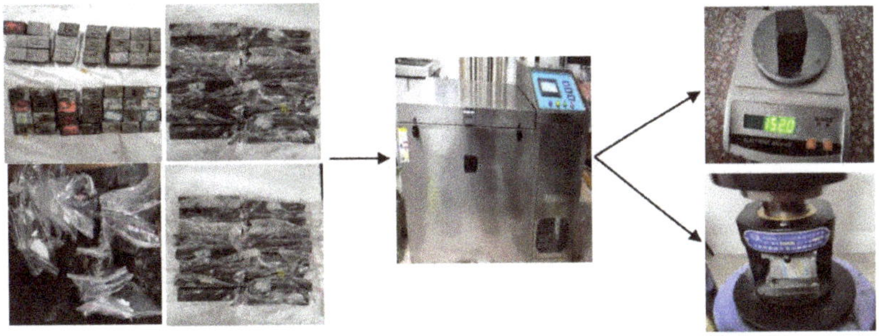

Figure 5. Apparatus and process of freeze-thaw cycle test.

2.3.4. Micro-Structure Analysis

X-ray diffractometer (XRD) tests were prepared by observation on sliced samples (2.5–5 mm). Sliced samples were firstly immersed in absolute alcohol for 24 h to stop the hydration, then ground into powders with a particle size of less than 0.08 mm; secondly, 20 g powders were took into a drying oven at (60 ± 2) °C for 48 h; finally, powders were tested at the Changsha Research Institute of Mining and Metallurgy. The working voltage used in the test was 32 kV. After the sliced samples were washed with ethanol and dried, the surface was sprayed with platinum to ensure conductivity. The Jim-6490lv scanning electron microscope was used in this test at Changsha Research Institute of mining and metallurgy.

3. Experimental Results

3.1. Working Performances of Samples

3.1.1. Fluidity of Slurry

It can be seen in Table 8 that the method of replacing GBFS with silicon manganese slag had improved the fluidity of AASSC. The slurry with a larger replacement ratio tended to have a greater fluidity. The reason may be that the alkali-activated concrete with less slag has fewer silicon aluminum products, and silicomanganese slag with low activity slows down the early reaction. With the increase of modulus, it was found that the fluidity of slurry with a modulus 2.0 was 7.9% and 2.2% higher than slurries with modulus 1.0 and modulus 1.5, respectively. This is because the mixture with a higher modulus has a better fluidizing effect on extra silicate content [35] and a higher reduction in solid concentration [36]. When the content of steel fiber increased from 1% to 3%, the fluidity of concrete changed from 225 mm to 200 mm. The reason can be attributed to the fact that steel fibers are disorderly distributed in the mixture, which increases the friction and shearing force of the mixture [37]. In addition, steel fibers adsorb a certain amount of moisture on their surface, which reduces the free water content of the material, resulting in a decrease in the fluidity of the mixture [38].

Table 8. Fluidities of slurries (mm).

-	S0	S10	S20	S30	S10-M1	S10-M1.5	S10-M2	S10-SF1	S10-SF2	S10-SF3	S60-SF2	S100-SF2
Fluidity	230	235	238	245	215	227	232	225	210	200	235	242

3.1.2. Setting Times

It can be seen in Table 9 that the initial setting times of AASSC were between 20 min and 66 min, and the final setting times were between 32 min and 74 min. When the content of silicon manganese slag increased, the setting time of slurries increased, which alleviated the problem of the fast setting time of alkali-activated concrete. The rapid condensation of alkali-activated concrete was due to the internal C-S-H composites accelerated by $[SO_4]^{4-}$ [39].

The reason that the addition of silicomanganese slag increased the setting time of alkali-activated concrete is that silicomanganese slag is less active than GBFS. The setting time of the material decreased with the increase of modulus M_S; this is because the activators with higher modulus M_S have higher $[SO_4]^{4-}$ contents and PH values, which can enhance the hydration of slurry [40]. The addition of steel fiber had a small effect on the initial and final setting times of the alkali-activated system; the difference in initial setting time was 7 min, and the difference in final setting times was 5 min between slurries with different fiber contents. The mixture with a higher fiber content had a longer setting time because the steel fibers in the mixture diluted the binders then slowed down the hydration.

Table 9. Setting times of samples (min).

-	S0	S10	S20	S30	S10-M1	S10-M1.5	S10-M2	S10-SF1	S10-SF2	S10-SF3	S60-SF2	S100-SF2
Initial	20	25	28	35	35	28	24	28	30	35	48	66
Final	32	38	39	45	42	36	30	34	35	41	55	74

3.2. *Mechanical Strength of Mixtures*

3.2.1. Effects of the Replacement Ratio of Silicomanganese Slag

Figure 6a shows the influence of the content of silicon manganese slag on the compressive strength of AASSC. The strength of alkali-activated concrete mixed with silicomanganese slag increased rapidly in the early stage and gradually slowed down in the later stage, which is consistent with previous research [25]. The reason can be attributed to the early formation of nano-crystalline or amorphous C-A-S-H gels produced through early activation of slag in the alkaline solution. The additional C-A-S-H gels can be generated through the reaction between the unreacted CaO and the SiO_2 from fine amorphous quartz [41]. The 28 day compressive strengths of alkali-activated concrete containing 0, 10%, 20% and 30% substitution ratios of silicomanganese slag were 91.5 MPa, 81.4 MPa, 63.5 MPa and 60.8 MPa, respectively. The development of mixtures at 56 days also exhibited the same tendency. The reason is that the silicon manganese slag is less active than the original cementitious materials [42], and the mixture containing a larger content of silicon manganese slag has a lower development speed, which affects the early and final strength of concrete.

It can be seen in Figure 6b that the development of flexural strength of AASSC was similar to its compressive strength, which developed particularly fast in the early stage and increased slowly in the later stage. The 28 day flexural strengths of mixtures with 0, 10%, 20%, and 30% substitution ratios were 12.9 MPa, 10.5 MPa, 8.1 MPa and 7.7 MPa, respectively. The 56 day flexural strengths of the samples also decreased with the increase in silicomanganese slag content. In summary, The sample S10, with a 10% substitution ratio of silicomanganese slag, had 2.2% increased fluidity in comparison to sample S0 without silicomanganese slag, as well as 1.3%, 4.2% decreased fluidities compared to samples S20 (with 20% substitution ratio) and S30 (30% substitution ratio), respectively; the difference is very small and could be neglected. To the compressive strength at 28 days, sample S10 had an 11% decreased value compared to sample S0, and 28.2% and 33.9% higher values than samples S20 and S30, respectively. To the flexural strength at 28 days, sample S10 had an 18.6% decreased value compared to sample S0, and 29.6% and 36.4% higher values than samples S20 and S30, respectively. Therefore, the substitution ratio of silicomanganese slag hardly affected the fluidity of slurry, but seriously influenced the compressive and flexural strengths, so sample S10 had limited strength induction when compared to S0, and was the best among other mixtures.

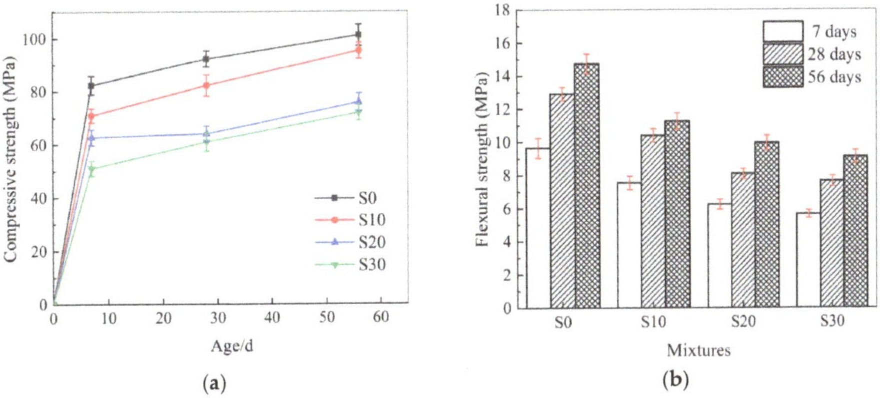

Figure 6. Effects of the content of silicomanganese slag (replacing GBFS) on compressive strength (**a**) and flexural strength (**b**) of AASSC.

3.2.2. Effect of Modulus M_S

As seen in Figure 7a, the 28 day compressive strengths of AASSC with modulus M_S of 1, 1.5 and 2 reached 71.4 MPa, 73.8 MPa and 79.4 MPa, respectively, and the 56 day compressive strengths of AASSC reached 80.5 MPa, 79.9 MPa and 90.4 MPa, respectively. The compressive strength increased with the modulus M_S. The previous study [43] had proven that the higher M_S value can provide more active silicon and generate more silicon–oxygen bonds under the same alkali content, and silicon oxygen bonds are more stable than other bonds, which is conducive to improving their strength. Moreover, water glass has more C-S-H gel than other activating materials in the case of alkali excitation, and C-S-H gel is likely to form a dense structure [44], which indicates that the high modulus can improve the strength of concrete. In general, the strength of the alkali-activated slag system increases with the modulus. However, it had a limit. The modulus of 2.0 is suitable for AASSC in this study.

As shown in Figure 7b, the 28 day flexural strengths of AASSC with 1, 1.5 and 2 modulus reached 9.4 MPa, 10.9 MPa and 11.3 MPa, respectively. When the modulus of water glass was two, the flexural strength of AASSC reached the maximum value. In addition, the flexural strength of AASSC increased with the M_S value, which is consistent with a previous study [45].

3.2.3. Effect of Steel Fiber

As seen in Figure 8a, The 28 day compressive strengths of samples with 0, 1%, 2% and 3% steel fiber content reached 79.4 MPa, 85.5 MPa, 99.4 MPa and 116.6 MPa, respectively, which indicates that the compressive strengths of samples with 1%, 2% and 3% fiber contents increased 7.7%, 25.2% and 46.9% than sample without fibers, respectively. The reason is that the steel fibers effectively change the direction of crack propagation in the process of composite cracking, reduce the stress concentration in the crack, and increase the interfacial bonding force in the concrete composite to improve the compressive strength [46].

As seen in Figure 8b, the 28 day flexural strengths of samples with 0, 1%, 2% and 3% steel fiber content reached 11.3 MPa, 12.5 MPa, 14.6 MPa and 16.5 6 MPa, respectively, which indicated that the flexural strengths of samples with 1%, 2% and 3% fiber content increased 10.6%, 29.2% and 46.0% than the sample without fibers, respectively. Furthermore, it was found that the sample without fiber addition had great brittleness, which was characterized by the sudden break when the sample reached its peak load. However, it was different when the concrete incorporated steel fibers, which showed a ductile failure with small cracks in the middle of the AASSC, and the sample did not completely break at the end of the test. Both the compressive and flexural strengths increased with the dosage of steel fibers [19]. Considering the effects of fluidity, strength and economic factors, the steel fiber content of 2% is more suitable than other samples.

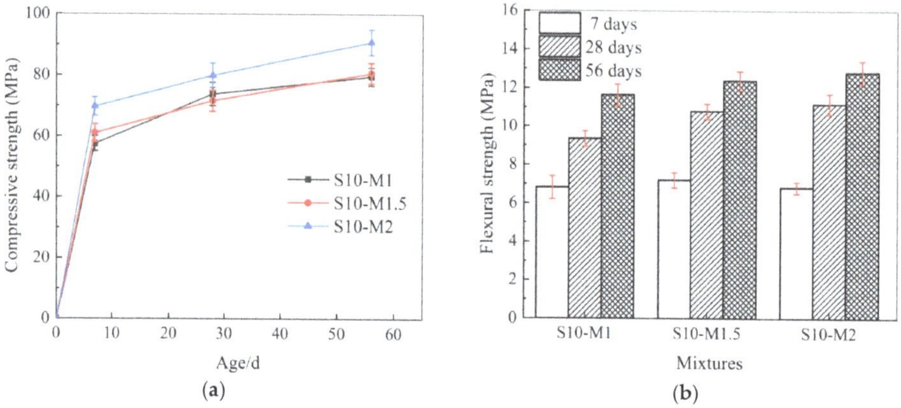

Figure 7. Effects of modulus M_S on compressive strength (**a**) and flexural strength (**b**) of AASSC.

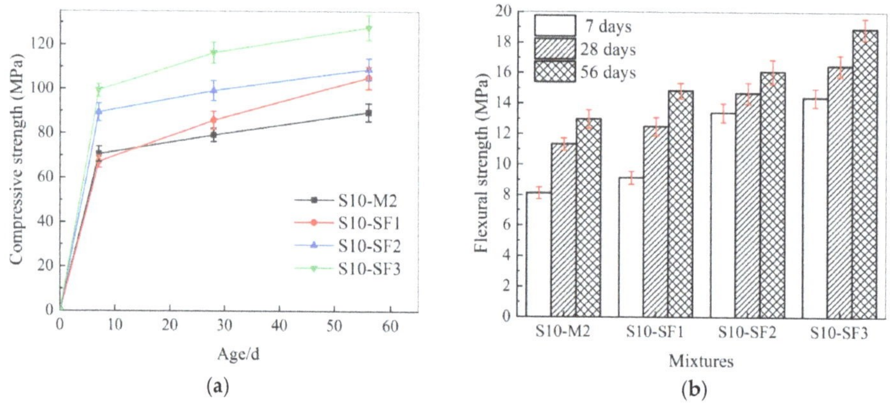

Figure 8. Effects of steel fiber content on compressive strength (**a**) and flexural strength (**b**) of AASSC.

3.2.4. Massive Substitution Ratio

As seen in Figure 9, with the same content of steel fiber, The 28 day compressive strengths of samples with 60% and 100% substitution ratio of silicomanganese slag to GBFS reached 83.4 MPa and 44.7 MPa, respectively, and the flexural strengths of these samples were 10.1 MPa and 4.1 MPa, respectively. This indicates that considerable silicomanganese slag content is unfavorable to the strength development of AASSC, which also reflects that the excessive silicomanganese slag hardly participates in the alkali-activated polymerization reaction, and eventually affects the mechanical properties of concrete [1]. Therefore, the amount of silicomanganese slag can directly control the final mechanical properties of the AASSC. The test shows that the suitable proportion of silicon manganese slag is 10% when considering high-strength properties and 60% for economic effects.

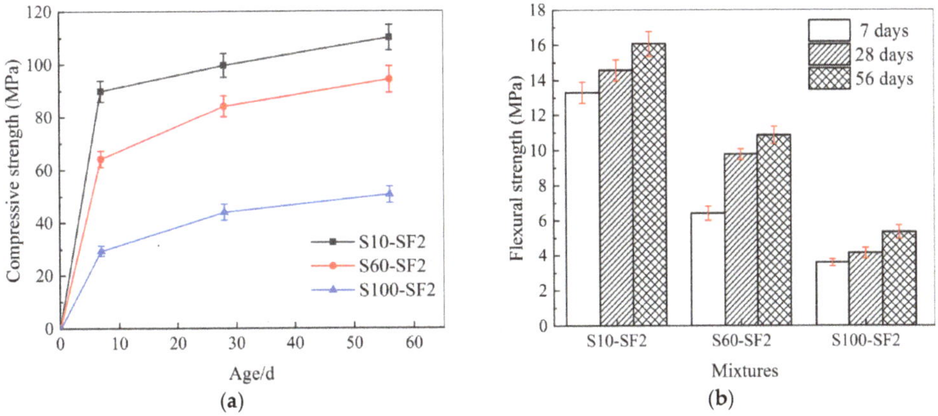

Figure 9. Effects of a massive content of silicomanganese slag on compressive strength (**a**) and flexural strength (**b**) of AASSC.

3.3. Freeze-Thaw Resistance

3.3.1. Effect of Steel Fiber

As seen in Figure 10, the change rule of compressive strength of GSCC with different fiber contents was the same during the freeze-thaw cycle test. For example, after 100 freeze-thaw cycles, the loss rates of compressive strengths of samples with 1%, 2% and 3% were 25%, 15% and 13%, respectively, which implies that the addition of steel fiber can reduce the loss rate of compressive strength of AASSC. However, the difference in the loss rate of AASSC between 2% and 3% fiber contents was small and could be neglected. The loss rate was the same as the strength loss rate; the mixture with a higher fiber content had less mass lost after the test, mixtures with 2% and 3% both achieved steady improvements compared to other mixtures and there was limited difference between them. Moreover, the mixture S10-M2 without steel fiber addition had more than a 5% mass loss ratio at 700 cycles, which indicates that the sample was broken. The reasons that steel fiber addition decreased the strength and mass loss rate in the test are that the steel fibers restrained the micro-cracks of the matrix under tension [47], which were caused by the freeze-thaw test, and optimized the micro-structure of the matrix [48], so the sample with steel fibers had improved resistance to freeze-thaw cycling. However, a high fiber content could generate flaws in the matrix and induce the resistance to the freeze-thaw test. In summary, 2% steel fiber is suitable for AASSC in the study by considering the strength of the freeze-thaw cycle test and economy.

3.3.2. Effect of Substitution Ratio

As seen in Figure 11, the mass and compressive strength of the samples decreased when the freeze-thaw cycle increased, and there were some large differences between each line, which indicates that a massive substitution ratio has remarkable influence on the concrete resistance to freeze-thaw test. When the cycle reached 300 times, the relative compressive strength ratios of AASSC with the replacement ratios of 10%, 60%, and 100% were 0.84, 0.74, and 0.51, respectively, with 0.4%, 0.9% and 2% mass loss ratios, respectively. Therefore, the AASSC with a higher substitution ratio has larger mass and strength losses. This is because a sample with a higher content of silicomanganese slag tends to have more unreacted components in the matrix [5], which generates a certain amount of holes inside the matrix and make the matrix expand when moisture invades, eventually causing tension and improving the development of micro-cracks. After 300 freeze-thaw cycles, the mass loss rate of AASSC with a 100% substitution ratio of silicon manganese slag reached 6.374%, which indicates the sample was damaged. After 600 freeze-thaw cycles, the mass loss rate of

AASSC with a 60% substitution ratio of silicon manganese slag reached 6.978%; the sample was damaged. In summary, AASSC with 2% steel fiber content and a 10% substitution ratio of silicomanganese slag had the best resistance to the freeze-thaw test, which could still have 75% compressive strength left and only 3.5% mass lost after 800 cycles. With the consideration of massive substitution, AASSC with 2% steel fiber content and a 60% substitution ratio can be applied to construction; it has 3.4% mass lost and 85% strength left after 300 cycles and is destroyed at 600 cycle times.

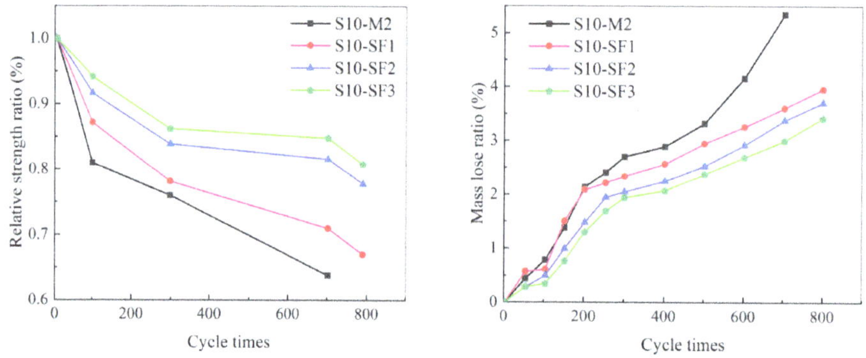

(**a**) Strength lose rate of AASSC (**b**) Mass loserate of AASSC

Figure 10. The influences of different steel fiber content on the freeze-thaw resistance of AASSC where relative strength ratio is that the compressive strength of sample before test versus the compressive strength of sample after test.

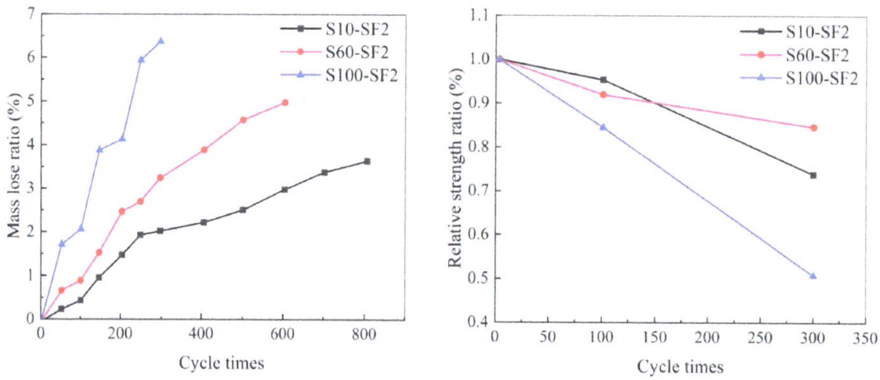

(**a**) Mass lose rate of AASSC (**b**) Strength lose rate of AASSC

Figure 11. The effects of different substitution ratios of silicomanganese slag on sample's resistance to freeze-thaw test.

3.4. Scanning Electron Microscope (SEM) Test

3.4.1. Effect of Substitution Ratio on Micro-Structure

It can be seen from in Figure 12(a1–d1) that when the content of silicomanganese slag increased, there were more sporadic holes on the surface of the matrix, which made a looser structure for concrete composite. When the substitution ratio was 30%, the matrix had the largest number of holes with a loose structure on the surface (as shown in (d1), (d2) and (d3)), which implied the compressive strength of that AASSC had a small value. This is consistent with the existing literature [49]. However, the sample S10 with a 10%

substitution ratio had a better surface than S0 without silicomanganese slag. This was maybe because the silicomanganese slag contains hydrated calcium silicate [50] and fine particles that fill the internal gaps of the mortar [6], which improve the adhesion of the particle interfaces. When compared with the micro-structure of S10, S20, and S30, S0 generally had a more compacted micro-structure, which promoted a higher strength in S0. This could explain why compressive and flexural strengths decreased with the increase of the silicomanganese slag content and the phenomenon of freeze-thaw resistance.In addition, as seen in Figure 12(a3–d3), the sample with less silicon manganese slag content had more hydration products. This is because the activity of silicomanganese slag isless than GBFS [49], and mixtures with a higher percentage of GBFS have higher hydration.

The content of silicomanganese slag also directly affected the morphology of hydration products; the hydrated products of the sample S0 were in the form of flakes and knits, crisscrossed and evenly distributed on the surface, and the sample S10 had a small amount of flake hydrated products, but a large amount of knits densely distributed on the surface. The sample S30 had more knits products than S10. This reflected that silicomanganese slag not only affected the hydration process of alkali-activated concrete, but also caused changes in the morphology of hydrated products, which was related to different chemical species, mineralogy and particle sizes, as well as the shape of fly ash and silicomanganese slag, and this ultimately led to differences in reaction and reaction product formation [25]. The silicomanganese slag addition could promote the development of knitted substances in the hydrated products and generate calcium zeolite in the matrix. When compared with the sample S0, there was a denser structure of the net and flocculent gel in the matrix with silicomanganese slag, but the hydration products of S0 had a denser and more even structure in comparison.

3.4.2. Effect of Different Modulus on Micro-Structure

The morphology of alkali-silica acid gel generated by the alkali–silica reaction was generally rose petal-type and lamellar-type in Figure 13, which can be presumed to be alkali–silica acid gel by previous research [51–53]. From the comparison of (a1), (b1), and (c1), it can be seen that the collective surfaces of alkali-activated concrete prepared with different modulus M_s were dense and plate-shaped. The interface was more continuous and smoother with the increase of the modulus, and there were few holes in the matrix. It can be seen from the microscopic morphology of the hydrated products in Figure 13(a2–c2) that the hydrated products of AASSC prepared with different modulus were identical and consisted of flocculent gel of hydrated calcium silicate (C-S-H) and some crystals such as calcite, which means that the alkali-activated mortars were hydrated well with applicable strengths and densities. The sample with a higher modulus had more hydrated products and a denser matrix, which is consistent with its strength development. This is because that the activator with a higher modulus has more alkali (Na_2O), which plays a vital role during alkali-activation; the system with a higher content of Na_2O had higher pH, which promoted more dissolution of reactive products (SiO_2, Al_2O_3) [54]. Therefore, the modulus can influence the gel product of the alkali-activated concrete and change the structure and morphology, which affects the flexural and compressive strengths in return, resulting in a sample with a higher modulus that has more uniform and denser hydration products and better compressive and flexural strengths.

As seen from Figure 13(a3–c3), S10–M1 with the modulus 1.0 had many gel substances staggering each other, and some particles were not tightly wrapped; S10–M1.5 with the modulus 1.5 had more sheet products and denser structures than S10–M1; S20–M2 with the modulus 2.0 had quartz and matrix connected, and its dense structure consists of crystals and gels. This indicates that the hydration products in the matrix increase with the modulus and the sample has better-completed hydration with a uniform distribution of hydration products, which eventually improve the mechanical properties of the sample.

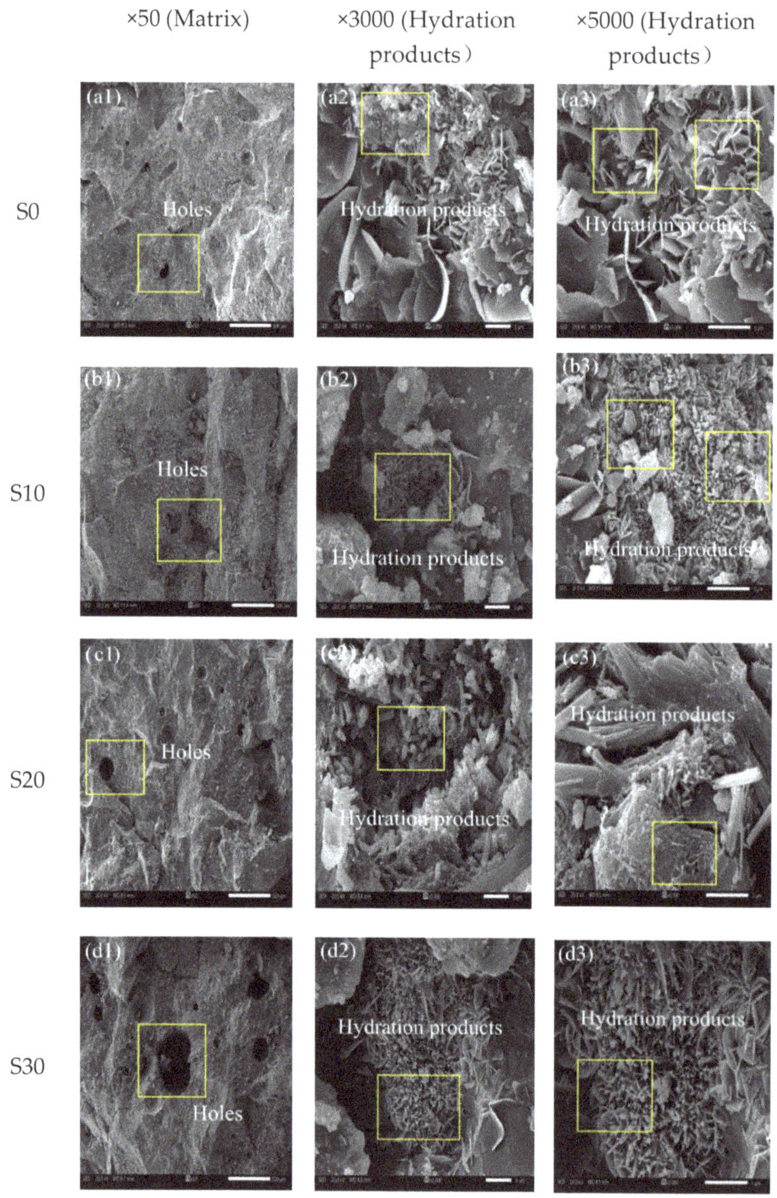

Figure 12. SEM images of AASSC with different substitution ratios of silicomanganese slag.

3.4.3. Effect of Steel Fiber on Micro-Structure

It can be seen from in Figure 14(a1–d1) that the bond between steel fibers and the matrix were connected closer when the steel fiber content was higher. When the content of steel fibers reached 3%, there were no obvious micro-cracks and pores at the bonding interface. When the content of steel fibers was 1%, the steel fibers and the matrix were separated, the surrounding holes were obvious with the uneven surface formed the matrix, and the loose connection between steel fibers and the matrix in the interface with obvious debris indicated the poor bonding performance and mechanical strength. This indicates that the steel fiber content optimizes the bonding performance between steel fibers and the matrix. The reason for this is that the steel fiber is a hydrophilic material; its polar molecules absorb water molecules to form a hydrated product film on the fiber surface, which improves the adhesion between steel fibers and matrix [37]. In summary, the research found that the fibers in the matrix promote the distribution of hydration products to improve the micro-structure of the matrix and enhance the mechanical performance of concrete.

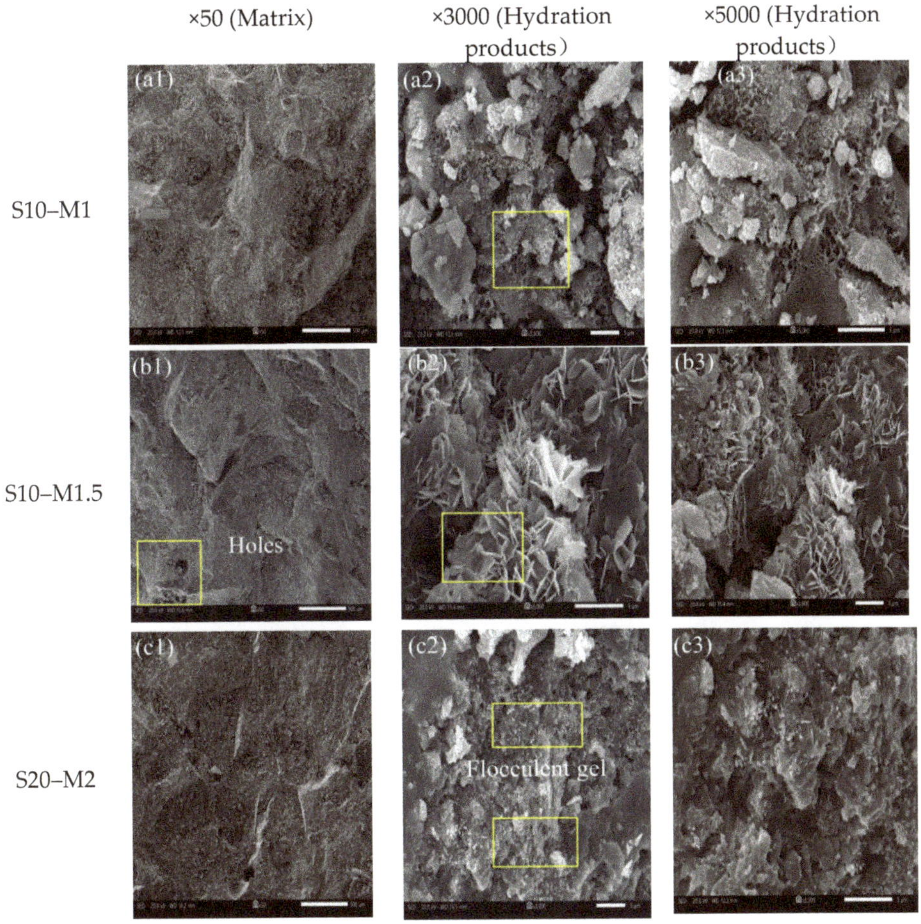

Figure 13. SEM images of AASSC with different modulus.

It can be seen from in Figure 14(a2–d2) and Figure 14(a3–d3) that the distribution of hydration products was more and more uniform and became more dense with this sequence; flocculent gel and knitted crystals were vertically and horizontally distributed on the matrix. For example, when compared with the S10 without steel fiber sample, the matrix with the higher steel fiber content had more hydration products with better hydration. Moreover, In the matrix, the steel fibers were tightly wound to form a mesh structure, which improved the bearing capacity of the matrix. The steel fibers traversed and bridged the micro-cracks to bear the tension between micro-cracks, and caused an obvious toughening effect, which effectively improved the mechanical properties of concrete, consistent with a previous study [55]. This explained the phenomenon that AASSC with steel fibers had a higher strength and toughness than the alkali-activated mortar without steel fibers.

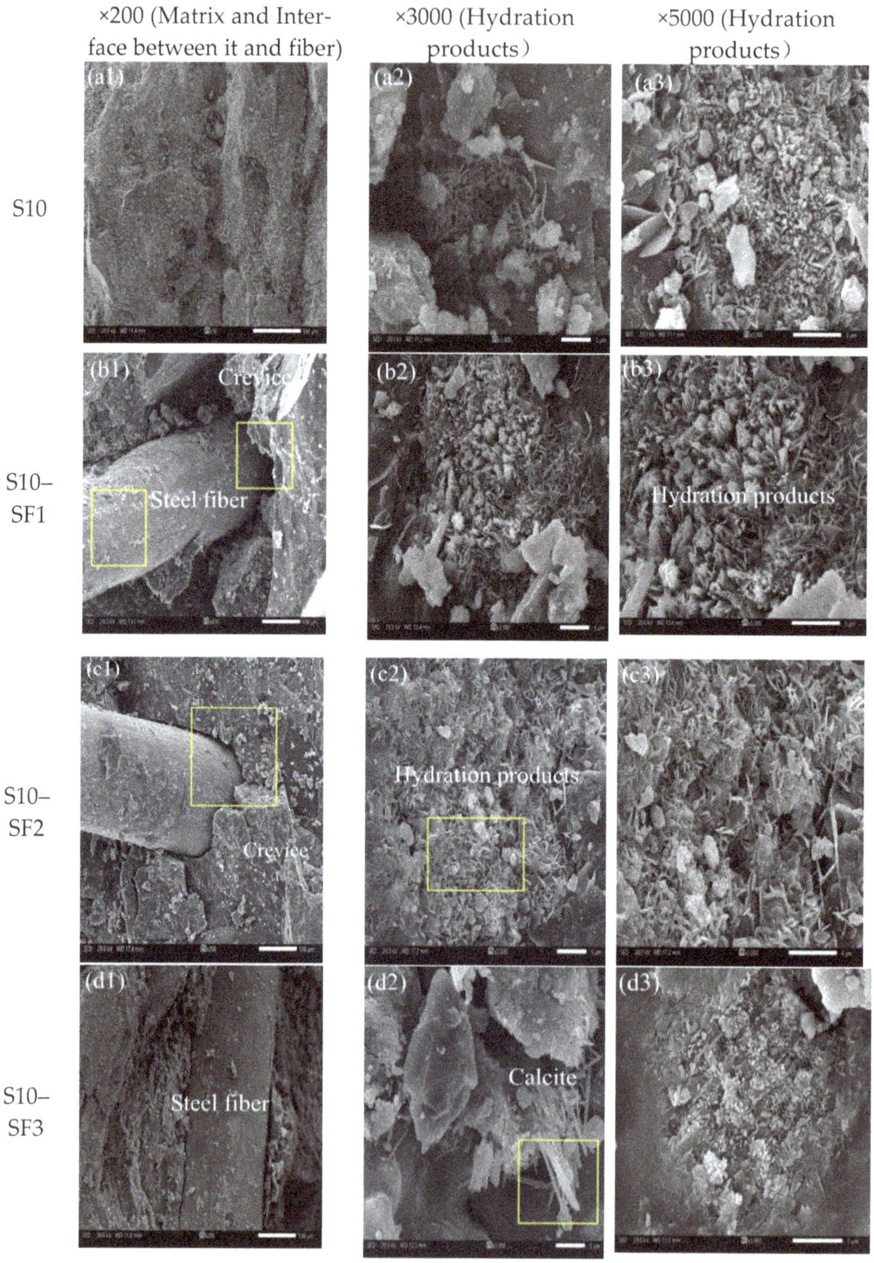

Figure 14. SEM images of AASSC with different steel fiber contents.

3.4.4. Effect of Massive Silicomanganese Slag Addition on Micro-Structure

As seen in Figure 15(a1–a3), the matrix of S10–SF2 was flat with a dense structure accompanied by considerable matrix debris on the surface of the steel fiber after being pulled out. The steel fibers were closely connected to the matrix on the surface. This indicates that there was a strong adhesive force between steel fibers and the matrix, which had a toughening effect on concrete composite. Moreover, there were many hydrated products, and the crystals distributed on the matrix, which shows a completed hydration happened in the matrix; as seen in Figure 15(b1–b3), S60-SF2 with 60% silicomanganese slag substitution ratio to GBFS had an irregular matrix surface with few holes and micro-cracks, and the steel fiber was closely connected with the matrix, which indicates that the interface between steel fibers and the matrix was closely bonded. Moreover, there were fewer hydration products than S10–SF2, which indicates S60-SF2 was less hydrated than S60-SF2; as seen in Figure 15(c1–c3), S100-SF2 with 100% silicomanganese slag substitution ratio to GBFS had an uneven matrix with a large number of pores and micro-cracks, which implied the matrix had poor compactness; the steel fiber and the matrix were separated from each other and accompanied by obvious micro-cracks. Moreover, there were few hydration products, with unreacted quartz particles on the surface, which indicated that the matrix had poor mechanical performance and durability. In conclusion, the micro-structure results can vividly express the mechanical properties and durability of AASSC.

3.5. XRD Analysis

3.5.1. Effect of Different Contents of Silicomanganese Slag

As shown in Figure 16, the main peaks of the samples from 10 degrees to 80 degrees were quartz (SiO_2), calcite ($CaCO_3$) and hydrated calcium silicate (C-S-H), among which the peak of SiO_2 was the largest. There was a "convex hull" phenomenon at 25–35° in the figure, which indicates the formation of C-S-H gel generated in the matrix [56]. From the number of diffraction peaks, the number of hydration products and calcites decreased with the increased content of silicomanganese slag. The peak of cobalt chromite (Ca_2CoSiO_7) may be caused by silicomanganese slag, which was observed in the XRD test of silicomanganese slag. Calcite may be produced by the reaction of hydrated products with carbon dioxide [57]. As seen in Figure 17, S10 with a 10% content of silicon manganese slag, had a small relative reduction of each main diffraction peak, which indicates that the crystal structure of the material was relatively completed. Therefore, the mechanical properties of the materials prepared with the small content of silicon manganese slag were better than those with a high substitution ratio, which is consistent with the analysis of the micro-structure.

3.5.2. Effect of Different Modulus

It can be seen in Figure 18 that the XRD phases of these three samples were mainly quartz (SiO_2), calcite ($CaCO_3$) and hydrated calcium silicate (C-S-H), and there was also a small amount of cobalt chromite. From Figure 19, it can be seen from the diffraction peak that the hydration products and calcite of samples increased with the modulus, but the quartz (SiO_2) of each sample had no obvious difference between different modulus. S10-M2 with modulus 2.0 resulted in an increased relative intensity and quantity of each main diffraction peak when compared to other samples, which indicates that the crystal structure of the hydrated product was better completed and explains the reason why S10-M2 had higher strength than other samples. There was no $Ca(OH)_2$ detected in the samples, which may be due to the pozzolanic nature of silica fume and fly ash or the formation of calcite with carbon dioxide.

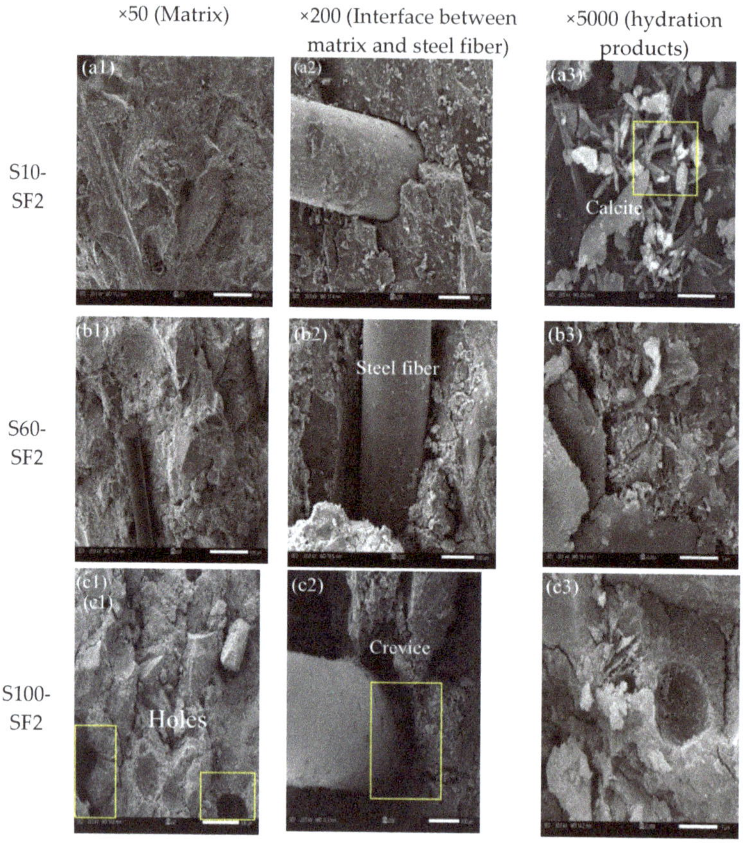

Figure 15. SEM images of AASSC with different substitution ratios of silicomanganese slag.

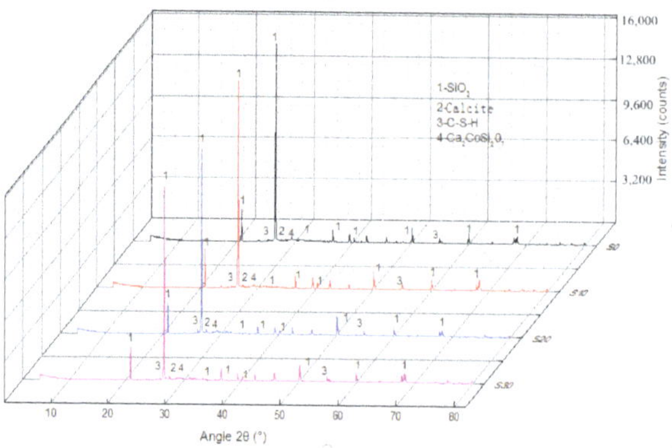

Figure 16. XRD patterns of AASSC with different contents of silico-manganese slag.

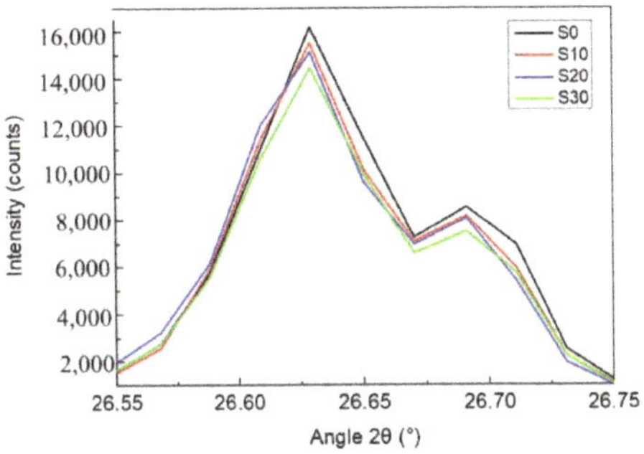

(a) SiO_2 peak values in different mortars

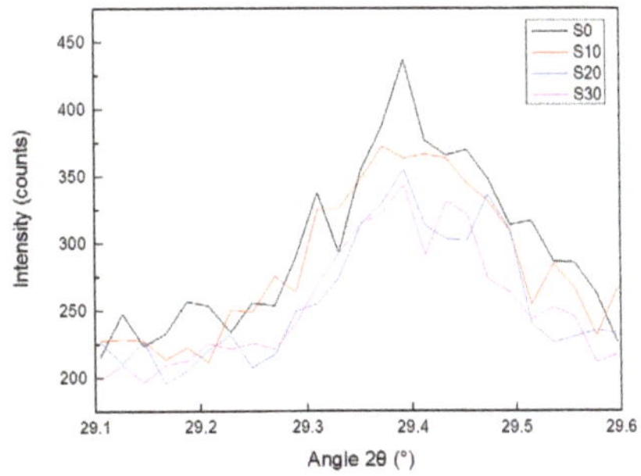

(b) C-S-H peak values in different mortars

Figure 17. Effect of different contents of silicomanganese slag on the comparison of peak values of SiO_2 and C-S-H of samples.

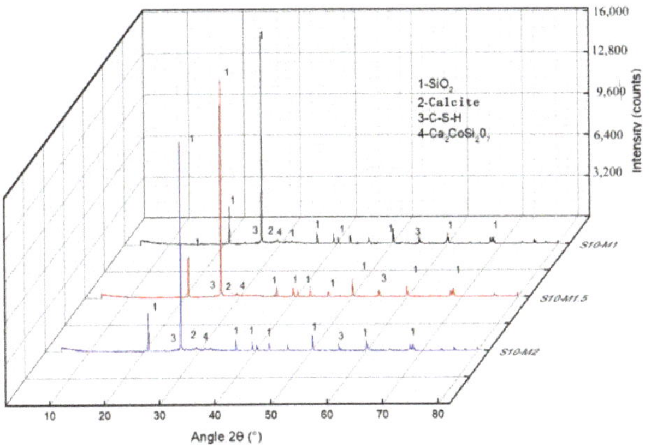

Figure 18. XRD patterns of AASSC with different modulus.

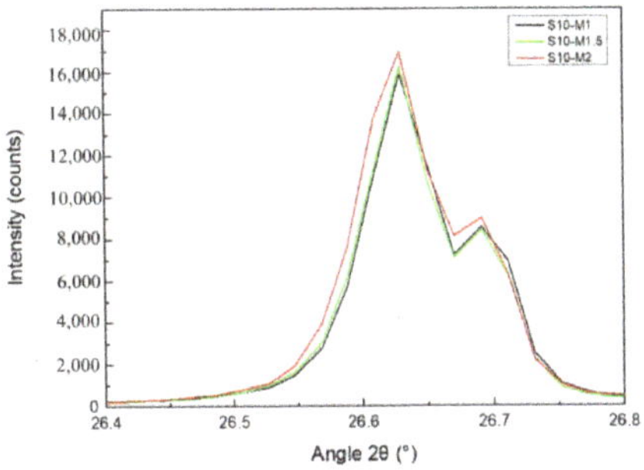

(a) SiO_2 peak values in different mortars

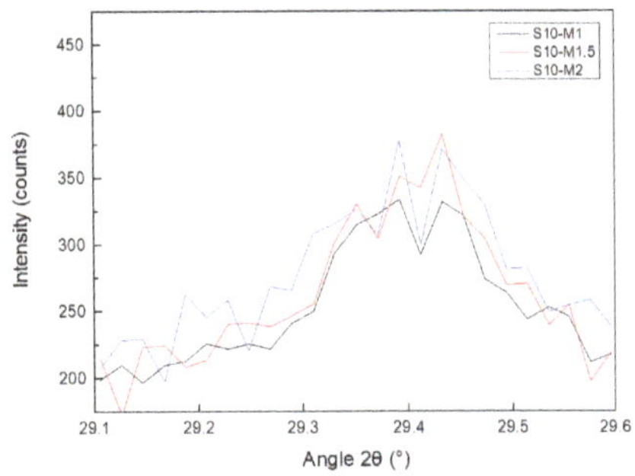

(b) C-S-H peak values in different mortars

Figure 19. Effect of different modulus on comparison of peak values of SiO_2 and C-S-H of samples.

4. Conclusions

In this paper, the method of GBFS of alkali-activated concrete partly replaced with silicomanganese slag was studied. The effects of the content of silicomanganese slag, alkali-activated modulus (M_S) and steel fiber content on the mechanical properties, freeze-thaw resistance and micro-structure of alkali-activated silicomanganese slag concrete (AASSC) were deeply investigated. The conclusions can be drawn:

(1) AASSC with a 10% substitution ratio of silicomanganese slag can reach over 80 MPa and over 132 MPa with the 3% steel fiber dosage; it can retain 80% compressive strength left with 3.5% mass lost after 800 freeze-thaw cycles by adding 2% steel fibers, which can allow the AASSC to be successfully applied to practical needs.
(2) ASSC with the higher silicomanganese slag content has lower compressive and flexural strengths, but higher fluidity and setting times. When compared to pure alkali-activated concrete, ASSC with a 10% substitution ratio of silicomanganese slag has a limited strength reduction and good working performance, which is preferable to practical applications.
(3) AASSC with a higher steel fiber content has better mechanical performance and resistance to freeze-thaw cycling. The steel fibers reduce the fluidity and setting times of slurry but can be neglected. The steel fiber content of 2% volume fraction is preferred for real applications in consideration of workability and strength.
(4) The steel fiber addition is important to the freeze-thaw resistance of AASSC, and AASSC with the higher fiber dosage has better resistance. Moreover, despite the harmful effect of large amount of silicomanganese slag on AASSC, AASSC with a 60% substitution ratio of silicomanganese slag was damaged at 600 freeze-thaw cycles and had 3.3% mass lost with 60% strength left; it could also applied to construction considering a massive substitution to GBFS.

5. Recommendation

The future research should focus on investigating other properties such as carbonization, projectile strength of AASSC. The application of AASSC on concrete structures is also worth to study in the future. Moreover, preplaced aggregate concrete is a special method that reduces the mortar content with increased economic benefits and is an interest for future AASSC research to solve disposal of recycled waste.

Author Contributions: B.L. contributed to the conceptualization, funding acquisition, supervision, and resources of this study; D.W. contributed to the investigation, formal analysis, methodology, writing—original draft of this study; E.M. contributed to the validation, writing—review & editing, visualization of this study. All authors have read and agreed to the published version of the manuscript.

Funding: Natural Science Foundation of Hunan Province, China (2020JJ4579).

Data Availability Statement: This paper is new. Neither the data nor any part of its content has been published or has been accepted elsewhere.

Acknowledgments: This study has been financially supported by the Natural Science Foundation of Hunan Province, China (2020JJ4579).

Conflicts of Interest: The authors declared that they have no conflict of interest to this work. We declare that we do not have any commercial or associative interest that represents a conflict of interest in connection with the work submitted.

References

1. Susanta, K.N.; Navneet, S.R.; Sanjay, K. A review on characteristics of silico-manganese slag and its utilization into construction materials. *Resour. Conserv. Recycl.* **2022**, *176*, 105946. [CrossRef]
2. Navarro, R.; Alcoce, E.G.; Sánchez, I.; Garcés, P.; Zornoza, E. Mechanical properties of alkali activated ground SiMn slag mortars with different types of aggregates. *Construct. Build. Mater.* **2018**, *186*, 79–89. [CrossRef]
3. Gao, H.L.; Yan, W.S. Sustainable development of manganese ferroalloys in post crisis period. *China Manganese Ind.* **2010**, *28*, 6–11. (In Chinese)
4. Liu, Q.; Li, J.; Lu, Z.Y.; Li, X.; Jiang, J.; Niu, Y.; Xiang, Y. Silicomanganese slag: Hydration mechanism and leaching behavior of heavy metal ions. *Construct. Build. Mater.* **2022**, *326*, 126857. [CrossRef]
5. Yun, C.N.; Bakri, M.K.B.; Rahman, M.R.; Kuok, K.K.; Khui, P.L.N.; Huda, D. Effect of Chemical Treatment on Silicon Manganese: Its Morphological, Elemental and Spectral Properties and Its Usage in Concrete. *Fash. Texit.* **2022**, *14*, 8081–8096. [CrossRef]

6. Kumar, S.; García-Triñanes, P.; Teixeira-Pinto, A.; Bao, M. Development of alkali activated cement from mechanically activated silico-manganese (SiMn) slag. *Cem. Concr. Compos.* **2013**, *40*, 7–13. [CrossRef]
7. Syed, K.N.; Megat, A.M.J.; Ashraf, A.B. Room Temperature Cured Silicomanganese Fume-Based Alkali Activated Mortar: Experimental, Statistical and Environmental Impact Studies. *Stat. Environ. Impact Stud.* **2022**, *2*, 31.
8. Li, X.M.; Yang, H.B.; Ruan, J.; Li, Y. Effect of mechanical activation on enhancement of carbon thermal reduction of nickel slag. *J. Iron Steel Res. Int.* **2020**, *27*, 1311–1321. [CrossRef]
9. Vavrenyuk, V.E.; Yu, S.V.; Vavrenyuk, V.G. Mechanical Activation, Mineral Powders Chemical Modification, Solid-Phaseeactivity, Polyorganosiloxanes. *IOP Conf. Ser. Mater. Sci. Eng.* **2021**, *1079*, 072043. [CrossRef]
10. Wang, F.; Long, G.; Bai, M.; Wang, J.; Zhou, J.L.; Zhou, X. Application of electrolytic manganese residues in cement products through pozzolanic activity motivation and calcination. *J. Clean. Prod.* **2022**, *338*, 130629. [CrossRef]
11. Bian, W.; Ma, K.; Long, G.; Liu, W.; Zhang, C. Study on properties of alkali activated slag fly ash permeable concrete. *J. Railw. Sci. Eng.* **2020**, *119*, 83–91. (In Chinese)
12. Huang, L.; Ye, H. Factors Affecting Kinetics and Gel Composition of Alkali–Silica Reaction in Alkali-Activated Slag Mortars. *Int. J. Civ. Eng.* **2021**, *19*, 453–462. [CrossRef]
13. Lv, Q.M.; Zhang, R.L.; Wang, Z. Effects of calcination temperature and cooling conditions on pozzolanic activity of paper mill sludge ash. *J. Environ. Eng.* **2018**, *12*, 213–219. (In Chinese)
14. Tao, X.W. Study on the Polymerization, Structure and Properties of SSBR at High Temperature. Master's Thesis, Dalian University of Technology, Dalian, China, 2009. (In Chinese).
15. He, W.; Li, R.; Zhang, Y.; Nie, D. Synergistic use of electrolytic manganese residue and barium slag to prepare belite-sulphoaluminate cement study. *Constr. Build. Mater.* **2022**, *326*, 126672. [CrossRef]
16. Wang, X.J.; Ren, S.M.; Ding, H. Research and application of micro powder C60 concrete with silicon manganese slag. *Silic. Build. Prod.* **1994**, *5*, 13–15. (In Chinese)
17. Srikavya, M.; Rajesh, T. Experimental Study on Concrete Using Silico Manganese Slag as A Partial Replacement of Coarse Aggregate and Fine Aggregate. *Turk. Online J. Qual. Inq.* **2021**, *12*, 4697–4720.
18. Zhang, J.; Shi, C.; Zhang, Z.; Ou, Z. Durability of alkali-activated materials in aggressive environments: A review on recent studies. *Constr. Build. Mater.* **2017**, *152*, 598–613. [CrossRef]
19. Amran, M.; Fediuk, R.; Abdelgader, H.S.; Murali, G.; Ozbakkaloglu, T.; Lee, Y.H. Fiber-reinforced alkali-activated concrete: A review. *J. Build. Eng.* **2022**, *45*, 103638. [CrossRef]
20. Ambily, P.S.; Ravisankar, K.; Umarani, C.; Dattatreya, J.K.; Iyer, N.R. Development of ultra-high performance geopolymer concrete. *Mag. Concr. Res.* **2013**, *66*, 82–89. [CrossRef]
21. You, W.J.; Mark, A.B.; Liu, H.N.; Zhao, W.; Yang, G.T. Steel-alkali activated cement based ultra-high performance concretelight-weight composite bridge decks: Flexural behavior. *Eng. Struct.* **2022**, *266*, 114639. [CrossRef]
22. Qaidi, S.M.; Atrushi, D.S.; Mohammed, A.S.; Ahmed, H.U.; Faraj, R.H.; Emad, W.; Tayeh, B.A.; Najm, H.M. Ultra-high-performance geopolymer concrete: A review. *Constr. Build. Mater.* **2022**, *346*, 128495. [CrossRef]
23. Su, Z.; Liu, Z.; Wang, H.; Xu, S.; Wang, D.; Han, F. Reaction kinetics, microstructure and phase evolution of alkali-activated Si-Mn slag during early age. *Constr. Build. Mater.* **2022**, *333*, 127437. [CrossRef]
24. Huang, T.Z. Study on the Utilization of Silicomanganese Slag in Building Materials. Master's Thesis, Chongqing University, Chongqin, China, 2012. (In Chinese).
25. Nath, S.K.; Kumar, S. Reaction kinetics, microstructure and strength behavior of alkali activated silico-manganese (SiMn) slag–Fly ash blends. *Constr. Build. Mater.* **2017**, *147*, 371–377. [CrossRef]
26. Navarro, R.; Zornoza, E.; Sánchez, I.; Alcocel, E.G. Influence of the type and concentration of the activator on the microstructure of alkali activated SiMn slag pastes. *Constr. Build. Mater.* **2022**, *342*, 128067. [CrossRef]
27. Liu, Y.; Zhan, Z.; Shi, C. Development of ultra-high performance geopolymer concrete (UHPGC): Influence of steel fiber on mechanical properties. *Cem. Concr. Compos.* **2020**, *112*, 103670. [CrossRef]
28. Li, N.; Shi, C.J.; Zhang, Z.; Zhu, D.; Hwang, H.; Zhu, Y.; Sun, T. A mixture proportioning method for the development of performance-based alkali-activated slag-based concrete. *Cem. Concr. Compos.* **2018**, *93*, 163. [CrossRef]
29. Wetzel, A.; Middendorf, B. Influence of silica fume on properties of fresh and hardened ultra-high performance concrete based on alkali-activated slag. *Cem. Concr. Compos.* **2019**, *100*, 53–59. [CrossRef]
30. *GB/T 2419-2016*; Determination of Fluidity of Cement Mortar. Ministry of Housing and Urban Rural Development: Beijing, China, 2016.
31. Wu, Z.; Shi, C.; Khayat, K.H. Investigation of mechanical properties and shrinkage of ultra-high performance concrete: Influence of steel fiber content and shape. *Compos. Part B* **2019**, *174*, 107201. [CrossRef]
32. *DL/T5144-2001*; Code for Construction of Hydraulic Concrete. Ministry of Housing and Urban Rural Development: Beijing, China, 2016.
33. *GB/T 17671-2020*; Inspection Method of Cement Mortar. Ministry of Housing and Urban Rural Development: Beijing, China, 2020.
34. *GB/T 50082-2009*; Standard for Test Methods of Long-Term Performance and Durability of Ordinary Concrete. Ministry of Housing and Urban Rural Development: Beijing, China, 2009.
35. Alonso, M.M.; Gismera, S.; Blanco, M.T.; Lanzón, M.; Puertas, F. Alkali-activated mortars: Workability and rheological behaviour. *Constr. Build. Mater.* **2017**, *145*, 576–587. [CrossRef]
36. Struble, L.; Sun, G.-K. Viscosity of Portland cement paste as a function of concentration. *Adv. Cem. Base Mater.* **1995**, *2*, 62–69. [CrossRef]

37. Wu, Z.; Shi, C.; He, W.; Wu, L. Effects of steel fiber content and shape on mechanical properties of ultra high performance concrete. *Constr. Build. Mater.* **2016**, *103*, 8–14. [CrossRef]
38. Yu, R.; Spiesz, P.; Brouwers, H.J.H. Mix design and properties assessment of ultra-high performance fibre-reinforced concrete (UHPFRC). *Cem. Concr. Res.* **2014**, *56*, 29–39. [CrossRef]
39. Serdar, A.; Baradan, B. Effect of activator type and content on properties of alkali-activated slag mortars. *Compos. Part B Eng.* **2014**, *57*, 166–172. [CrossRef]
40. Jiao, Z.Z. Study on Basic Mechanical Properties of Alkali Activated Slag Cementitious Block Masonry. Master's Thesis, Harbin Institute of Technology, Harbin, China, 2019. (In Chinese).
41. Manjunath, M.C.; Narasimhan, K.M. Studies on development of high performance, self-compacting alkali activated slag concrete mixes using industrial wastes. *Constr. Build. Mater.* **2019**, *198*, 33–147. [CrossRef]
42. Nath, S.K.; Kumar, S. Influence of Granulated Silico-Manganese Slag on Compressive Strength and Microstructure of Ambient Cured Alkali-Activated Fly Ash Binder. *Waste Biomass Valorization* **2018**, *10*, 2045–2055. [CrossRef]
43. Leong, H.Y.; Ong, D.E.L.; Sanjayan, J.G. The effect of different Na_2O and K_2O ratios of alkali activator on compressive strength of fly ash based-geopolymer. *Constr. Build. Mater.* **2016**, *106*, 500–511. [CrossRef]
44. Nicolas, R.S.; Bernal, S.A.; De Gutierrez, R.M.; van Deventer, J.S.; Provis, J. Distinctive microstructural features of aged sodium silicate-activated slag concretes. *Cem. Concr. Res.* **2014**, *65*, 41–51. [CrossRef]
45. Barbosa, V.F.F.; Mackenzie, K.J.D.; Thaumaturgo, C. Synthesis and characterisation of materials based on inorganic polymers of alumina and silica: Sodium polysialate polymers. *Int. J. Inorg. Mater.* **2000**, *2*, 309–317. [CrossRef]
46. Cai, R.J.; Tian, Z.S.; Ye, H.L. Durability characteristics and quantification of ultra-high strength alkali-activated concrete. *Cem. Concr. Compos.* **2022**, *134*, 104743. [CrossRef]
47. Aghaee, K.; Khayat, K.H. Benefits and drawbacks of using multiple shrinkage mitigating strategies on performance of fiber-reinforced mortar. *Cem. Concr. Compos.* **2022**, *133*, 104714. [CrossRef]
48. Kuranlı, Ö.F.; Uysal, M.; Canpolat, O. Mechanical and durability properties of slag/fly ash based alkali-activated concrete reinforced with steel, polypropylene and polyamide fibers. *Eur. J. Environ. Civ. Eng.* **2022**, *26*, 1–24. [CrossRef]
49. Nath, S.K.; Kumar, S. Evaluation of the suitability of ground granulated silico-manganese slag in Portland slag cement. *Constr. Build. Mater.* **2016**, *125*, 127–134. [CrossRef]
50. Navarro, R.; Zornoza, E.; Garces, P.; Sanchez, I.; Alcocel, E.G. Optimization of the alkali activation conditions of ground granulated SiMn slag. *Constr. Build. Mater.* **2017**, *150*, 781–791. [CrossRef]
51. Fernández-Jiménez, A.; Puertas, F. The alkali–silica reaction in alkali-activated granulated slag mortars with reactive aggregate. *Cem. Concr. Res.* **2002**, *32*, 1019–1024. [CrossRef]
52. Sarkar, S. *Handbook for Identification of Alkali-Silica Reactivity in Airfield Pavements*; AC No150/5380-8 Appendix2004; US Department of Transportation, Federal Aviation Administration: Washington, DC, USA, 2004; pp. 1–12.
53. García-Lodeiro, A.; Palomo, A.; Fernández-Jiménez, A. Alkali–aggregate reaction inactivated fly ash systems. *Cem. Concr. Res.* **2007**, *37*, 175–183. [CrossRef]
54. Ahmed, F.A.; Fei, J.; Al-Tabbaa, A. Development of greener alkali-activated cement: Utilization of sodium carbonate for activating slag and fly ash mixtures. *J. Clean. Prod.* **2016**, *113*, 66–75. [CrossRef]
55. Anıl, N.; Necip, A.E.; Abdulkadir, Ç. Effects of nanosilica and steel fibers on the impact resistance of slag based self-compacting alkali-activated concrete. *Ceram. Int.* **2021**, *47*, 23905–23918. [CrossRef]
56. Chen, F.S.; Lv, X.J.; Zhu, X.; Liu, X.B.; Fu, L.; Shi, L. Research progress of alkali slag cementitious materials and their retarders. *J. Fly Ash* **2012**, *24*, 8–10. (In Chinese)
57. Yang, G.Y. Study on the Interface Effect of Fiber in Ultra High Performance Concrete and Its Compatibility with Cement Substrate 636. Master's Thesis, South China University of Technology, Changsha, China, 2020. (In Chinese).

buildings

MDPI

Article

Flexural Behavior of RC Beams Strengthened with GFRP Laminate and Retrofitting with Novelty of Adhesive Material

Basem O. Rageh [1,*], Mahmoud A. El-Mandouh [2], Ahmed H. Elmasry [1] and Mohammed M. Attia [3]

[1] Civil Engineering Department, Delta Higher Institute for Engineering and Technology, Mansoura 35511, Egypt
[2] Civil Construction Technology Department, Faculty of Technology and Education, Beni-Suef University, Beni-Suef 62521, Egypt
[3] Civil & Architectural Construction Department, Faculty of Technology and Education, Suez University, Suez 43511, Egypt
* Correspondence: bassem@dhiet.edu.eg

Abstract: Two unique approaches were proposed to strengthen the bond between the glass fiber reinforcement polymer (GFRP) and the RC concrete surfaces. The two bonding materials are epoxy (EP) and geopolymer (GPP) with different ratios of short glass fibers (SGF). The experimental program includes seven reinforced concrete (RC) beams that have the same cross-section (150 mm × 200 mm) and are 1500 mm in length. The first beam is the control beam (B0-Control). The next three beams B1-0-GPP, B2-0.6-GPP, and B3-1.2-GPP have GPP with SGF ratios of 0%, 0.6%, and 1.2%, respectively. The last three beams B4-0-EP, B5-0.6-EP, and B6-1.2-EP have EP with SGF ratios of 0%, 0.6%, and 1.2%, respectively. The results show that the failure loads of beams B1-0-GPP, B2-0.60-GPP, and B3-1.2-GPP are greater than the control beam B0-Control by approximately 20.80%, 25.60%, and 31.40%, respectively, whereas the failure loads of beams B4-0-EP, B5-0.6-EP, and B6-1.2-EP are greater than the B0-Control by approximately 16.90%, 26.90%, and 26.10%, respectively; it is also noted that debonding occurs. In addition to the adhesive material, GPP has a great effect on increasing the beam's failure load capacity due to the enhanced interfacial bond shear strength. Additionally, a finite-element program ABAQUS is performed to verify the experimental results.

Keywords: geopolymer paste; flexural behavior; GFRP; epoxy; short glass fibers; strengthening

1. Introduction

Among the biggest problems that contemporary civil engineering is currently facing are the strengthening, updating, and retrofitting of old structures. One of the most promising answers to these needs is the use of carbon fiber reinforcement polymer (CFRP) laminates bonded to the tensile face of the member. In addition, the type of adhesive material that is used is important to prevent CFRP debonding.

The need for reinforced concrete (RC) structures strengthening is becoming more important, especially when there is an increase in ultimate load, degradation trouble, and/or design construction shortcomings [1–3]. One of the more conventional approaches to strengthening is using epoxy to bond structural elements with steel plates. There are significant drawbacks to using steel plates as a strengthening material [4–6], such as corrosion, the heavy weight of the steel plate, the difficulty of forming the steel plate, and the good preparation of the RC beam surface.

A. Jahami et al. [7] studied numerically the effect of using CFRP as a strengthening technique for RC beams subjected to blast loading. Three different configurations of CFRP were considered: bottom CFRP strips for flexural strengthening, diagonal side strips for shear strengthening, and U-shaped strips for both shear and flexural strengthening. Results showed that using CFRP in the bottom tensile face of RC beams helped in absorbing blast energy. In addition, using CFRP has shown a reduction in the tensile strain of the beam reinforcements.

To create a new technique that can replace steels and alloys with one that is dependable, affordable, and easy to use, extensive research has been done in this area [8,9]. Therefore, using externally bonded (EB) fiber-reinforced polymer (FRP) sheets in concrete structures as a means of strengthening and retrofitting has received widespread acceptance in the industries [10–12]. This method involves using epoxy to bond the glass fiber reinforcement polymer (GFRP) laminates to the tension surface of RC beams. This EB technique, on the other hand, results in greater ultimate flexural strength and stiffness [13].

According to several studies, FRP has many advantages including high tensile resistance. Furthermore, when FRP is applied to RC members, it improves rehabilitation performance, durability, and cost-effectiveness [14,15]. Glass-fiber-reinforced polymer (GFRP) is the most suited FRP reinforcement for a variety of civil engineering applications because glass fibers offer the highest temperature resistance, the highest strength, and the lowest cost [16]. The process of debonding between GFRP laminates and RC concrete surface due to the weak cohesion of the epoxy as an adhesive material is one of the problems that limit the development of the use of GFRP laminates as an effective system for strengthening concrete structures [17]. Numerous types of FRP debonding failure have been documented in prior studies [18–20], including concrete cover separation, intermediate crack debonding, and plate end debonding. The effectiveness of the strengthening mechanism is greatly reduced when FRP debonding fails [21,22].

Current and previous studies have developed various anchorage systems, such as FRP anchors, FRP U-jacket anchors, and mechanical anchors to postpone the debonding process between GFRP laminates and RC concrete surfaces [23,24]. The enhanced concrete beam with anchorage had a bond strength that was eight times greater than the specimen without an anchor, according to the experiment results. Additionally, it was shown that the interfacial bond's performance was dependent on the site of the anchorage. Anchoring FRP laminates at their ends with anchor bolts can effectively prevent brittle debonding failure while also improving bond shear resistance and ductility of the reinforced structures [25,26]. Such anchorage systems have the potential to significantly improve the interfacial bond between FRP and RC concrete surfaces [27].

On the other hand, the preparation and installation of these anchors greatly increase the difficulty of implementation and the demand for craftsmanship [28,29]. To increase interfacial shear resistance, several indirect anchorage techniques have been created [30,31]. By expanding the adhesive area between the FRP and the concrete, the near-surface mounted (NSM) approach increased the indirect anchoring of FRP composites [32,33]. The American Code Institute (ACI) [34] has suggested the concrete grooving method (GM) as an easy version of the NSM due to its important development in concrete surface roughness [35]. The interfacial bond strength with the adhesive region between FRP and concrete can be greatly enhanced by grooving the concrete before applying epoxy resin [36]. The preparation of concrete surface grooves, on the other hand, would significantly increase construction work [37]. As a result, a variety of substances and techniques have been researched to identify viable alternatives that might be utilized as a partial or full replacement for Portland cement to create ecologically friendly concrete [38].

Geopolymer is typically synthesized by activating aluminosilicate source material with an alkaline solution (such as NaOH or KOH) [39]. Metakaolin (MK) with a lower calcination temperature or industrial waste products such as fly ash (FA), slag, and rice husk ash are frequently employed as an aluminosilicate source material. Because of this, producing geopolymer uses less energy and generates less carbon dioxide. Geopolymer is regarded as a possible substitute for cement due to its exceptional qualities including low shrinkage, fast concretion, high early compressive strength, and superior fire resistance [40]. The bond property of geopolymer has recently piqued the interest of researchers [41–43].

Composite materials produce a combination property of two or more materials that neither fiber nor matrix can achieve when acting alone [44]. Short-fiber-reinforced polymer

(SFRP) composites are particularly attractive due to their simplicity in manufacture, low cost, and outstanding mechanical properties [45]. High fiber content is generally required to achieve a high-performance SFRP composite. As a result, the effect of fiber content on the mechanical properties of SFRP composites is of particular interest and significance [46]. Glass-fiber-reinforced polymeric composites were the most used materials in the production of composite materials [47]. Finally, adequate bonding between the FRP laminates and the RC concrete surfaces is critical for successful strengthening; failure to do so would result in undesirable premature failure before the ultimate load.

The main objectives of this study are (1) to suggest two adhesive materials to improve the bond between the GFRP laminates and the concrete surface and (2) to study the impact of the adhesive materials on the tested beams' structural behavior. The adhesive materials used are Sikadur®−330 epoxy (EP) and geopolymer paste (GPP) with different short glass fibers (SGF) ratios. In addition, an analytical examination using the non-linear finite element computer software ABAQUS [48] has been performed to validate the experimental consequences of the tested beams.

2. Experimental Program

2.1. Materials

All the tested beams were constructed using ordinary Portland cement (OPC) 42.5 grade, which has a specific gravity of 3.10 g/cm^3 and an average particle size of 20.67 µm. The chemical composition of the OPC is shown in Table 1. In this investigation, natural river sand was used as the fine aggregate. The sand had a specific gravity of 2.42 g/cm^3, water absorption of 1.92%, and a fineness modulus of 2.25. As a coarse aggregate, natural crushed stone with a nominal maximum size of 12.5 mm was employed. The coarse aggregate had a specific gravity of 2.78 g/cm^3 and water absorption of 1.56%.

Table 1. Chemical and physical properties: Composition, OPC, and FA.

Composition	OPC	FA
	Chemical properties	
SiO_2	19.39	57.2
Al_2O_3	4.13	24.4
CaO	55.66	2.2
Fe_2O_3	4.70	7.1
MgO	1.70	2.4
K_2O	0.28	3.4
Na_2O	0.31	0.4
SO_3	3.90	0.3
	Physical properties	
Specific gravity	3.10	2.8

Concrete was made using tap water, with a water-to-cement ratio of 0.42 in this investigation. It was necessary to utilize a superplasticizer to ensure the required workability. To improve the workability of concrete, this superplasticizer, which complies with ASTM C494 type F [49,50], is made of synthetic resins and added at a dose of 1.5% by weight of cement. The main bottom and top longitudinal deformed bars of all tested beams with a diameter of 12 mm and 10 mm, respectively, and stirrups with diameters of 8 mm. Sika Wrap®-430 G of Sika Egypt, as shown in Figure 1a, was used as unidirectional GFRP laminates, which have a density of 2.56 g/cm^3, a thickness of 0.168 mm, a tensile strength of 2500 N/mm^2, and modulus of elasticity equal to 72,000 N/mm^2, according to the manufacturing sheet.

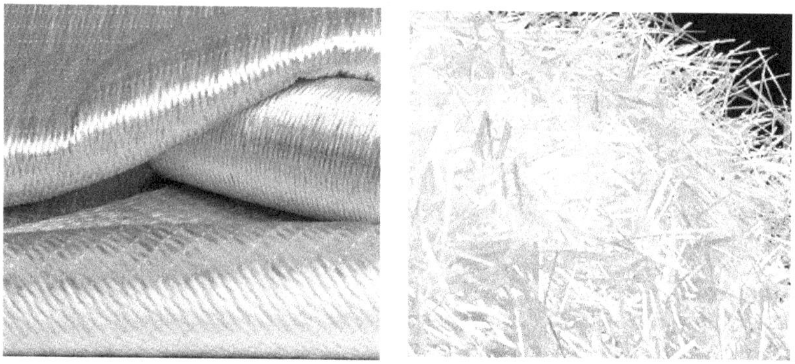

(a) GFRP (b) SGF

Figure 1. (a) Glass fiber sheets; (b) Short glass fiber.

The GFRP installation uses the wet hand-layup system as per ACI [51]. Short glass fibers (SGF), acquired from Sika Egypt, as shown in Figure 1b, have a length of 6 mm, a specific gravity of 0.91, a diameter of 13 microns, and tensile strength of 480 Mpa, according to the manufacturing sheet.

The EP has two components of resin and hardener. The mixing ratio of resin to hardener is 4:1 by weight. The mechanical properties of EP with and without SGF are presented in Table 2. Two percentages of SGF, 0.6% and 1.20%, from the volume of the EP were used. The second bonding material is GPP which includes fly ash (FA) and a mix of sodium silicate (Na_2SiO_3) solution with sodium hydroxide (NaOH) solution as an alkaline activator. The chemical composition of FA is illustrated in Table 1. The specific surface area and the specific gravity of FA are 379 m^2/kg and 2.25, respectively.

Table 2. The proportion of the beam's concrete mix.

Cement (kg/m^3)	Water (Liter)	Fine Aggregate (kg/m^3)	Coarse Aggregate (kg/m^3)	Super Plasticizer (Liter)
350	147	870	1270	5.25

The utilized sodium hydroxide (NaOH) is available as flakes with a particle size of approximately 3 mm and a specific gravity of 2.13, and a pH of 14. 40 g/mol was utilized as the molar mass. By dissolving sodium hydroxide flakes in water, a sodium hydroxide (NaOH) solution was created. The mass of NaOH solids in a solution disperses according to the solution concentration expressed in terms of the molar. In this research, NaOH solution with 12 M (480 g/L) was used. To provide this molarity, a volume-measured beaker was used. Firstly, NaOH flakes were added, and then water was added until the volume was completed to one liter. After the stirring process was completed, the solution was left to cool down, and then the beaker was refilled to one liter.

From the market, a sodium silicate solution in a ready-to-use form was purchased. The chemical formula for the sodium silicate solution is SiO_2 = 29.4%, Na_2O = 14.7%, and water of approximately 55.9% by mass. The specific gravity of sodium hydroxide solution was 1.48 g/cc and the viscosity was 400 cp at 20 °C. Inappropriately mixed mortar may cause failures such as not hardening, flash setting, or both, which cause inapplicability. To avoid these drawbacks, the selection of mixture ingredients was made based on the previous research that was summarized in previous works and the results of various trial mixtures in the preliminary study. The selection steps are summarized as follows: FA was used as a base material with 700 kg/m^3, and a mixture of sodium hydroxide (NaOH) solution and sodium silicate solution (Na_2SiO_3) was used as an alkaline liquid by a ratio of

1 to 2.5. The molarity of sodium hydroxide was 12 M, the ratio of alkaline solution to FA was 1:2 by mass, and SGF was added with 0.6 and 1.2% of the total volume of the mixture.

Six direct shear tests were performed under pull static load to calculate the ultimate shear strength between the concrete and examined adhesive materials [GPP without SGF (0-GPP), GPP with 0.60% ratio of SGF (0.6-GPP), GPP with 1.20% ratio of SGF (1.2-GPP), EP without SGF (0-EP), EP with 0.60% ratio of SGF (0.6-EP), and EP with 1.20% ratio of SGF (1.2-EP)]. Six concrete blocks were cast with dimensions (length L = 300 mm, width b = 125 mm, and thickness h = 125 mm). Using an abrasive dry grinder, the concrete block's surface was prepared by grinding down to a depth of roughly 0.50 mm. Six specimens of GFRP laminates were prepared with dimensions of 152 mm bonded length and 35 mm unbounded length. Each one of the GFRP laminates was pasted to one concrete block along the centerline on one side of the concrete block. The tensile load was applied to the GFRP laminate while the concrete block was restrained from the bottom. Figure 2 shows the test setup. LVDTs that were installed on the concrete surface near the upper corners of the bonded area were used to measure the initial slip and final slip. The typical applied load-slip response is illustrated in Figure 3.

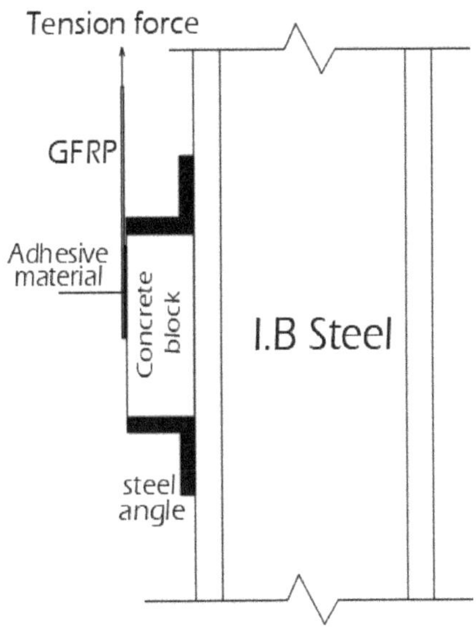

Figure 2. Pull out static test.

2.2. *Mix Proportioning and Details of Tested Beams*

The experimental program included seven simply supported RC beams. All the beams had an equal section of 200 mm depth, 150 mm width, and 1500 mm length. The reinforcement bars, dimensions, and geometry information of the tested beams are illustrated in Figure 4. The main bottom reinforcement was two bars of 12 mm diameter with a design yield stress of 410 MPa, and the top reinforcement bars were two bars of 10 mm diameter with the same yield strength. Vertical stirrups were constructed using 125 mm-spaced, 8 mm-diameter bars with a yield strength of 240 Mpa to ensure that all tested beams failed in the flexural mode according to ACI-19 [52].

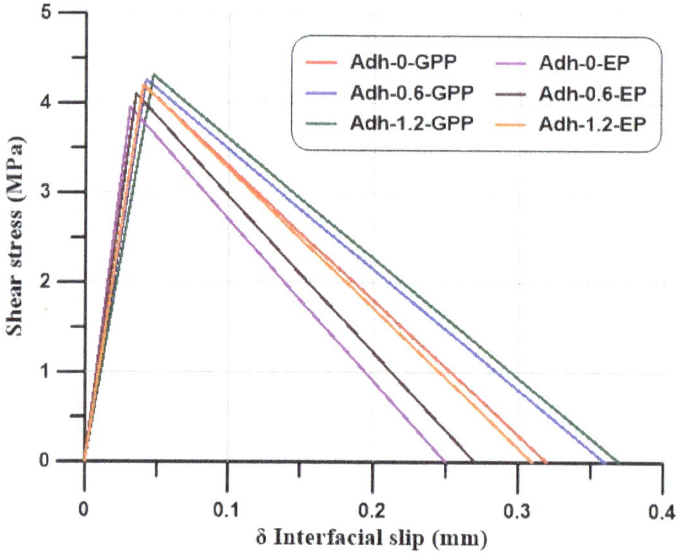

Figure 3. The interface bond shear stress between the GFRP laminate and concrete surface experimentally.

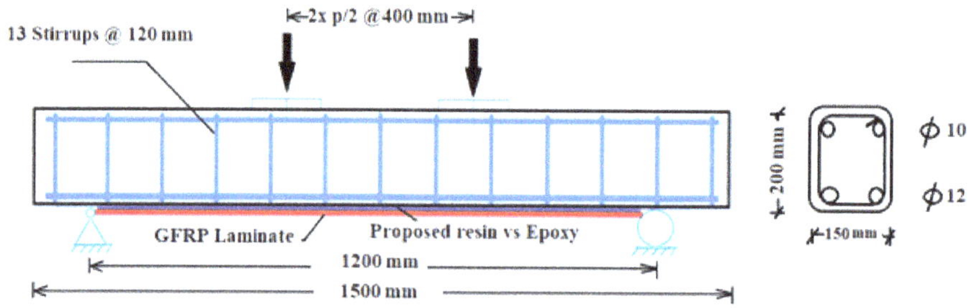

Figure 4. GFRP externally strengthened RC beam under two-point loads.

The proportioning of the concrete mix of the tested beams is given in Table 2. Table 3 provides a summary of the information and the tested beams' designation. The beams' designation is arranged as follows: the beam number, numbers 0, 0.6, and 1.2 referring to the SGF ratio, and GPP and EP referring to the geopolymer and Sikadur®-330 epoxy paste of Sika Egypt, respectively. Three groups of beams were used. The first group included the beam B0-control, which referred to the control beam without strengthening. The second group included three tested beams, which were marked as B1-0-GPP, B2-0.6-GPP, and B3-1.2-GPP, and had bonding material GPP with SGF by the percentages of 0%, 0.6%, and 1.2%, respectively. The third group included three beams, which were marked as B4-0-EP, B5-0.6-EP, and B6-1.2-EP, and had EP with SGF percentages of 0%, 0.6%, and 1.2%, respectively.

Table 3. The names and specifics of different beams.

Group	Bonding Material	Beam ID	SGF %
A	-	B0-Control	0
B	GPP	B1-0-GPP B2-0.6-GPP B3-1.2-GPP	0 0.6 1.2
C	EP	B4-0-EP B5-0.6-EP B6-1.2-EP	0 0.6 1.2

2.3. Beam Preparation and GFRP Strengthening

2.3.1. Beam Preparation

Seven plywood forms were prepared for fabricated RC beams, and then steel reinforcement cages were prepared as shown in Figure 5. The concrete was mixed until the mixture became homogenous, then fresh mixtures were cast in forms at 23 ± 2 °C using an electrical poker vibrator for two minutes; specimens were then covered with polyethylene sheets started immediately after casting and were cured with pure water after 24 h from casting until the test. Three concrete cylinders (ϕ 150 × 300 mm) and three concrete cubes (150 mm × 150 mm ×150 mm) were taken to determine the average cylinder compressive strength (f_c') and the average cube compressive strength (f_{cu}), respectively. The average cylinder compressive strength (f_c') and the cube compressive strength (f_{cu}) were equal to 26.3 Mpa, and 32.60 Mpa, respectively. The average experimental compressive stress–strain curve of the tested beams concrete is shown in Figure 6. The average experimental tensile strength was equal to 2.1 Mpa.

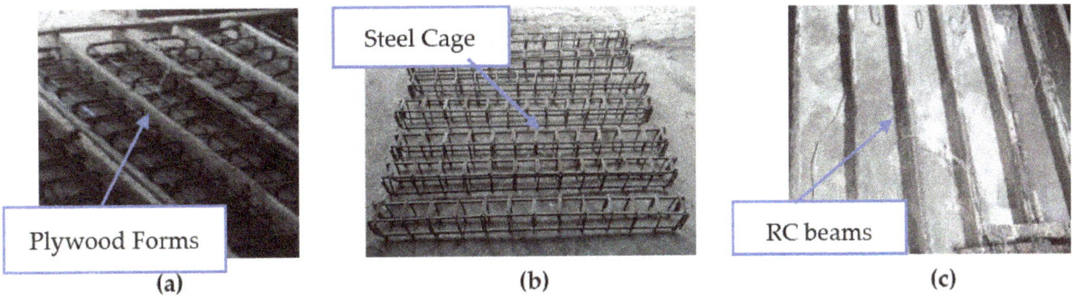

Figure 5. (**a**) plywood forms; (**b**) steel cages; and (**c**) casting of RC beams.

2.3.2. GFRP Strengthening

All the beams were cast and cured for 28 days, and then ACI 440.2R-2017's "wet layup" technology was used to enhance the tension surface of the RC beams [50]. Before the laminates were bonded, all the beams' tension surfaces were cleaned, dried, and greased, and contaminates, if any, removed. The International Concrete Repair Institute's recommended minimum rough profile (CSP3) was used to prepare the tension side of the beams (ICRI 03730) [34]. Any loose cement paste layer was removed using an abrasion wheel so that the exposed aggregate particles could be seen. Figure 7 displays the various surface preparation phases. When using epoxy, the resin and hardener were mixed in the predetermined ratio of 4:1, and the GPP was prepared by the method mentioned in Section 2.1. The GFRP was then cleaned to eliminate contaminants and cut to the required size. The prepared surface was uniformly coated with EP and GPP, and the GFRP was then pressed onto the top surface while employing light pressure and a rubber roller. To ensure that the GFRP adhered well to the surface, released any trapped air, and removed any excess epoxy that might have been present between the GFRP and concrete surface, these steps were taken. Thusly produced laminated beams were cured for a week before being tested. Figure 7a,b illustrate the process for strengthening beams.

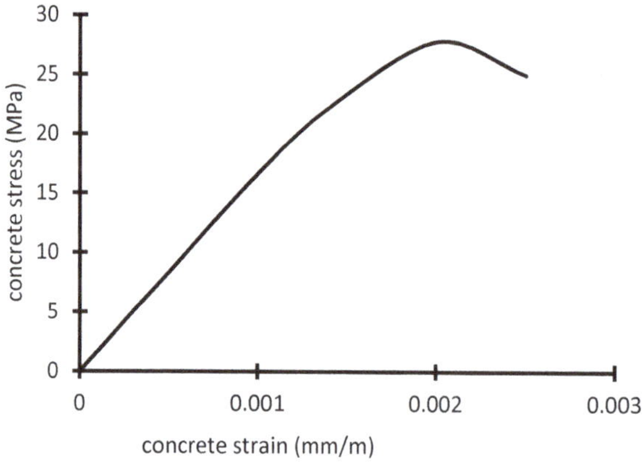

Figure 6. Stress–strain curve of concrete in compression.

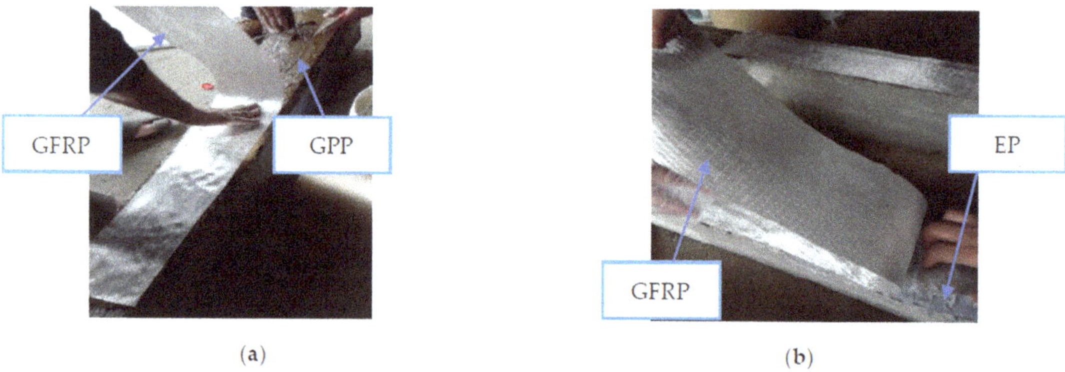

Figure 7. Strengthening RC beams with (**a**) GFRP and (**b**) EP.

2.4. Test Setup and Instrumentation

Every RC beam underwent testing with two points of loading. The distance between two-point loads is 400 mm and the distance between two beams supports is 1200 mm. On the right and left ends of the examined beams, roller and hinged supports were employed, respectively. Flexural testing was carried out using load-controlled mode loading at 0.02 KN/s until failure. A steel I-section spreader beam was used to transfer the load from the load cell to the beam. This load from the spreader beam got transferred onto the two loading points on the beam. A linear voltage displacement transducer (LVDT) with an accuracy of 0.01 mm was used to measure the defections at the mid-span of beams. Electrical resistance was used to measure the strains at the midpoint of the bottom longitudinal reinforcing bars. Strain gauges 6 mm long were used. Figure 8 displays an image of the test setup and the associated equipment. Each load interval produced a record of the load and the greatest defection.

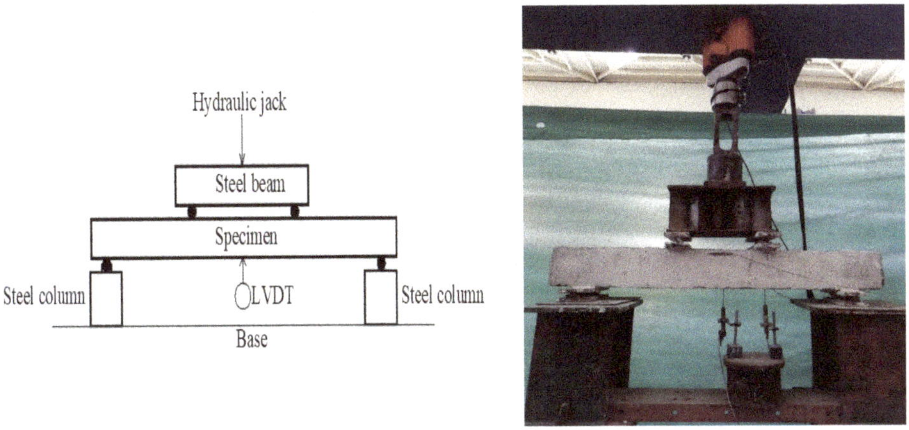

Figure 8. Test setup of RC beams.

3. Results and Discussion

At each loading increment up to failure, deflections and crack pattern propagation were observed. The impact of material paste type on the tested beams' structural behavior was assessed. Results from beams with stronger ones were compared with those from the control beam. Compared with the control beam, all beams had higher load-carrying capacities.

3.1. Failure Loads

The loads at first crack P_{cr}, beam longitudinal reinforcement bars yield P_y, and the failure load P_u are summarized for all the examined RC beams in Table 4. To assess the effectiveness of the different types of the adhesive material paste, the maximum loads for the tested beams were recorded and compared, as shown in Figure 9. The failure loads of beams B1-0-GPP, B2-0.60-GPP, and B3-1.2-GPP were greater than control beam B0-Control by approximately 20.80%, 25.60%, and 31.40%, respectively. The results showed that increasing the SGF ratio in the GPP had a clear effect on raising the capacity of the beam due to the enhanced interfacial bond shear strength, as illustrated in Figure 3.

The failure loads of beams B4-0-EP, B5-0.6-EP, and B6-1.2-EP were greater than the control beam B0-Control by approximately 16.90%, 26.90%, and 26.10%, respectively. The results showed that the GPP as the adhesive material paste had a great effect on increasing the failure load capacity of the tested beams. Compared with the costs of the bonding materials EP and GPP, we found that bonding material GPP is much less expensive than EP. The GPP is a novel material to increase the bond between the GFRP laminates and the concrete surface. The results showed that the use of GFRP laminates increases the flexural capacity of RC beams [53].

3.2. Failure Behavior and Crack Patterns

Figures 10–16 depict the crack patterns of the tested beams that were tested until they failed. For beam B0-Control, the first crack was noticed at a load of approximately 31.20% of the ultimate load. The first crack appeared at a load of approximately 26.90%, 26.90%, 25.70%, 28.90%, 27.70%, and 25.30% of the ultimate load for beams B1-0-GPP, B2-0.60-GPP, B3-1.20-GPP, B4-0-EP, B5-0.6-EP, and B6-1.20-EP, respectively. Most cracks appeared initially under the loads on the bottom surface of all tested beams. Based on the outcomes of the experiment, no debonding occurred for laminates when using GPP with different SGF ratios. On the contrary, there was a debonding when using EP for beams B5-0.6-EP and B6-1.2-EP; however, for the beam B4-0-EP, the failure mode was laminate debonding and concrete cover separation. GFRP reduced the tensile damage that occurred in concrete due to the increase in energy absorption [54,55].

Table 4. Test results synopsis.

Beam	P_{cr} (kN)	P_y (kN)	P_u (kN)	Δ_y (mm)	Δ_u (mm)	Δ_u/Δ_y	Failure Mode
B0-Control	24.0	70.14	76.90	5.50	17.70	3.21	Concrete crushing
B1-0-GPP	25.0	76.98	92.90	4.97	13.60	2.74	Concrete crushing
B2-0.6-GPP	26.0	78.68	96.58	4.64	15.93	3.43	Concrete crushing
B3-1.2-GPP	26.0	79.71	101.03	4.37	15.30	3.23	Concrete crushing
B4-0-EP	26.0	75.6	89.87	4.77	12.36	2.59	Debonding-C.C. S *
B5-0.6-EP	27.0	76.1	97.58	4.65	14.06	3.02	GFRP rupture
B6-1.2-EP	26.0	76.9	102.96	4.55	12.10	2.66	GFRP rupture

* C.C. S: Concrete Cover Separation.

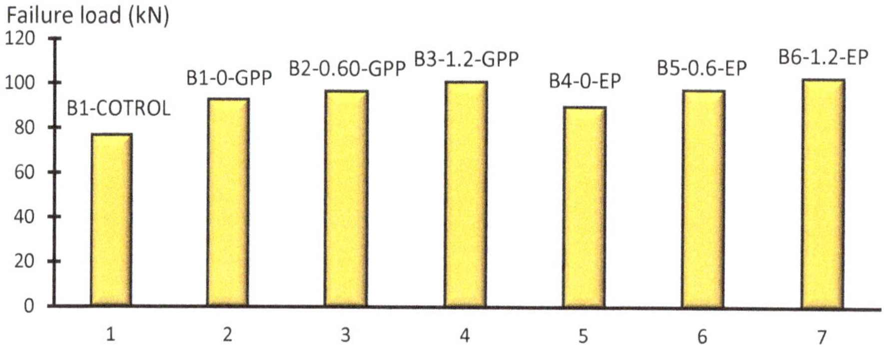

Figure 9. Load–deflection relationship for tested beams.

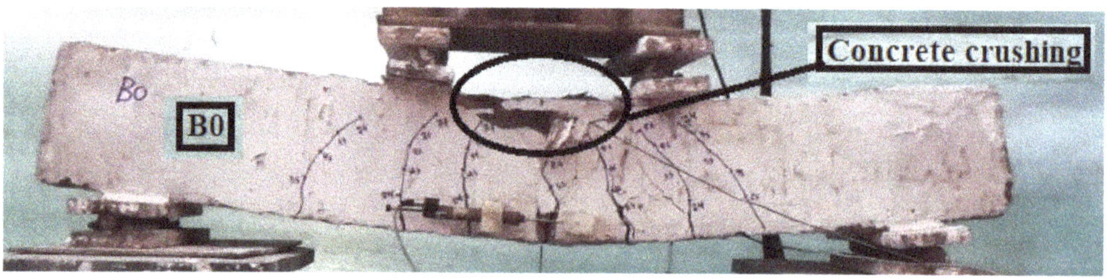

Figure 10. The final crack of beam B0-Control.

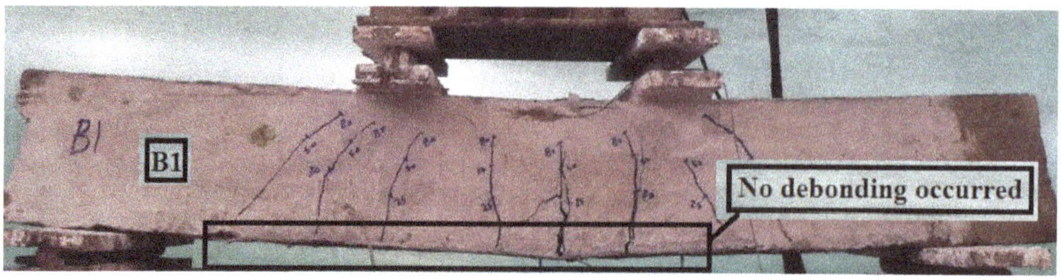

Figure 11. The final crack of beam B1-0-GPP.

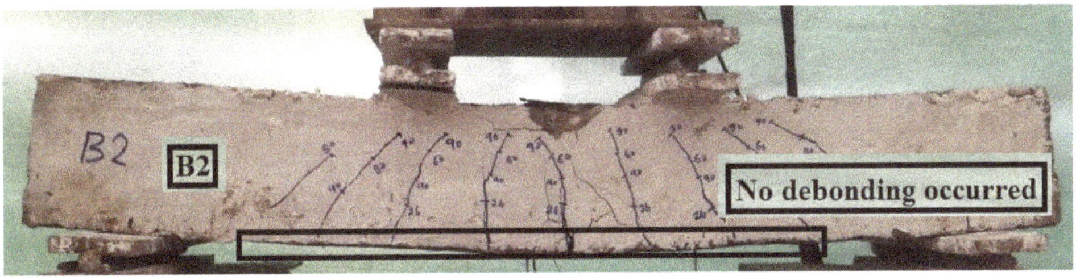

Figure 12. The final crack of beam B2-0.6-GPP.

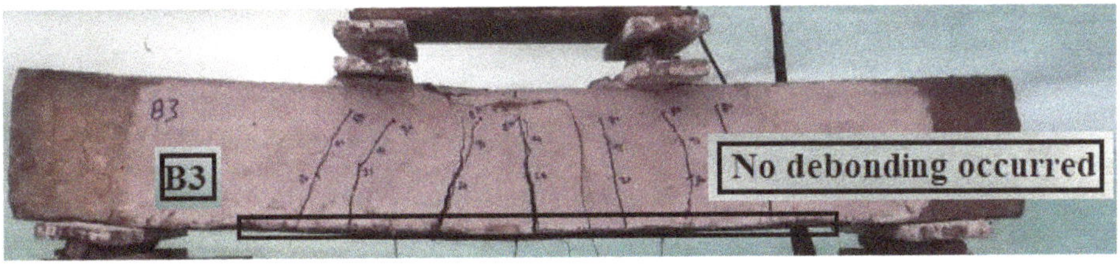

Figure 13. The final crack of beam B3-1.2-GPP.

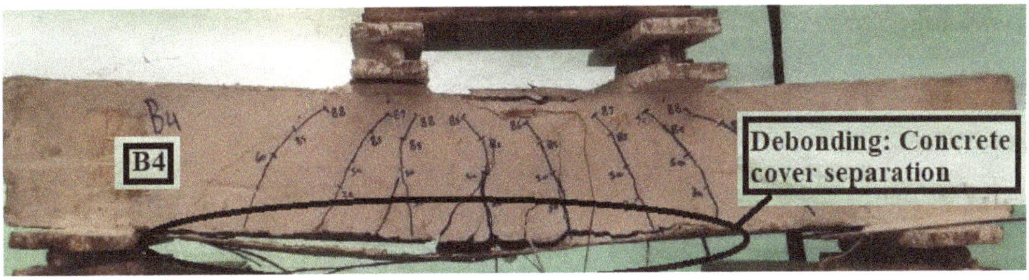

Figure 14. The final crack of beam B4-0-EP.

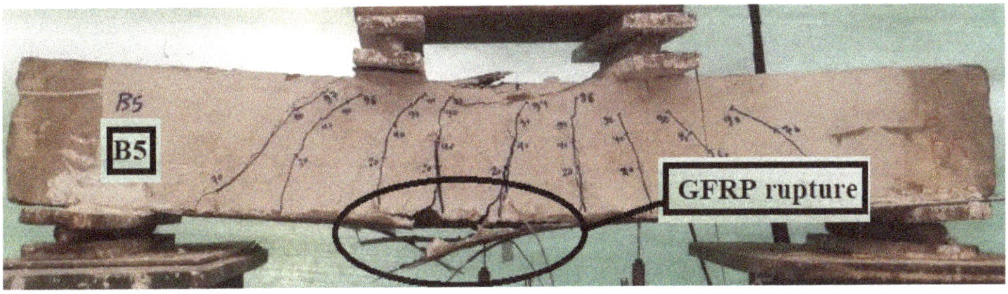

Figure 15. The final crack of beam B5-0.6-EP.

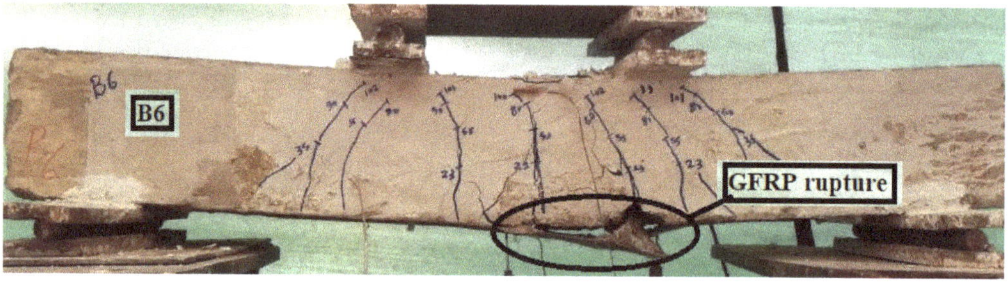

Figure 16. The final crack of beam B6-1.2-EP.

3.3. Load–Deflection Curves of Tested Beams

Figure 17 illustrates the load–deflection relationship for tested beams with GPP and EP as adhesive material. All beams displayed linear behavior up to the cracking load. The maximum deflection of tested beams B1-0-GPP, B2-0.6-GPP, B3-1.2-GPP, B4-0-EP, B5-0.6-EP, and B6-1.2-EP was less than the control beam by approximately 14.60%, 6.90%, 6.20%, 19.30%, 5.92%, and 17.10%, respectively. The results showed that GPP as adhesive material paste enhanced the maximum beam's deflections better than EP. Also, it was found that using GFRP with EP reduced the mid-span displacement of the tested beams [56–58].

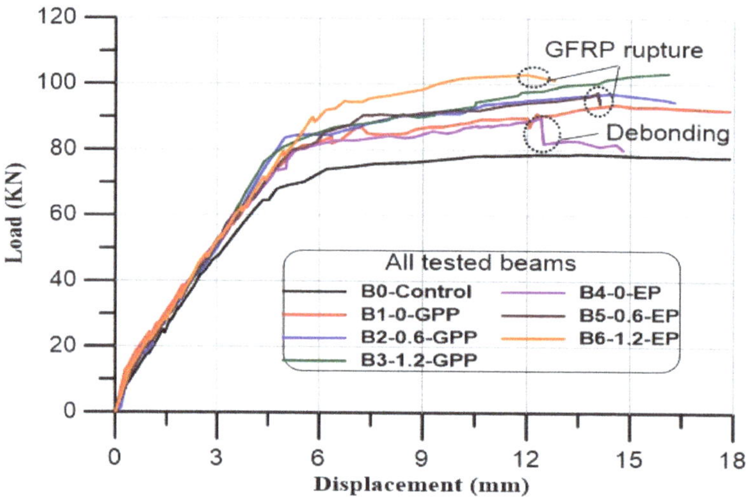

Figure 17. Load–deflection relationship for tested beams.

3.4. Ductility of Specimens

The ductility factor can be determined as the ratio of the maximum displacement at failure load (Δ_u) to the maximum displacement at yield load (Δ_y). The ductility factor of beams B1-0-GPP, B2-0.6-GPP, and B3-1.2-GPP was larger than corresponding beams B4-0-EP, B5-0.6-EP, and B6-1.2-EP by approximately 5.80%, 13.60%, and 21.40%, respectively. Table 4 shows that the beam B3-1.2-GPP had a similar ductility factor to the control beam due to the large ratio of SGF. In addition, the results showed that the ductility factors of beams B2-0.6-GPP and B3-1.2-GPP had a little more ductility than the control beam.

4. Finite Element Analysis

In this section, finite element analysis (FE) was carried out using the ABAQUS software [47]. The FE program can accurately simulate the GFRP/concrete interface, taking into consideration both geometric and material nonlinearity. A series of vertical displacement increments were applied until the maximum experimental deflection value was reached for each beam. The Newton–Raphson technique was utilized in the nonlinear analysis to allow convergence within tolerance limits at the end of each deflection increment. To further prevent a divergent solution, automatic stabilization and short time increments were applied.

4.1. Modeling of Material Properties

4.1.1. Concrete

The concrete was modeled using concrete damage plasticity (CDP). Due to the model's combination of compressive plasticity and isotropic tensile plasticity, it can simulate the concrete in the inelastic zone. The elastic modulus values, the Poisson's ratio, characterization of compressive and tensile behavior, and the plastic damage parameters, are necessary for the plastic damage model. In the concrete damage plasticity model (CDP), the plastic damage parameters are the dilation angle, the flow potential eccentricity, the ratio of initial biaxial compressive yield stress to initial uniaxial compressive yield stress, the ratio of the second stress invariant on the tensile meridian to that on the compressive meridian, and the viscosity parameter which defines viscoplastic regularization. The final four parameters for defining concrete material damage are described in Table 5.

Table 5. Concrete damage parameters (CDP).

Dilation Angle (Ψ)	35°
Eccentricity (e)	0.10
f_{bo}/f_{co}	1.16
K	0.66
Viscosity Parameter	0.00005

The experimental compressive stress–strain curve for concrete as illustrated in Figure 4 was used in the FE program. The concrete tensile strength was equal to 2.1 MPa. The modulus of elasticity of concrete (E_c) was 26,985 MPa according to the compressive stress–strain curve (See Figure 4, Section 2.3.1). Two primary failure modes are considered for concrete according to (CDP): compressive crushing and tensile cracking. The stress–strain response under uniaxial tension followed a linear elastic relationship up until a value of the failure stress was attained.

The initiation of microcracking in the concrete material is correlated with failure stress. A softening stress–strain response was used to illustrate the propagation of microcracks beyond the failure stress. Therefore, elastic modulus, E_c, and tensile strength, f_t, are required to define the first part of the relationship. For descending branch, the fracture energy G_f approach was used to define the post-peak tension failure behavior of concrete, as shown in Figure 18. The area beneath the concrete's tensile curve following the peak value was the fracture energy G_f.

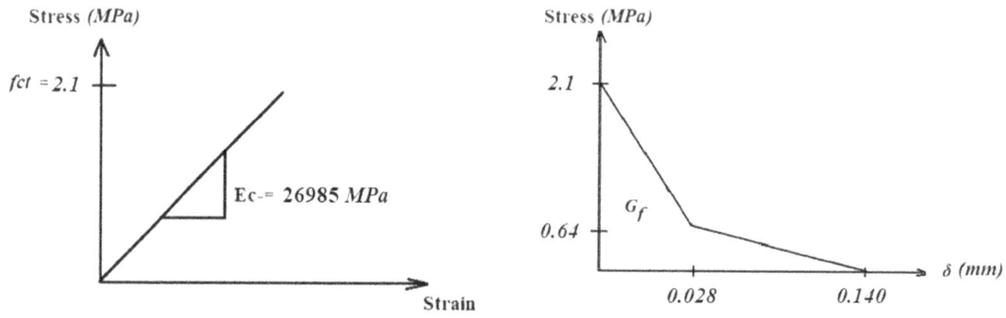

Figure 18. The tensile curve of the concrete (Simplified curve).

4.1.2. Steel Reinforcement Bars and GFRP Laminate

Figure 19 illustrates a bilinear elastic–plastic stress–strain curve for reinforcement bars. The embedded region was modeled for the reinforcement in this investigation. Table 6 illustrates the mechanical properties of reinforcement bars. The interaction surface between reinforcement bars and concrete was assumed to be a perfect bond. In the current study, the GFRP was assumed to be a linear elastic isotropic material. Table 7 illustrates the properties of CFRP laminates as specified by the manufacturer.

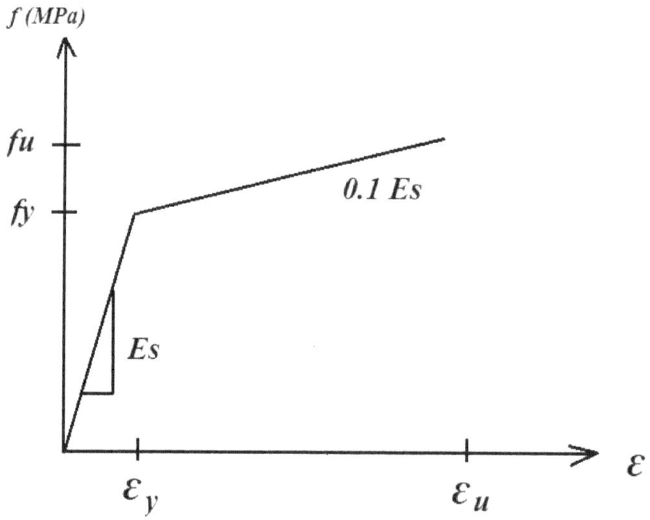

Figure 19. Stress–strain curve for steel in finite element analysis.

Table 6. The mechanical properties of reinforcement bars.

Modulus of Elasticity, E_s (GPa)	202
Yield strength, f_y (MPa)	410
Ultimate strength f_u (Mpa)	520
Poisson's ratio	0.20

Table 7. The properties of GFRP as specified by the manufacturer.

Modulus of Elasticity, E (GPa)	72
Tensile strength (GPa)	2.5

4.1.3. GFRP Laminate–Concrete Interface

The GFRP laminate–concrete interface can be more effectively modeled using the cohesive model in ABAQUS. When dealing with debonding issues of strengthening materials, cohesive elements and surface-based cohesive behavior are preferable alternatives for this type of modeling. Surface-based cohesive behavior requires a linear elastic traction separation as well as damage initiation and evolution laws, and it has very comparable formulas and constitution laws to cohesive elements. A graphic representation of a standard bilinear traction–separation rule expressed in terms of effective traction τ and effective opening displacement (Interfacial slip) δ is shown in Figure 20. The initial stiffness K_0 is specified as [56] and is modeled as a rich zone of small thickness at the interface. The initial stiffness can be expressed as follows:

$$K_0 = \frac{1}{\frac{t_i}{G_i} + \frac{t_c}{G_c}} \quad (1)$$

where t_i and t_c represent the thickness of adhesive and the concrete thickness, respectively. G_i and G_c are the shear modulus of the adhesive and concrete, respectively. The values of shear stress of each resin $\tau_{\max}$ according to the experimental results used for B1-0-GPP, B2-0.6-GPP, B3-1.2-GPP, B4-0-EP, B5-0.6-EP, and B6-1.2-EP are 4.2 MPa, 4.26 MPa, 4.31 MPa, 3.95 MPa, 4.1 MPa, and 4.2 MPa, respectively (See Figure 3). The damage initiation criterion was described by using the quadratic nominal stress function:

$$\left\{\frac{\sigma_n}{\sigma_n^0}\right\} + \left\{\frac{\tau_n}{\tau_s^0}\right\} + \left\{\frac{\tau_t}{\tau_t^0}\right\} = 1 \quad (2)$$

where σ_n is the cohesive tensile, τ_s and τ_t are the interface's shear stresses, and n, s and t denote the direction of the stress component. Damage was considered to initiate when the traction function reached one. The maximum normal stress σ_n^0 in comparable studies is equal to 2.1 MPa, while the maximum shear stresses τ_s^0, τ_t^0 are taken according to the experimental results (see Figure 3). Energy release was used to assess the development of interface damage. The description of this model can be found in the ABAQUS material library. Previous investigations have found values for the fracture energy, G_{cr}, ranging from 300 J/m^2 to 1500 J/m^2 [57]; for this investigation, the values taken were calculated based on experimental results.

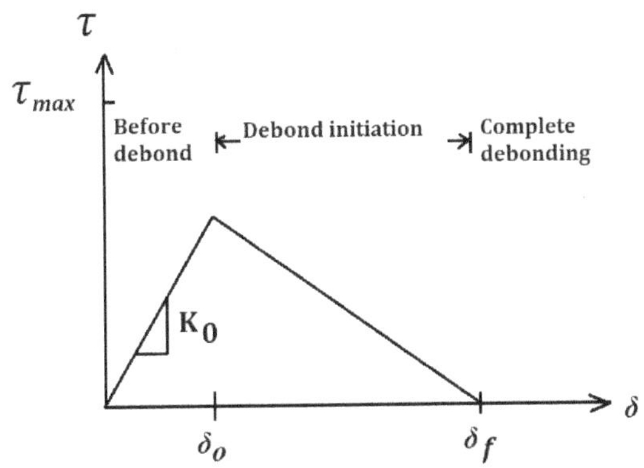

Figure 20. Bilinear traction–separation constitutive law.

4.2. Boundary Conditions and Meshing

As shown in Figure 21, the geometry of the beams, loads, and boundary conditions were all symmetrical, so just one-half of a beam was modelled using a standard finite element mesh. This approach significantly decreased the amount of computing time and the computer storage space, especially in the case of the GFRP/Concrete interface which is not modeled as a full bond. To choose an appropriate mesh with acceptable accuracy in terms of ultimate load and ultimate deflection, three models with various mesh sizes—fine, medium, and coarse of 15, 25, and 35 mm, respectively—were examined. Figure 21 depicts the concrete volume medium meshes and reinforcing medium meshes for tested beams. The solid three-dimensional eight-node bricks (C3D8R) were adopted in the FE study to model the concrete. The main reinforcement bars were modeled using a bar truss element (T2D3) with the same mesh size as the concrete. GFRP laminate was modeled as a four-nodded doubly thin curved shell, reduced integration, and hourglass control (S4R). The applied load and supports were taken as illustrated in Figure 21a to simulate the experimental program.

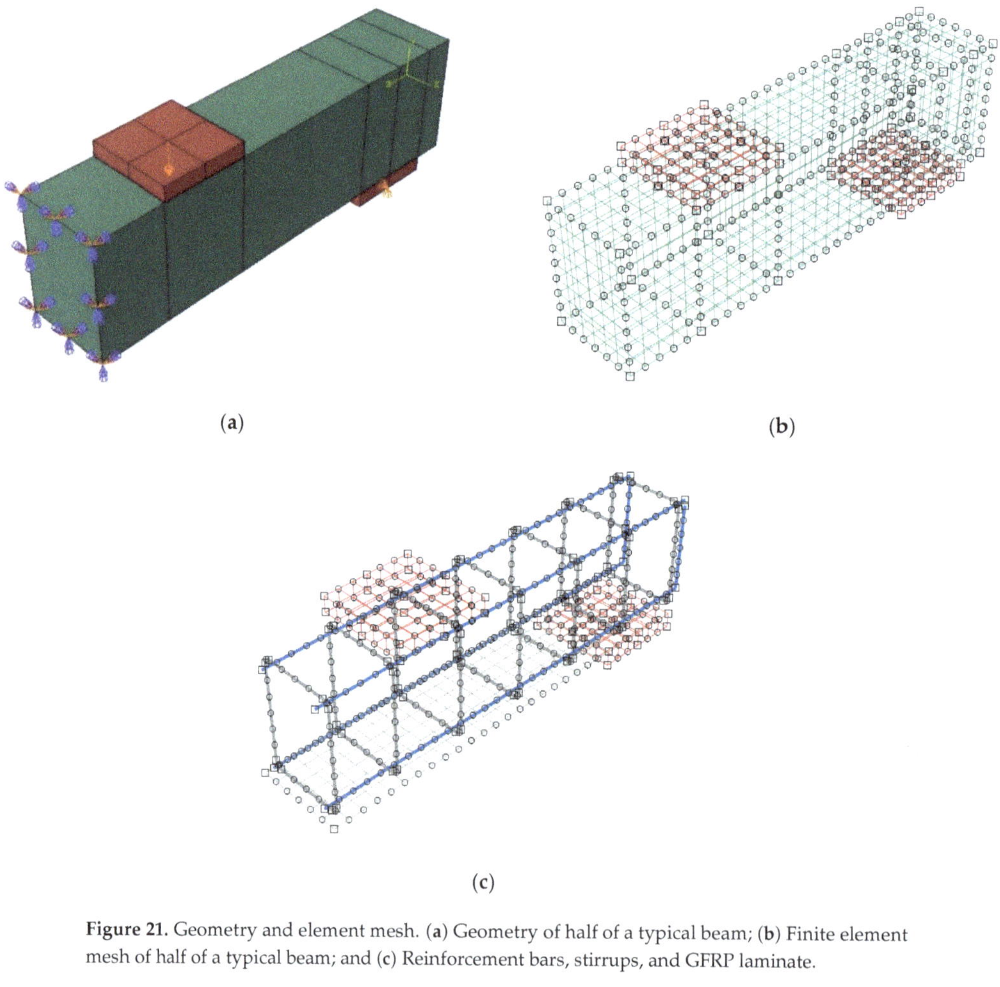

Figure 21. Geometry and element mesh. (**a**) Geometry of half of a typical beam; (**b**) Finite element mesh of half of a typical beam; and (**c**) Reinforcement bars, stirrups, and GFRP laminate.

5. Comparison of the Experimental and Finite Element Results

To confirm the experimental findings, the ABAQUS program was used to mimic each tested beam with the same information and the same material characteristics as the experiment. It was found that the medium mesh gives the best results in the ultimate load and deflection values, as shown in Table 8. Additionally, it had an acceptable computational time compared to the other meshes. The load–deflection curves for both experimental and numerical were presented for all tested beams, as illustrated in Figure 22. Compared with the experimental results, the load–deflection curves from numerical simulations behaved more stiffly, which was likely caused by a sizable scatter in the tensile strength of the concrete. The range of 0.95 to 0.98 is the FE forecast for the experimental ultimate load ratio, as illustrated in Table 5. In addition, the range of 0.83 to 1.09 is the FE forecast for the experimental maximum beam displacement ratio.

Table 8. Comparison of the experimental and finite element findings.

	Failure Load, kN				Maximum Beam Displacement at the Ultimate Load (mm)			
	Finite Element			EXP./Finite Element (Medium)	Finite Element			EXP./Finite Element (Medium)
	Fine	Medium	Coarse		Fine	Medium	Coarse	
B0-Control	79.60	78.51	76.54	0.98	17.07	16.6	15.87	1.07
B1-0-GPP	97.11	95.85	93.53	0.97	16.86	16.4	15.67	0.83
B2-0.6-GPP	101.92	100.61	98.20	0.96	15.45	15.0	14.30	1.06
B3-1.2-GPP	104.43	103.1	100.64	0.98	14.64	14.2	13.52	1.07
B4-0-EP	95.85	94.6	92.31	0.95	13.43	13	12.34	0.95
B5-0.6-EP	103.42	102.1	99.66	0.95	13.33	12.9	12.24	1.09
B6-1.2-EP	106.45	105.1	102.60	0.98	12.93	12.5	11.85	0.97

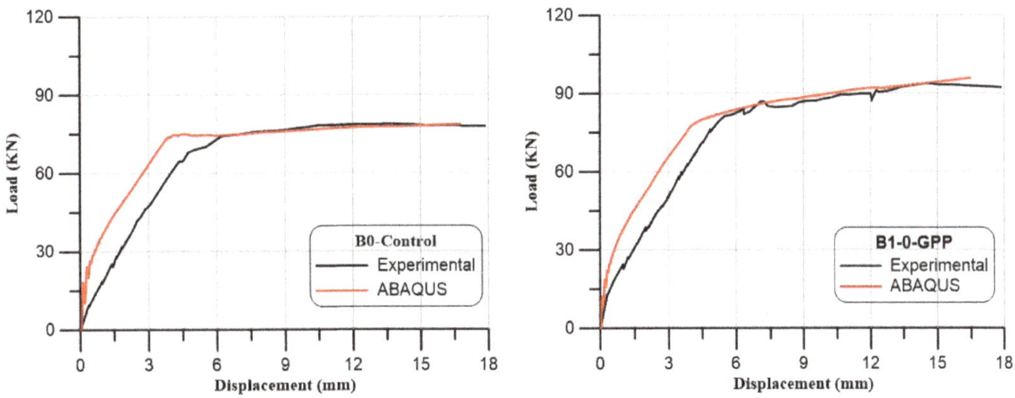

Figure 22. *Cont.*

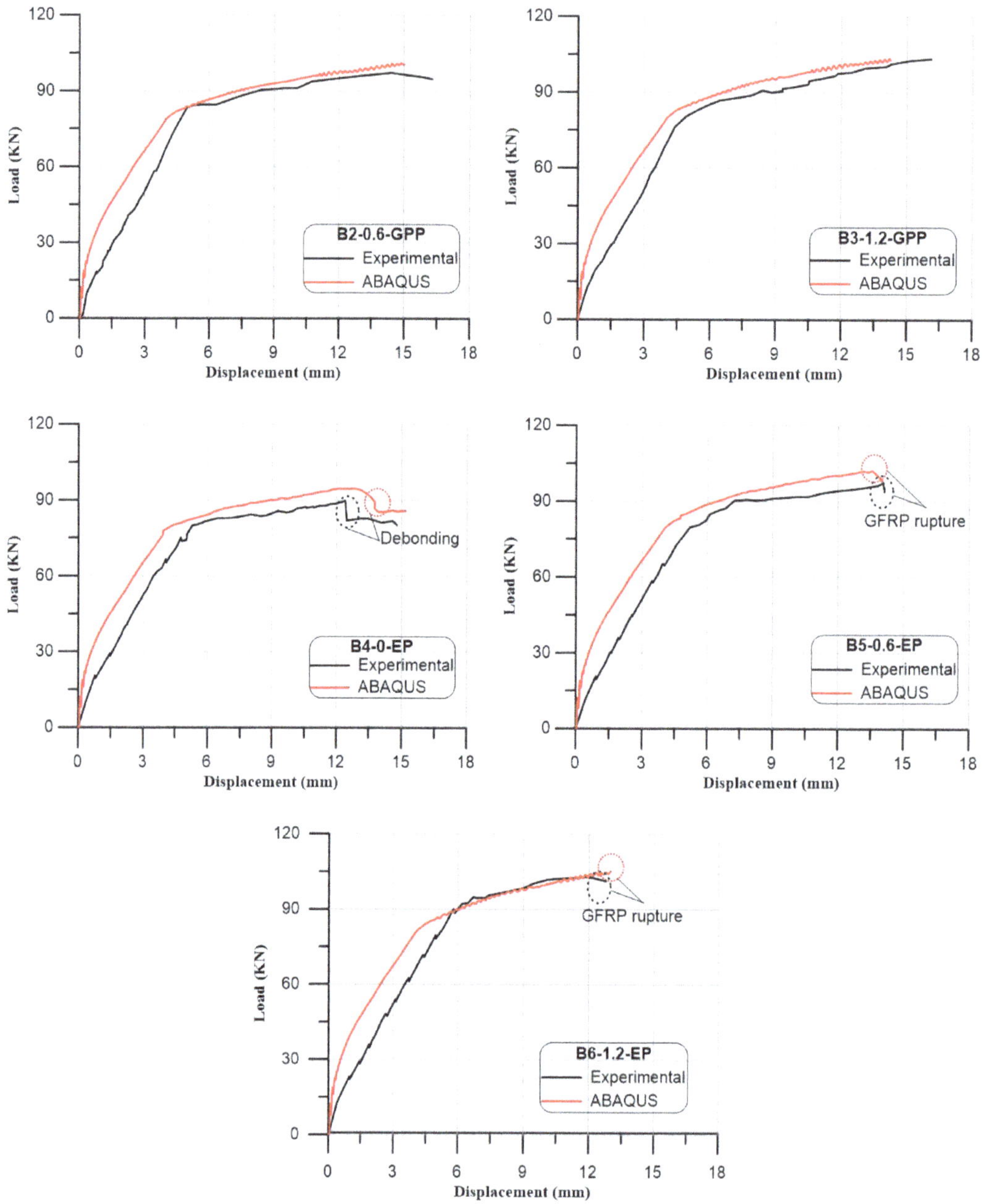

Figure 22. Load–deflection curves of the tested beams (numerical and experimental results).

Figure 23 shows the final finite element cracks of the examined beams. All failure modes obtained numerically were tension failures at the flexural zone. All cracks formed in the middle third of the beams verified the experimental results. The experimental findings and FE predictions can be seen to be in good agreement. Due to the complexity of using the cohesive model in modeling the GFRP laminate–concrete interface, there was a slight variation in the results.

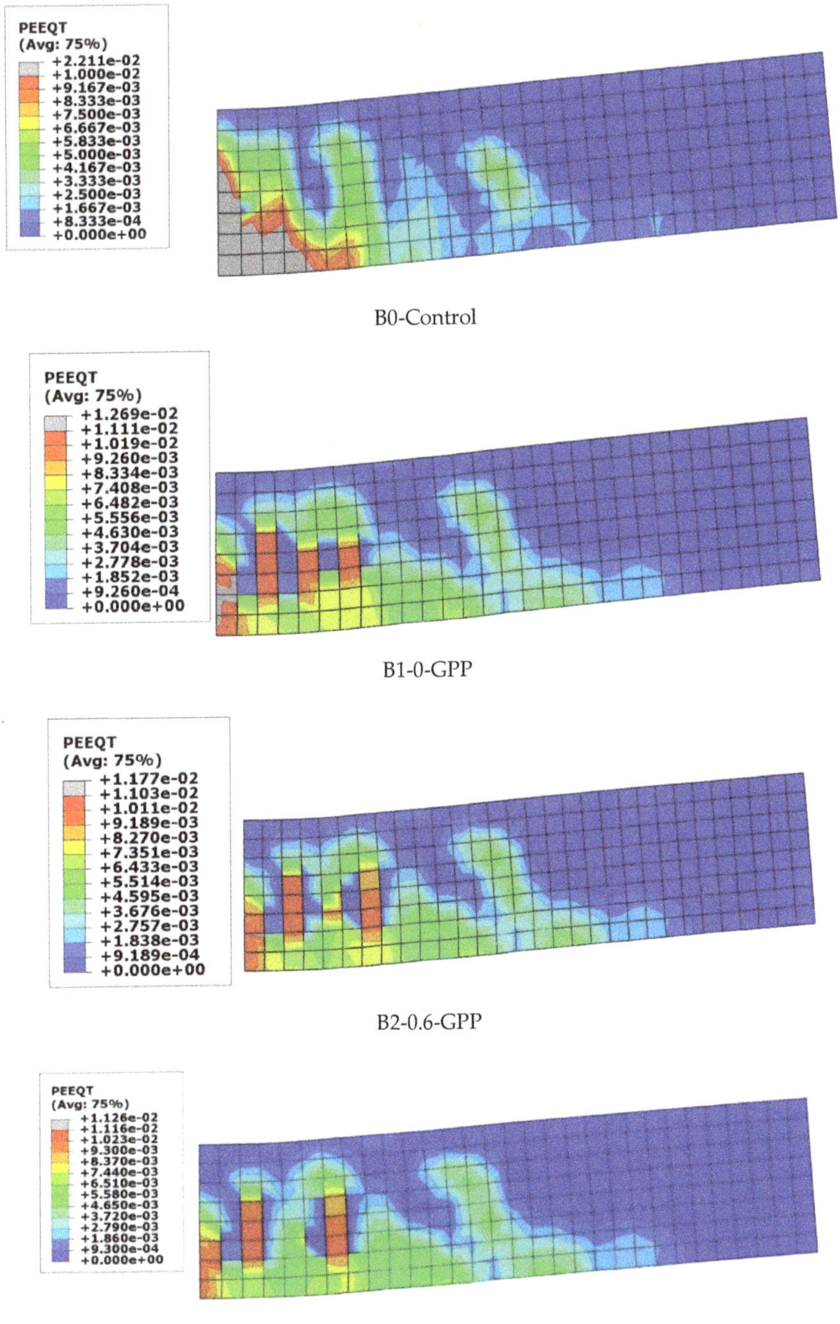

Figure 23. *Cont.*

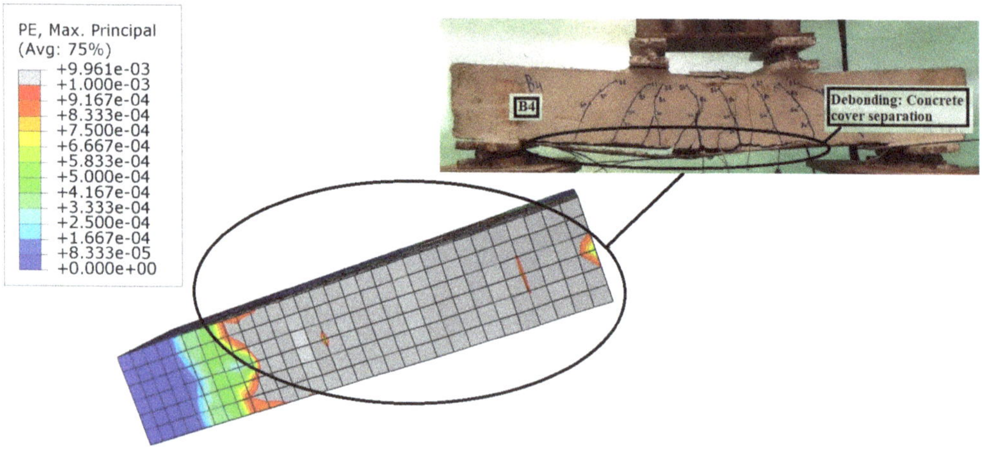

B4-0-EP

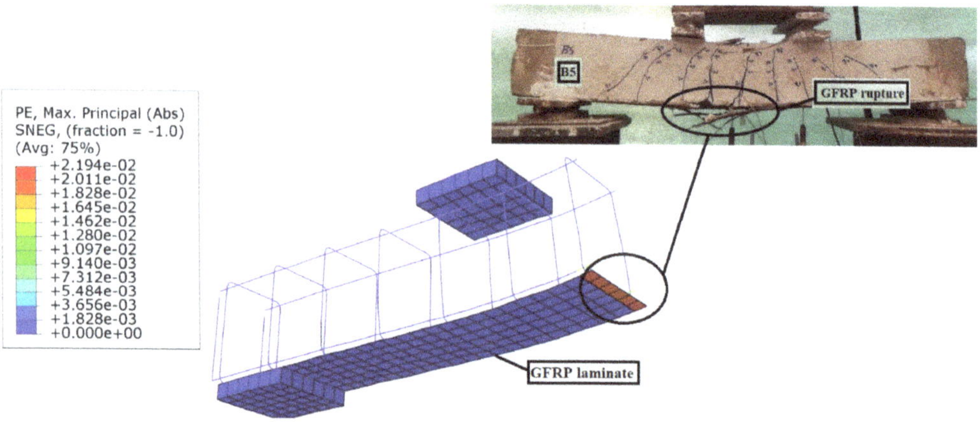

B5-0.6-EP

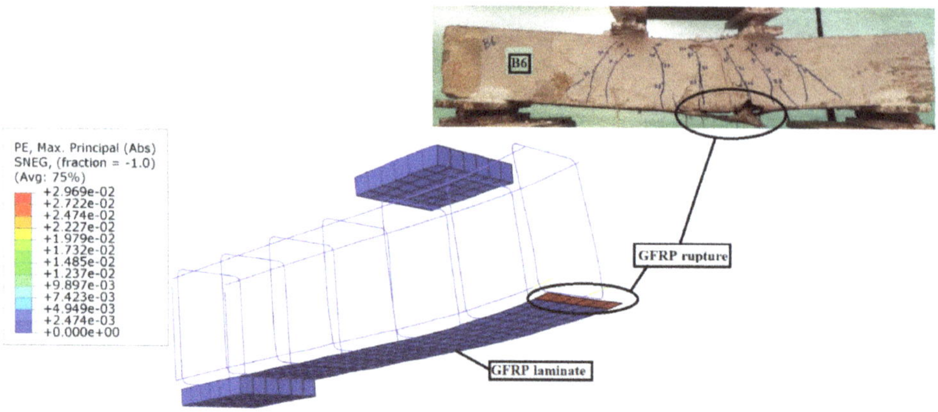

B6-1.2-EP

Figure 23. Finite element final cracks of studied beams.

6. Conclusions

The subsequent conclusions may be taken from the findings of the experimental and analytical investigations to suggest two novel ways to enhance the bond between the GFRP laminates and the concrete's surface.

1. The geopolymer paste with and without short glass fibers outperformed epoxy in the strengthened beams. No debonding was observed between the GFRP laminate and the concrete surface, which is the most significant issue that frequently limits the strengthening of structures using GFRP laminate. On the other hand, compared with epoxy, geopolymer paste with and without short glass fibers reduced adhesive materials costs by more than 90%.
2. The failure loads of beams B1-0-GPP, B2-0.60-GPP, and B3-1.2-GPP were greater than control beam B0-Control by approximately 20.80%, 25.60%, and 31.40%, respectively, whereas the failure loads of beams B4-0-EP, B5-0.6-EP, and B6-1.2-EP were greater than the control beam B0-Control by approximately 16.90%, 26.90%, and 26.10%, respectively. The results showed that increasing the SGF ratio in the GPP had a clear effect on raising the capacity of the beam due to the enhanced interfacial bond shear strength. In addition, the GPP had a great effect on increasing the failure load capacity of the tested beams.
3. Based on the experimental results, no debonding occurred for laminates when using GPP as adhesive material with different SGF ratios. On the contrary, there was a debonding when using EP as adhesive material.
4. The maximum deflection of the tested beams B1-0-GPP, B2-0.6-GPP, and B3-1.2-GPP was less than the control beam by approximately 14.60%, 6.90%, and 6.20%, respectively, whereas the maximum deflection of the tested beams B4-0-EP, B5-0.6-EP, and B6-1.2-EP was less than the control beam by approximately 19.30%, 5.92%, and 17.10%, respectively. This showed that the GPP as adhesive material pastes enhanced the maximum beam's deflections better than EP.
5. The ductility factor of beams B1-0-GPP, B2-0.6-GPP, and B3-1.2-GPP was larger than corresponding beams B4-0-EP, B5-0.6-EP, and B6-1.2-EP by approximately 5.80%, 13.60%, and 21.40%, respectively. This showed that the GPP as adhesive material pastes enhanced the beam's ductility factor more than EP.
6. The experimental findings and FE predictions are shown to be in good agreement. Due to the complexity of using the cohesive model in modeling the GFRP laminate–concrete interface, there was a slight variation in the results.

7. Recommendations for Future Research

The study will be continued to examine the efficiency of using the GPP as an adhesive material in strengthening beams in shear using GFRP and CFRP laminates.

Author Contributions: Conceptualization and methodology, M.M.A.; software, validation, formal analysis, data curation, writing—original draft preparation, writing—review and editing, and visualization, B.O.R., M.A.E.-M., A.H.E. and M.M.A.; investigation, B.O.R., M.A.E.-M. and M.M.A.; resources, supervision, and project administration, B.O.R., M.A.E.-M. and M.M.A. All authors have read and agreed to the published version of the manuscript.

Funding: This research received no external funding.

Institutional Review Board Statement: Not applicable.

Informed Consent Statement: Not applicable.

Data Availability Statement: Not applicable.

Conflicts of Interest: The authors declare no conflict of interest.

References

1. Attia, M.M.; El-Shaer, M.A.A.; Shawky, S.M.M.; Samaan, M.F. Replacement efficiency of steel reinforcement with FRB bars in RC beams under flexure load: Experimental and FE study. *Innov. Infrastruct. Solut.* **2022**, *7*, 281. [CrossRef]
2. El Ghadioui, R.; Wagner, J.; Klein, J.; Proske, T.; Curbach, M.; Graubner, C. RC members with a flexural-strengthening layer of CFRP textile-reinforced concrete under monotonic and cyclic long-term loading. *Struct. Concr.* **2022**, *23*, 939–953. [CrossRef]
3. Barros, J.; Figueiredo, F.P.; Costa, I.G.; Dourado, F. New type of CFRP reinforcement and technique for the flexural strengthening of RC balconies. *Compos. Struct.* **2022**, *280*, 114899. [CrossRef]
4. Prasertsri, T.; Lenwari, A.; Thepchatri, T. Flexural Response of CFRP-Strengthened Steel Beams with Initial Bond Defects. *Eng. J.* **2020**, *24*, 115–127. [CrossRef]
5. Yuan, P.; Xiao, L.; Wang, X.; Xu, G. Failure mechanism of corroded RC beams strengthened at shear and bending positions. *Eng. Struct.* **2021**, *240*, 112382. [CrossRef]
6. Song, L.; Li, L.G.; Fan, C. Analysis on bending resistance of RC beams reinforced with steel bonding plates under secondary stress. *Shenyang Gongye Daxue Xuebao/J. Shenyang Univ. Technol.* **2021**, *43*, 470–475. [CrossRef]
7. Jahami, A.; Temsah, Y.; Khatib, J.; Baalbaki, O.; Kenai, S. The behavior of CFRP strengthened RC beams subjected to blast loading. *Mag. Civ. Eng.* **2021**, *103*, 10309. [CrossRef]
8. Abuodeh, O.R.; Abdalla, J.A.; Hawileh, R.A. Flexural strengthening of RC beams using aluminum alloy plates with mechanically-fastened anchorage systems: An experimental investigation. *Eng. Struct.* **2021**, *234*, 111969. [CrossRef]
9. Abuodeh, O.R.; Hawileh, R.A.; Abdalla, J.A. Nonlinear finite element models of reinforced concrete beams strengthened in bending with mechanically fastened aluminum alloy plates. *Comput. Struct.* **2021**, *253*, 106573. [CrossRef]
10. Kumar, M.V.; Muthukannan, M. Investigation of HFRC beams retrofitted using GFRP for enhancement in flexural capacity. *Int. J. Manuf. Technol. Manag.* **2020**, *34*, 1–24. [CrossRef]
11. Yang, Y.; Fahmy, M.F.; Cui, J.; Pan, Z.; Shi, J. Nonlinear behavior analysis of flexural strengthening of RC beams with NSM FRP laminates. *Structures* **2019**, *20*, 374–384. [CrossRef]
12. Patil, S.; Dhongade, A.V. Literature review for use of FRP in Civil Engineering. *Int. Res. J. Eng. Technol.* **2021**, *8*, 446–449.
13. Ibrahim, S.S.; Kandasamy, S.; Pradeepkumar, S.; Bose, R.S.C. Effect of discrete steel fibres on strength and ductility of FRP laminated RC beams. *Ain Shams Eng. J.* **2021**, *12*, 1329–1337. [CrossRef]
14. Tahmouresi, B.; Momeninejad, K.; Mohseni, E. Flexural response of FRP-strengthened lightweight RC beams: Hybrid bond efficiency of L-shape ribbed bars and NSM technique. *Arch. Civ. Mech. Eng.* **2022**, *22*, 95. [CrossRef]
15. Attia, M.M.; Ahmed, O.; Kobesy, O.; Malek, A.S. Behavior of FRP rods under uniaxial tensile strength with multiple materials as an alternative to steel rebar. *Case Stud. Constr. Mater.* **2022**, *17*, e01241. [CrossRef]
16. Aravind, N.; Samanta, A.K.; Thanikal, J.V.; Roy, D.K.S. An experimental study on the effectiveness of externally bonded corrugated GFRP laminates for flexural cracks of RC beams. *Constr. Build. Mater.* **2017**, *136*, 348–360. [CrossRef]
17. Alam, A.; Onik, S.A.; Bin Mustapha, K.N. Crack based bond strength model of externally bonded steel plate and CFRP laminate to predict debonding failure of shear strengthened RC beams. *J. Build. Eng.* **2020**, *27*, 100943. [CrossRef]
18. Radfar, S.; Foret, G.; Saeedi, N.; Sab, K. Simulation of concrete cover separation failure in FRP plated RC beams. *Constr. Build. Mater.* **2012**, *37*, 791–800. [CrossRef]
19. De Maio, U.; Fabbrocino, F.; Greco, F.; Leonetti, L.; Lonetti, P. A study of concrete cover separation failure in FRP-plated RC beams via an inter-element fracture approach. *Compos. Struct.* **2019**, *212*, 625–636. [CrossRef]
20. Fu, B.; Teng, J.G.; Chen, J.F.; Chen, G.M.; Guo, Y.C. Concrete Cover Separation in FRP-Plated RC Beams: Mitigation Using FRP U-Jackets. *J. Compos. Constr.* **2017**, *21*, 04016077. [CrossRef]
21. Zhang, W.; Tang, Z.; Yang, Y.; Wei, J. Assessment of FRP–Concrete Interfacial Debonding with Coupled Mixed-Mode Cohesive Zone Model. *J. Compos. Constr.* **2021**, *25*, 04021002. [CrossRef]
22. Zhang, F.; Gao, L.; Wu, Y.-F.; Liu, J. Flexural design of reinforced concrete structures strengthened with hybrid bonded FRP. *Compos. Struct.* **2021**, *269*, 113996. [CrossRef]
23. Yazdani, N.; Aljaafreh, T.; Beneberu, E. Concrete beam flexural strengthening with anchored pre-saturated CFRP laminates. *Compos. Struct.* **2020**, *235*, 111733. [CrossRef]
24. Mostofinejad, D.; Hosseini, S.M.; Tehrani, B.N.; Eftekhar, M.R.; Dyari, M. Innovative warp and woof strap (WWS) method to anchor the FRP sheets in strengthened concrete beams. *Constr. Build. Mater.* **2019**, *218*, 351–364. [CrossRef]
25. Ababneh, A.N.; Al-Rousan, R.Z.; Ghaith, I.M. Experimental study on anchoring of FRP-strengthened concrete beams. *Structures* **2020**, *23*, 26–33. [CrossRef]
26. Sun, W.; Liu, S.; Zhang, C. An effective improvement for enhancing the strength and feasibility of FRP spike anchors. *Compos. Struct.* **2020**, *247*, 112449. [CrossRef]
27. Grelle, S.V.; Sneed, L.H. Review of Anchorage Systems for Externally Bonded FRP Laminates. *Int. J. Concr. Struct. Mater.* **2013**, *7*, 17–33. [CrossRef]
28. Ibrahim, W.; Ali, W. Comparison of Anchorage Systems for Externally Bonded FRP Laminates. *Int. J. Eng. Technol.* **2019**, *11*, 1128–1135. [CrossRef]
29. Al-Saoudi, A.; Kalfat, R.; Al-Mahaidi, R.; Cervenka, J.; Pryl, D. Numerical and experimental investigation into the fatigue life of FRP bonded to concrete and anchored with bidirectional fabric patches. *Eng. Struct.* **2021**, *239*, 112335. [CrossRef]

30. Godat, A.; Hammad, F.; Chaallal, O. State-of-the-art review of anchored FRP shear-strengthened RC beams: A study of influencing factors. *Compos. Struct.* **2020**, *254*, 112767. [CrossRef]
31. Eslami, A.; Moghavem, A.; Shayegh, H.R.; Ronagh, H.R. Effect of FRP stitching anchors on ductile performance of shear-deficient RC beams retrofitted using FRP U-wraps. *Structures* **2020**, *23*, 407–414. [CrossRef]
32. Zhang, S.; Ke, Y.; Smith, S.; Zhu, H.; Wang, Z. Effect of FRP U-jackets on the behaviour of RC beams strengthened in flexure with NSM CFRP strips. *Compos. Struct.* **2021**, *256*, 113095. [CrossRef]
33. Abdallah, M.; Al Mahmoud, F.; Khelil, A.; Mercier, J.; Almassri, B. Assessment of the flexural behavior of continuous RC beams strengthened with NSM-FRP bars, experimental and analytical study. *Compos. Struct.* **2020**, *242*, 112127. [CrossRef]
34. ACI. ACI SP 275. In *Fiber-Reinforced Polymer Reinforcement for Concrete Structures 10th International Symposium, Tampa, FL, USA, 2–4 April 2011*; ACI: Farmington Hills, MI, USA, 2011.
35. Yuan, C.; Chen, W.; Pham, T.M.; Hao, H.; Chen, L.; Zhang, M. New epoxy anchor for better bonding between FRP sheets and concrete. *Constr. Build. Mater.* **2020**, *248*, 118628. [CrossRef]
36. Mostofinejad, D.; Mofrad, M.H.; Hosseini, A.; Mofrad, H.H. Investigating the effects of concrete compressive strength, CFRP thickness and groove depth on CFRP-concrete bond strength of EBROG joints. *Constr. Build. Mater.* **2018**, *189*, 323–337. [CrossRef]
37. Jiang, C.; Wan, B.; Wu, Y.F.; Omboko, J. Epoxy interlocking: A novel approach to enhance FRP-to-concrete bond behavior. *Constr. Build. Mater.* **2018**, *193*, 643–653. [CrossRef]
38. Almutairi, A.L.; Tayeh, B.A.; Adesina, A.; Isleem, H.F.; Zeyad, A.M. Potential applications of geopolymer concrete in construction: A review. *Case Stud. Constr. Mater.* **2021**, *15*, e00733. [CrossRef]
39. Mackenzie, K.; Welter, M. Geopolymer (aluminosilicate) composites: Synthesis, properties and applications. In *Advances in Ceramic Matrix Composites*; Victoria University of Wellington: Wellington, New Zealand, 2014; pp. 445–470. [CrossRef]
40. Singh, N.; Middendorf, B. Geopolymers as an alternative to Portland cement: An overview. *Constr. Build. Mater.* **2020**, *237*, 117455. [CrossRef]
41. Dong, M.; Feng, W.; Elchalakani, M.; Li, G.; Karrech, A.; Sheikh, N.M. Material and glass-fibre-reinforced polymer bond properties of geopolymer concrete. *Mag. Concr. Res.* **2020**, *72*, 509–525. [CrossRef]
42. Abdulrahman, H.; Muhamad, R.; Visintin, P.; Shukri, A.A. Mechanical properties and bond stress-slip behaviour of fly ash geopolymer concrete. *Constr. Build. Mater.* **2022**, *327*, 126909. [CrossRef]
43. Saranya, P.; Nagarajan, P.; Shashikala, A.P. Experimental Investigation on Bond Strength Properties of Geopolymer Concrete. In *Advances in Civil Engineering*; Lecture Notes in Civil Engineering; Singh, R.M., Sudheer, K.P., Kurian, B., Eds.; Springer: Singapore, 2021; Volume 83. [CrossRef]
44. Rajak, D.K.; Pagar, D.D.; Kumar, R.; Pruncu, C.I. Recent progress of reinforcement materials: A comprehensive overview of composite materials. *J. Mater. Res. Technol.* **2019**, *8*, 6354–6374. [CrossRef]
45. Mohamedou, M.; Zulueta, K.; Chung, C.N.; Rappel, H.; Beex, L.; Adam, L.; Arriaga, A.; Major, Z.; Wu, L.; Noels, L. Bayesian identification of Mean-Field Homogenization model parameters and uncertain matrix behavior in non-aligned short fiber composites. *Compos. Struct.* **2019**, *220*, 64–80. [CrossRef]
46. Li, Z.; Liu, Z.; Lei, Z.; Zhu, P. An innovative computational framework for the analysis of complex mechanical behaviors of short fiber reinforced polymer composites. *Compos. Struct.* **2021**, *277*, 114594. [CrossRef]
47. Sharma, K.K.; Kushwaha, J.; Kumar, K.; Singh, H.; Shrivastava, Y. Fabrication and testing of hybrid fibre reinforced composite: A comprehensive review. *Aust. J. Mech. Eng.* **2022**, 1–17. [CrossRef]
48. Abaqus, S. *CAE 6.14 User's Manual Dassault*; Systémes Inc. Provid.: Johnston, RI, USA, 2014; Volume IV.
49. *ASTM C494*; Standard Specification for Chemical Admixtures for Concrete. ASTM Int.: West Conshohocken, PA, USA, 2019; Volume 04.
50. Attia, M.M.; Abdelsalam, B.A.; Amin, M.; Agwa, I.S.; Abdelmagied, M.F. Metal-Nails Waste and Steel Slag Aggregate as Alternative and Eco-Friendly Radiation Shielding Composites. *Buildings* **2022**, *12*, 1120. [CrossRef]
51. ACI Committee 440. ACI440.2R-17 Guide for the Design and Construction of Externally Bonded FRP Systems for Strengthening Concrete Structures Reported by ACI Committee 440. *Am. Concr. Inst.* **2017**, *5*, 45.
52. American Concrete Institute. *ACI 318 Building Code Requirements for Structure Concrete and Commentary*; American Concrete Institute ACI Committee: Farmington Hills, MI, USA, 2019.
53. Hamrat, M.; Bouziadi, F.; Boulekbache, B.; Daouadji, T.H.; Chergui, S.; Labed, A.; Amziane, S. Experimental and numerical investigation on the deflection behavior of pre-cracked and repaired reinforced concrete beams with fiber-reinforced polymer. *Constr. Build. Mater.* **2020**, *249*, 118745. [CrossRef]
54. Almusallam, T.; Al-Salloum, Y.; Alsayed, S.; Iqbal, R.; Abbas, H. Effect of CFRP strengthening on the response of RC slabs to hard projectile impact. *Nucl. Eng. Des.* **2015**, *286*, 211–226. [CrossRef]
55. Radnić, J.; Matešan, D.; Grgić, N.; Baloević, G. Impact testing of RC slabs strengthened with CFRP strips. *Compos. Struct.* **2015**, *121*, 90–103. [CrossRef]
56. Guo, Z.; Cao, S.; Sun, W.; Lin, X. Experimental study on bond stress-slip behaviour between FRP sheets and concrete. In Proceedings of the International Symposium on Bond Behaviour of FRP in Structures, Hong Kong, China, 7–9 December 2005.

57. Obaidat, Y.T.; Heyden, S.; Dahlblom, O. The effect of CFRP and CFRP/concrete interface models when modelling retrofitted RC beams with FEM. *Compos. Struct.* **2010**, *92*, 1391–1398. [CrossRef]
58. Jahami, A.; Temsah, Y.; Khatib, J.; Sonebi, M.; Ali, J. Numerical study for the effect of carbon fiber reinforced polymers (CFRP) sheets on structural behavior of posttensioned slab subjected to impact loading. In Proceedings of the Symposium on Concrete Modelling–CONMOD2018, Delft, The Netherlands, 27–30 August 2018; pp. 259–267.

MDPI AG
Grosspeteranlage 5
4052 Basel
Switzerland
Tel.: +41 61 683 77 34

Buildings Editorial Office
E-mail: buildings@mdpi.com
www.mdpi.com/journal/buildings

www.ingramcontent.com/pod-product-compliance
Lightning Source LLC
Chambersburg PA
CBHW040802250926
51850CB00007BA/1055
* 9 7 8 3 7 2 5 8 6 9 2 6 8 *